Theory and Methods of Quantification Design on System-Level Electromagnetic Compatibility

Donglin Su · Shuguo Xie ·
Fei Dai · Yan Liu · Yunfeng Jia

Theory and Methods of Quantification Design on System-Level Electromagnetic Compatibility

Donglin Su
Beihang University
Beijing, China

Shuguo Xie
Beihang University
Beijing, China

Fei Dai
Beihang University
Beijing, China

Yan Liu
Beihang University
Beijing, China

Yunfeng Jia
Beihang University
Beijing, China

ISBN 978-981-13-3689-8 ISBN 978-981-13-3690-4 (eBook)
https://doi.org/10.1007/978-981-13-3690-4

Jointly published with National Defense Industry Press, Beijing, China
The print edition is not for sale in China Mainland. Customers from China Mainland please order the print book from: National Defense Industry Press.

Library of Congress Control Number: 2018963290

© National Defense Industry Press and Springer Nature Singapore Pte Ltd. 2019
This work is subject to copyright. All rights are reserved by the Publisher, whether the whole or part of the material is concerned, specifically the rights of translation, reprinting, reuse of illustrations, recitation, broadcasting, reproduction on microfilms or in any other physical way, and transmission or information storage and retrieval, electronic adaptation, computer software, or by similar or dissimilar methodology now known or hereafter developed.
The use of general descriptive names, registered names, trademarks, service marks, etc. in this publication does not imply, even in the absence of a specific statement, that such names are exempt from the relevant protective laws and regulations and therefore free for general use.
The publisher, the authors and the editors are safe to assume that the advice and information in this book are believed to be true and accurate at the date of publication. Neither the publisher nor the authors or the editors give a warranty, express or implied, with respect to the material contained herein or for any errors or omissions that may have been made. The publisher remains neutral with regard to jurisdictional claims in published maps and institutional affiliations.

This Springer imprint is published by the registered company Springer Nature Singapore Pte Ltd.
The registered company address is: 152 Beach Road, #21-01/04 Gateway East, Singapore 189721, Singapore

Foreword

System-level electromagnetic compatibility (EMC) design is one of the core technologies in the research and development of large-scale platform. With the increasingly harsh electromagnetic environment and increasingly complex electronic systems on large platforms, system-level EMC quantitative design becomes a very important methodology in great demand.

This book is based on the research and engineering practice of Prof. Donglin Su and the EMC Research Team in Beihang University which she has led for about 30 years. The book starts with an introduction of the basic EMC theories and the basic concept of system-level EMC quantitative design. Then, combined with the experience in aircraft EMC performance analysis, engineering design, and troubleshooting, the book discusses the key technologies in the quantitative design of system-level EMC. Next, taking the aircraft system-level EMC design as an example, the authors introduce the method of quantitative design and evaluation of system-level EMC. Evaluation and quality control methodologies of EMC are also discussed from the perspective of lifecycle EMC performance. The authors then discuss the common problems and solutions of CE102, RE102, and RS103 tests. This book is the first academic monograph about system-level EMC quantitative design theory and method in China. I hope the book will provide good guidance to the readers in both theory and applications.

Professor Donglin Su has a great enthusiasm for aviation and EMC. She has devoted herself to the theoretical research and engineering practice in system-level quantitative design of EMC for 30 years, and she has made a significant achievement in this field. She is the leading expert in EMC quantitative design with outstanding contributions to the development of the EMC industry. She is the recipient of the Second Prize of the State Science and Technology Progress Awards of China with her theory and method of "top-down quantitative design of system-level EMC for aircraft." This achievement is a major innovation to solve the system-level EMC issues and plays an important role in the successful development of aircraft.

I believe the publication of this book will greatly promote the development of China's EMC field.

Shijiazhuang, Hebei, China

Liu Shanghe
Academician of Chinese Engineering Academy

Preface

Aircraft, satellites, ships, and electronic information systems are large and complex information platforms. Their EMC demonstration and design at system level (also known as platform level) have always been a universal problem. There are two main causes of the problem: One is the lack of design standard; the other is the lack of demonstration and design methodologies. These two might also be the major reasons why EMC is difficult to include in the demonstration and design of most large-scale complex information systems and platforms.

High quality of EMC is a result of design.

"System-level EMC quantitative technology" was proposed by the EMC Research Team of Beihang University, China. There are more than a hundred of teachers, students, and engineering technicians in the team. This book covers our analysis of the advanced experience and achievements of EMC research both at home and abroad, as well as our own research on theories and methodologies of EMC and relevant disciplines. The book also reflects our experience from more than 20 cases of system-level EMC design and troubleshooting and our unremitting efforts in the field of EMC in the past 30 years. I hope this book will serve as a good reference to researchers and engineers in EMC and relevant field in China.

We are sincerely grateful to the great opportunity created by China's national defense and aviation industry. We would also like to thank Beihang University for the excellent platform it provides for our careers.

The EMC Research Team of Beihang University has been supported by authorities at all levels and by other research institutes and manufacturers. We have also benefited enormously from ideas and discussions with experts in EMC and related fields at home and abroad. We would like to gratefully acknowledge all of the people who helped us write this book.

The research work covered in this book has been supported by the Key Program, General Program, and Young Scientist Fund of the National Natural Science Foundation of China as well as the National Defense Fund and Aviation Fund.

This book is divided into two parts: Part I introduces the basic theories of system-level EMC quantitative design, including electromagnetic field and electromagnetic wave, microwave technology, and antenna principle. Part II focuses on the system-level EMC quantitative design, including basic concepts, theoretical methods, major technological solutions, EMC quality control and evaluation methods, software, and application cases.

Professor Donglin Su is the corresponding author of this book; Prof. Shuguo Xie wrote the chapter of antenna theories; the part of test requirements and indicators was written by Assoc. Prof. Fei Dai; the EMC quantitative design application cases were written by Dr. Yan Liu; EMC quality control application cases were written by Dr. Yunfeng Jia. The preparation of this book has also been assisted by other members of the EMC Research Team of Beihang University, to whom we would like to express our gratitude for their hard work.

Special thanks to Academician Liu Shanghe, Academician Zhang Minggao, and Prof. Wang Junhong who have recommended the book.

Finally, I would like to express my heartfelt thanks to the National Defense Industry Press, which supports the publication of this book.

We welcome our readers to point out if there are any errors or mistakes in this book.

Beijing, China

Donglin Su
Shuguo Xie
Fei Dai
Yan Liu
Yunfeng Jia

Contents

Symbols

B	System bandwidth (Hz)
B_R	Transmitter bandwidth (Hz)
B_T	Receiver bandwidth (Hz)
C	Capacitance (F)
D_r	Maximum size of the receiving antenna (equivalent diameter) (m)
D_t	Maximum size of the transmitting antenna (equivalent diameter) (m)
$D(\theta, \phi)$	Power gain (dB)
$\boldsymbol{E}$	Electric field strength (V/m)
E_{mn}	Electric field strength of mode m, n (V/m)
$EMC(s)$	EMC condition
$\tilde{\mathbf{E}}$	Complex electric field vector (V/m)
$E(\Theta, \Phi, t, f)$	Environmental electromagnetic interference source model
$\mathbf{F}$	Force (N)
F	Total noise figure (dB)
FIM	Fundamental inference margin (dB)
$F(\theta, \varphi)$	Directivity function
G	Conductance (S)
G	Total gain (dB)
$G_r(\theta_r, \varphi_r)$	Receiving antenna gain in the transmitting direction (dB)
$G_r(f_{in})$	Power gain of the receiving antenna (dB)
G_{sat}	Gain compression at the saturation point (dB)
$G_t(\theta_t, \varphi_t)$	Transmitting antenna gain in the receiving direction (dB)
$G_t(f_{in})$	Power gain of the transmitting antenna (dB)
$\boldsymbol{H}$	Magnetic field strength (A/m)
$H_{m,n}$	Magnetic field strength of mode m, n (A/m)
$\tilde{\mathbf{H}}$	Complex magnetic field vector (A/m)
$H(\Theta, \Phi, t, f)$	Interference coupling path model
$\boldsymbol{I}$	Linear current (A)
$I(E, \mathrm{R}; f)$	Isolation matrix
$\mathbf{I}_{\mathrm{limit}}(\mathrm{dB})$	Isolation safety margin (dB)

$I(t,f)$	Safety margin function
$\dot{I}$	Current vector (A)
$\boldsymbol{J}$	Volume current (A/m^2)
J_A	Interference power matrix (dBm)
$\boldsymbol{K}$	Surface current (A/m)
L	Inductance (H)
L	Isolation (dB)
L_a	Antenna isolation (dB)
L_d	Spatial isolation (dB)
L_P	Loss caused by the polarization mismatch (dB)
$\boldsymbol{L_{rB}}$	Reception suppression matrix of the receivers at the analysis frequency point (dB)
$\boldsymbol{L_{rf}}$	Receiving feeder loss matrix of the receivers (dB)
$\boldsymbol{L_{tB}}$	Emission attenuation matrix of the transmitters at the analysis frequency (dB)
$\boldsymbol{L_{tf}}$	Transmitting feeder loss matrix of the transmitter (dB)
M	Mutual inductance (H)
N_0	Noise power (dBm)
NMSE	Normalized mean square error (NMSE)
$O(\Theta, \Phi, t, f)$	Interference output model
P	Particle
P_{1dB}	Output power of the 1 dB gain compression point (dBm)
P_D	Desired signal level (dBm)
P_{IIP3}, P_{OIP3}	Input or output power of the TOI point (dBm)
P_{in}	Input power (dBm)
P_{out}	Output power (dBm)
P_{REF}	Reference signal level (dBm)
P_r	Receiving power (dBm)
P_{sat}	Output power of the saturation point (dBm)
$\boldsymbol{P_{smin}}$	Sensitivity matrix (dBm)
$\boldsymbol{P_t}$	Transmitting power matrix (dBm)
P_t	Transmitting power (dBm)
$\dot{P}_\Sigma$	Radiation power (dBm)
Q	Total charge (C)
Q_{net}	Net charge (C)
R	Resistance (Ω)
RIM	Receiver inference margin (dB)
R_Σ	Radiation resistance (Ω)
$\mathbf{S}$	Poynting vector (W/m^2)
S	Surface (m^2)
SIM	Spurious inference margin (dB)
$\boldsymbol{S_m}$	EMC safety margin matrix (dBm)
S/N	SNR matrix
$(S/N)_{REF}$	SNR of reference signal level (dB)

$\tilde{\mathbf{S}}$	Poynting complex vector (W/m^2)
$S(\Theta, \Phi, t, f)$	Susceptive subject model
T	Temperature (K)
T_A	Equivalent noise temperature (K)
$\boldsymbol{T_E}$	Transmitting conversion matrix
T_o	Temperature(290 K) (K)
$\boldsymbol{T_R}$	Receiving conversion matrix
TE	Transversal electric wave
TEM	Transversal electromagnetic wave
TE_{mn}	Transversal electric wave of mode m, n
TIM	Transmitter inference margin (dB)
TM	Transverse magnetic wave
TM_{mn}	Transverse magnetic wave of mode m, n
$T(\Theta, \Phi, t, f)$	Inference source model
V	Volume (m^3)
V	Voltage (V)
$\dot{V}$	Voltage vector (V)
VSWR	Voltage standing wave ratio
W	Total electromagnetic field energy (J)
W_E	Total electric field energy (J)
W_e	Total electromagnetic field energy in capacitor (J)
W_H	Total magnetic field power (J)
W_m	Total magnetic field energy in inductor (J)
X	Reactance (J)
Y	Admittance (S)
Z	Impedance (Ω)
Z_c	Characteristic impedance (Ω)
f	Frequency (Hz)
f_c	Cutoff frequency (Hz)
f_E	Transmitting of power of transmitter (Hz)
f_0	Central operating frequency of equipment (Hz)
f_R	Receiver response frequency, receiver central frequency (Hz)
f_T	Transmitter central frequency (Hz)
i	Transient current (A)
i_n	Unit vector in the normal direction of the boundary
$\mathbf{i}_{r_c}$	Unit vector in r_c direction on column coordinate system
$\mathbf{i}_{r_s}$	Unit vector in r_s direction on spherical coordinate system
$\boldsymbol{i}_v$	Unit vector in the flowing direction
$\mathbf{i}_x$	Unit vector in x-axis of Cartesian coordinate system
$\mathbf{i}_y$	Unit vector in y-axis of Cartesian coordinate system
$\mathbf{i}_z$	Unit vector in z-axis of Cartesian or column coordinate system
$\mathbf{i}_\theta$	Unit vector in θ-axis of spherical coordinate system
$\mathbf{i}_\varphi$	Unit vector in φ-axis of column or spherical coordinate system
k	Free space phase constant (Rad/m)

k_c	Cutoff wave number (Rad/m)
$(k_c)_{mn}$	Cutoff wave number in m, n mode (Rad/m)
m, n	Various modes that can exist in the waveguide
p_d	Electromagnetic power density of the loss in the resistance bar (W/Ω)
q	Point charge (C)
$\boldsymbol{r}$	Radius vector of spatial point (M)
r_C	r_c-coordinate on column coordinate system (M)
$\boldsymbol{r}_P$	Radius vector of the position where charge P locates (M)
r_s	r_s coordinate on spherical coordinate system (M)
t	Time (s)
$\boldsymbol{\upsilon}$	Velocity (m/s)
v	Transient voltage (V)
w	Electromagnetic field energy density (J/m^3)
w_E	Electric field energy density (J/m^3)
w_H	Magnetic field energy density (J/m^3)
x	x-coordinate on Cartesian coordinate (M)
$x(k)$	System input
$x(t)$	System input
y	y-coordinate on Cartesian coordinate (M)
$y(k)$	System output
$y(t)$	System output
z	z-coordinate on Cartesian coordinate (M)
$d\boldsymbol{a}$	Surface element vector on surface S (m^2)
da	Surface element on surface S (m^2)
ds	Line element on curve C (M)
dV	Volume element in volume V (m^3)
α	Attenuation constant (dB/m)
β	Phase shift constant (Rad/m)
γ	Wave propagation constant
δ_e	Skin depth (M)
ε	Dielectric constant (F/m)
$\varepsilon_0.$	Vacuum dielectric constant: $1/36\pi \times 10^{-9}$ (F/m)
η	Surface charge (C/m^2)
η	Wave impedance (Ω)
η_0	Wave impedance in free space (Ω)
η_{TE}	TE wave impedance (Ω)
η_{TEM}	TEM wave impedance (Ω)
η_{TM}	TM wave impedance (Ω)
θ	θ-coordinate on spherical coordinate system (rad)
λ	Line charge (C/m)
λ	Wavelength (M)
λ_c	Cutoff wavelength (M)
$(\lambda_c)_{mn}$	Cutoff wavelength of mode m, n (M)

μ	Magnetic permeability (H/m)
μ_0	Magnetic permeability in vacuum: $4\pi \times 10^{-7}$ (H/m)
ρ	Volume charge (C/m^3)
ρ	Voltage standing wave ratio
σ	Conductivity (s/m)
Γ	Reflection coefficient
Φ	Potential (V)
φ	φ-coordinate on column coordinate or spherical coordinate system (rad)
ω	Angular frequency (rad/m)

Part I
Electromagnetic Compatibility Fundamental Theories

With the rapid development of electronic information technology, electromagnetic compatibility (EMC) involves more and more disciplines, such as electronic science and technology, information and communication engineering, control science, electrical appliances, power electronics, material science and engineering, and mechanical electronics. Especially with the wide application of radio frequency (RF) technology and high-rate digital technology, there are increasing number of EMC problems caused by the coupling channel composed of free space, the distributed parameter effect of metal conductor, the RF parasitic parameter of devices, the transmission line effect of metal apertures, etc. Therefore, in order to understand the basic principles of EMC, we first study the basic theories and principles of EMC including the theory of electromagnetic fields and waves, microwave engineering, and antenna theory.

This part introduces the fundamental theories and methods of EMC, namely electromagnetic fields and waves, microwave engineering, and antenna theory and engineering.

In the part of electromagnetic fields and waves, by introducing the overall physical meaning of Maxwell's equations, we explain that the characteristics of the electronic circuits under direct current (DC) or low frequency are essentially different from the characteristics under radio frequency or microwave. By analyzing the electromagnetic power flow, our readers will understand that the energy can be transmitted through the free space between the voltage source and the load even in the case of DC. By analyzing the reflection of electromagnetic waves, we illustrate that the tangential electric field of the ideal conductor surface is zero, which is called the electric wall. The electric wall does not have to be composed of ideal conductors, air can also be used for shielding instead (the grounded closed conductor shell, which can shield electric fields and electromagnetic fields, is a typical applications of metal electric walls. The high-speed digital connector is a typical application for air electric walls).

In the section of microwave engineering, by learning the transmission line theory, our readers will understand that the characteristics of the single-conductor and double-conductor transmission line involved in the case shielding, and the cross talk problem in the cable layout. We also explain that there is essential difference of electronic circuit characteristics between when the electronic circuit working in DC and when the linear degree of electronic circuit is comparable to the working wavelength.

In the section of antenna theory and engineering, by analyzing the field generated by the alternating electric dipole, we explain to our readers that after the airborne antenna being installed, its radiation characteristics may greatly change, which will further change the functional indicators of the airborne antenna, such as the working distance. Through this section, our readers will also understand that the system-level EMC design not only includes antenna layout design, but also involves the design of RF front-end part, feeder part, and baseband part.

Chapter 1
Electromagnetic Field and Wave

Field is a material form that exists objectively and may permeate the space, with a special law of motion. Field can vary with spatial position and time; i.e., the field parameters can be expressed as a function of space and time.

This chapter discusses the classical electromagnetic field theories [1], which only consider the macroscopic statistical electromagnetic field phenomenon, without consideration of the microscopic electromagnetic field and the quantum effects of the field. Therefore, the infinitesimal mentioned in the book is macroscopic, not mathematical.

This chapter is the basis of the entire book, and our readers need to understand the mathematical physical concepts in this chapter.

1.1 The Physical Meaning of Maxwell's Equations

Through the study of the overall physical meaning of Maxwell's equations, our readers will understand that the characteristics of the electronic circuits and system in DC and low frequency are essentially different than that in radio frequency (RF) and microwave.

To better explain the symbols of Maxwell's equations, we first define the source and field symbols used in this book, including the basic source parameters related to charge and current and the basic field parameters related to fields.

1.1.1 Basic Source Variables

The variables related to charge include point charge q, line charge λ, surface charge η, and volume charge ρ.

© National Defense Industry Press and Springer Nature Singapore Pte Ltd. 2019

D. Su et al., *Theory and Methods of Quantification Design on System-Level Electromagnetic Compatibility*, https://doi.org/10.1007/978-981-13-3690-4_1

(1) Point charge q, the unit is Coulomb (symbol: C): From a macroscopic point of view, if the charge distribution area is very small, it can be considered to be distributed only at one point. This kind of charge distribution is called a point charge distribution, and the electric quantity q is the point charge quantity. In general, the point charge is a function of time, i.e., $q(P) = q(\mathbf{r_P}, t)$, where $\mathbf{r_P}$ is the positional vector of the point where the point charge P is located.
(2) Line charge λ, the unit is Coulomb/meter (symbol: C/m): If the area of the charge distribution is very thin, its cross-sectional area can be considered to be zero from the macroscopic perspective. Such a charge distribution is called a line charge distribution, and the charge is distributed on a curve. For any point P on the curve, if the line element Δs containing point P has a charge amount Δq, when Δs shrinks to zero toward point P, the limit of the ratio of Δq to Δs is the line charge density of point P, i.e., $\lambda(\mathbf{P}) = \lim\limits_{\Delta s \to 0} \frac{\Delta q}{\Delta s}$. In general, the line charge is a function of time and space, i.e., $\lambda = \lambda(x, y, z, t) = \lambda(\mathbf{r}, t)$. For a curve C, the amount of charge on it should be $Q(t) = \int_c \lambda(\mathbf{r}, t)\mathrm{d}s$, where ds is the line element on curve C.
(3) Surface charge η, unit is Coulomb/meter^2 (symbol: C/m^2): If the region of the charge distribution is very thin, from a macroscopic point of view, the thickness is considered to be zero. Such a charge distribution is called a surface charge distribution, and these charges are distributed on a curved surface without volume. For any point P on the surface, if the area element Δa containing the point contains the charge amount Δq, when Δa shrinks toward point P and approaches zero, the limit of the ratio of Δq to Δa is the surface charge density $\eta(P)$ of point P, and $\eta(\mathbf{P}) = \lim\limits_{\Delta a \to 0(P)} \frac{\Delta q}{\Delta A}$. In general, the surface charge density is a function of spatial position and time, i.e., $\eta = \eta(x, y, z, t) = \eta(\mathbf{r}, t)$. For a surface S, the amount of charge on it should be $Q(t) = \int_S \eta(\mathbf{r}, t)\mathrm{d}a$, where d$a$ is the area element of surface S.
(4) Volume charge ρ, the unit is Coulomb/meter^3 (symbol: C/m^3): If the volume element ΔV containing any point P contains the charge amount Δq, when ΔV shrinks toward point P and approaches zero, the limit of the ratio of Δq to ΔV is the volume charge density, i.e., $\rho(P) = \lim\limits_{\Delta V \to 0(P)} \Delta q / \Delta V$. In general, the volume charge density can be a function of spatial position and time, i.e., $\rho = \rho(x, y, z, t) = \rho(\mathbf{r}, t)$, where r is the radius vector of the spatial point. For a known volume V, the amount of charge contained inside is $Q(t) = \int_V \rho(\mathbf{r}, t)\mathrm{d}V$, where d$V$ is the volume element of volume V.

The variables related to current are line current I, surface current K, and volume current J.

(1) Line current I, the unit is Ampere (symbol: A): If the area where the current goes through is very thin, from the macroscopic aspect, the area is considered to be a line with zero cross section. In this case, the current distribution can be considered as a line current with current I. In general, the line current is a function of space and time, i.e., $I = I(x, y, z, t) = I(\boldsymbol{r}, t)$.

(2) Surface current K, the unit is Ampere/meter (symbol: A/m): If the area which the current goes through is very thin, in a macroscopic ideal case, the current can be considered to flow on a curved surface. For a point P on the surface, if the flow direction of the current at point P is $\boldsymbol{i}_v$, the current flowing through the point P and the line element Δl perpendicular to $\boldsymbol{i}_v$ is ΔI, and the thickness of the surface $h \to 0$, then the surface current density at the point P is $\boldsymbol{K}(P) = i_v \lim_{\Delta l \to 0(P)} \Delta I/\Delta l$. In general, the surface current density is a function of spatial position and time, $\boldsymbol{K} = \boldsymbol{K}(x, y, z, t) = \boldsymbol{K}(\boldsymbol{r}, t)$. For a curve C on a surface with a surface current $K(\boldsymbol{r}, t)$, the current flowing through it should be $I(t) = \int_C \boldsymbol{K}(\boldsymbol{r}, t) \cdot i_{ns} ds$, where d$s$ is the line element on C and $\cdot\boldsymbol{i}_{ns}$ is the unit vector in the normal direction of the line element.

(3) Volume current $\boldsymbol{J}$, the unit is Ampere/meter^2 (symbol: A/m^2): For any point P in space, if the unit vector of the current in the flowing direction of point P is $\boldsymbol{i}_v$, and the current intensity flowing through the surface element Δa containing P point and perpendicular to $\boldsymbol{i}_v$ is ΔI, then the volume current density at point P is: $\boldsymbol{J}(P) = \boldsymbol{i}_v \underset{\Delta a \to 0(P)}{\boldsymbol{lim}} \Delta I/\Delta a$. In general, the volume current density is a function of spatial position and time, i.e., $\boldsymbol{J} = \boldsymbol{J}(x, y, z, t) = \boldsymbol{J}(\boldsymbol{r}, t)$. The total current flowing through a curved surface S is $\mathbf{I}(t) = \int_C J(\mathbf{r}, t) \cdot \mathrm{d}a$, where $\boldsymbol{da}$ is a vector surface element on S.

1.1.2 Basic Field Variables

(1) Lorentz force (F), the unit is Newton (symbol: N): Experiments have shown that a point charge q moving at velocity $\boldsymbol{v}$ is subjected to a force in the electromagnetic field in free space. The force can be written as $\boldsymbol{F} = q\boldsymbol{E} + q\boldsymbol{v} \times \mu_0 \boldsymbol{H}$. This formula is called Lorentz force formula, where μ_0 is the permeability of free space. The first part on the right side of the formula is independent of the speed of motion, and the second part is proportional to the speed and perpendicular to it.

(2) Electric field intensity E, the unit is Newton/Coulomb (symbol: N/C) or Volt/meter (symbol: V/m): The electric field intensity is defined by the portion of Lorentz force that is independent of speed, i.e., $\boldsymbol{E} = \boldsymbol{F}|_{v=0}/q$.

(3) Magnetic field intensity H, the unit is Ampere/meter (symbol: A/m): The magnetic field intensity is defined by the velocity-dependent part of Lorentz force formula. Let $\Delta \boldsymbol{F} = \boldsymbol{F} - q\boldsymbol{E} = \boldsymbol{F}|_{E=0}$, then $|\mathbf{H}| = |\Delta \mathbf{F}|/(\mu_0 |q||\mathbf{v}||\sin\alpha|)$, where α is the angle between $\boldsymbol{v}$ and $\boldsymbol{H}$. We can change the direction of the motion of q so that $|\Delta \boldsymbol{F}|$ reaches its maximum value; then, there is $\boldsymbol{H} = \Delta \boldsymbol{F} \times \boldsymbol{v}/(q\mu_0|\boldsymbol{v}|^2)$.

1.1.3 Maxwell's Equations in Free Space

There are five laws of electromagnetic fields in free space:

$$\oint_C \boldsymbol{E} \cdot d\boldsymbol{s} = -\frac{d}{dt}\int_S \mu_0 \boldsymbol{H} \cdot d\boldsymbol{a} \text{ (Faraday's Law of Electromagnetic Induction)} \tag{1.1a}$$

$$\oint_C \boldsymbol{H} \cdot d\boldsymbol{s} = \int_S \boldsymbol{J} \cdot d\boldsymbol{a} + \frac{d}{dt}\int_S \varepsilon_0 \boldsymbol{E} \cdot d\boldsymbol{a} \text{ (Modified Ampere's Circuital Law)} \tag{1.1b}$$

$$\oint_S \varepsilon_0 \boldsymbol{E} \cdot d\boldsymbol{a} = \int_V \rho dV = Q_{net} \text{ (Gauss's Law)} \tag{1.1c}$$

$$\oint_S \mu_0 \boldsymbol{H} \cdot d\boldsymbol{a} = 0 \text{ (Gauss's Law for Magnetism)} \tag{1.1d}$$

$$\oint_S \boldsymbol{J} \cdot d\boldsymbol{a} = -\frac{d}{dt}\int_V \rho dV = -\frac{dQ_{net}}{dt} \text{ (Law of Charge Conservation)} \tag{1.1e}$$

The first four formulas are often collectively referred to as Maxwell's equations. Since all of the five formulas are line, surface or volume integrals of the field quantities $\boldsymbol{E}$ and $\boldsymbol{H}$ and the source variables ρ and $\boldsymbol{J}$, the formulas are called the integral form of the field laws.

1.1.4 Physical Meaning of Maxwell's Equations

1. Physical meaning of Faraday's law of electromagnetic induction
   In free space, the electromotive force along a closed path is equal to the decreasing rate (the negative of the changing rate with time) of the magnetic flux interlinking with the path (the magnetic flux passing through any one of the curved surfaces bounded by the closed path). In other words, a time-varying magnetic field can generate a vortex electric field.
2. Physical meaning of the modified Ampere's circuital law
   In free space, the ring flow of a magnetic field intensity H along a closed curve (sometimes called magnetomotive force) is equal to the sum of the increasing rate of the cross-linking current and the electric flux. In other words, both the current and the time-varying electric field can generate a vortex magnetic field.
3. Physical meaning of the electric field Gauss's law
   In free space, the electrical flux (electric flux density flux) that passes through a closed curved surface is equal to the amount of net charge in the entire volume

enclosed by the curved surface. In other words, the charge is the source of the electric flux density vector.

4. Physical meaning of Gauss's law for magnetism
   In free space, the net magnetic flux that passes through any closed curved surface is zero; that is, there is no source magnetic charge of the magnetic flux density vector.
5. Physical meaning of the law of charge conservation
   For a system of volume V and external surface S, the net charge in the system changes only when there is charge in or out. If the system has no charge exchange with the outside world, that is, the system is a charge-closed system, then the net charge within the system is constant. In other words, the charge can only be transferred in the form of current, but cannot be generated or disappeared by itself.

1.1.5 The Overall Physical Meaning of Maxwell's Equations

The significance of Maxwell's equations in the electromagnetic field theory is the same as the significance of Newton's laws of mechanics in theoretical mechanics. Any real electromagnetic field behavior obeys Maxwell's equations. In the scope of nonrelativity, the behavior of electromagnetic fields must obey Maxwell's equations in integral form. From the form of the equations, Maxwell's equations describe the relationship between the electromagnetic field quantities $\boldsymbol{E}$ and $\boldsymbol{H}$ and their source quantities $\rho, \boldsymbol{J}$, which is illustrated in Fig. 1.1, where the arrow " $\rightarrow$" indicates direct relationship, and "~~ $\rightarrow$ " indicates a time-varying relationship.

Of all the relationships reflected in Maxwell's equations, there are two situations that need further discussion.

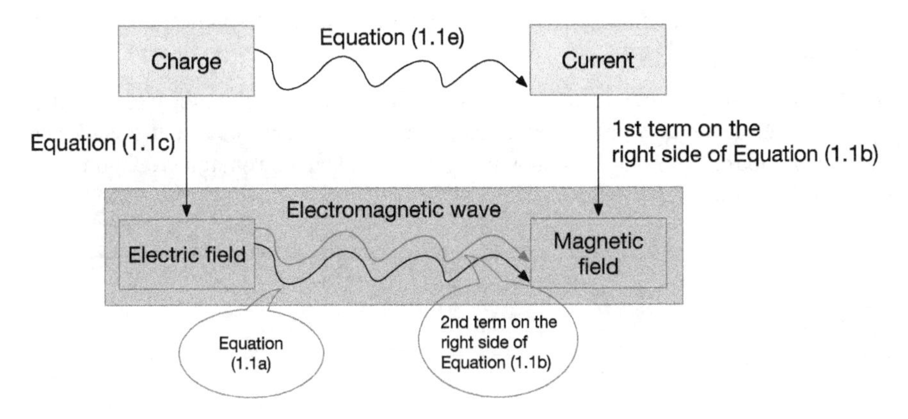

Fig. 1.1 Overall physical meaning of the electromagnetic field law

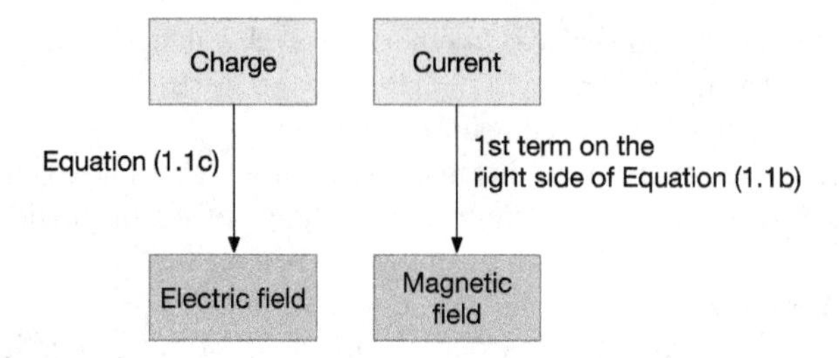

Fig. 1.2 Overall physical meaning of EMF laws when all physical variables are nontime varying

(1) All physical variables are time invariant. At this time, the relationship between the field quantities E and H and their source quantities ρ and J is shown as Fig. 1.2. The electromagnetic field law at this time is:

$$\oint_S \boldsymbol{E} \cdot d\boldsymbol{s} = 0 \tag{1.2a}$$

$$\oint_C \boldsymbol{H} \cdot d\boldsymbol{s} = \int_S \boldsymbol{J} \cdot d\boldsymbol{a} \tag{1.2b}$$

$$\oint_S \varepsilon_0 \boldsymbol{E} \cdot d\boldsymbol{a} = \int_V \rho dV \tag{1.2c}$$

$$\oint_S \mu_0 \boldsymbol{H} \cdot d\boldsymbol{a} = 0 \tag{1.2d}$$

$$\oint_S \boldsymbol{J} \cdot d\boldsymbol{a} = 0 \tag{1.2e}$$

...

In this case, there is no mutual coupling between $\boldsymbol{E}$ and $\boldsymbol{H}$. Only two sides of "→" can be retained in Fig. 1.1. This situation is a static (or nontime-varying) electromagnetic field issue.

2. Source quantities are zero, i.e., $\rho = 0$, $J = 0$. In this situation, the expression of the relationship between the electromagnetic field quantities E and H and their source quantities ρ and J is shown in Fig. 1.3. The electromagnetic field law in this situation can be written as

$$\oint_C \boldsymbol{E} \cdot d\boldsymbol{s} = -\frac{d}{dt} \int_S \mu_0 \boldsymbol{H} \cdot d\boldsymbol{a} \tag{1.3a}$$

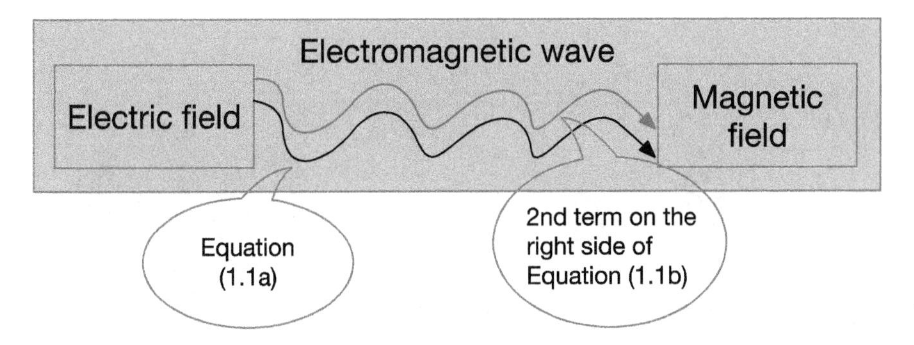

Fig. 1.3 Overall physical meaning of EMF law when the source quantities are zero

$$\oint_C \boldsymbol{H} \cdot d\boldsymbol{s} = \frac{d}{dt} \int_S \varepsilon_0 \boldsymbol{E} \cdot d\boldsymbol{a} \tag{1.3b}$$

$$\oint_S \varepsilon_0 \boldsymbol{E} \cdot d\boldsymbol{a} = 0 \tag{1.3c}$$

$$\oint_S \mu_0 \boldsymbol{H} \bullet d\boldsymbol{a} = 0 \tag{1.3d}$$

From Fig. 1.3, we see that in the region without charge and current, the time-varying electromagnetic field can still exist through mutual coupling, and this form of existence is called the electromagnetic wave. Free space is a typical medium for electromagnetic wave propagation.

Now, we further explain $\oint_C \boldsymbol{H} \cdot d\,\boldsymbol{s} = \frac{d}{dt} \int_S \varepsilon_0 \boldsymbol{E} \cdot d\boldsymbol{a}$. After transformation, the formula can be rewritten as $\oint_C \boldsymbol{H} \cdot d\,\boldsymbol{s} = \frac{\mathrm{d}}{\mathrm{d}t} \int_S \varepsilon_0 \boldsymbol{E} \cdot d\,\boldsymbol{a} = \int_S \frac{\partial}{\partial t}(\varepsilon_0 \boldsymbol{E}) \cdot d\,\boldsymbol{a}$. $\frac{\partial}{\partial t}\varepsilon_0 \boldsymbol{E}$ and the current $\boldsymbol{J}$ satisfies the same equation in generating magnetic field in form. $\frac{\partial}{\partial t}(\varepsilon_0 \boldsymbol{E})$ is added to electromagnetic field law by Maxwell for the mathematical integrity. This term is called the displacement current term. After adding this item to Ampere's law and combined with Faraday's law of electromagnetic induction, the existence of electromagnetic waves is theoretically proved. At first, people only thought that this was a mathematical treatment because there was no experimental evidence of the existence of electromagnetic waves. It was not until 1888, nine years after Maxwell's death, that Hertz's electromagnetic experiments proved the genius prophecy of Maxwell.

The purpose of revisiting the overall physical meaning of Maxwell's equations is to provide our reader systematical explanation that the characteristics of the electronic circuits in DC (frequency = 0 Hz) or low frequency (frequency < 100 kHz) and the characteristics of the electronic circuits in RF (frequency > 1 MHz) or microwave (frequency > 1 GHz) are essentially different.

Now, we explain the above with examples.

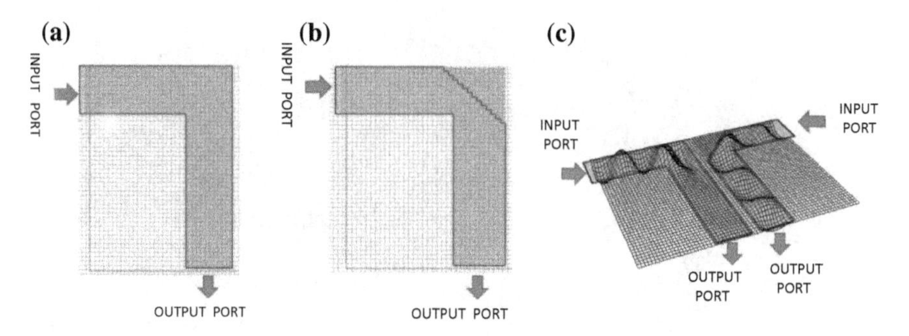

Fig. 1.4 Different transmission characteristics of the microwave transmission line under DC and time-varying conditions

Example 1.1 Connect a metal point to the "ground" with a metal wire; then, the potential of the metal point is the same as the "ground" potential. However, it should be noted that this conclusion can only be strictly established when the entire system is time invariant. In other words, when the frequency is high enough that the linearity (maximum length) of the metal wire can be compared with the wavelength, the phases of the entire metal wire will be unequal since the electrical size (ratio of the geometric size to the wavelength) of the metal wire is large. Therefore, the metal point is no longer equipotential to the "ground" potential.

Example 1.2 Figure 1.4 illustrates a microwave transmission line. The difference between Fig. 1.4a and b is their corners: Fig. 1.4a has a right angle and b has a chamfer angle.

There are input signals at the input port respectively. If the input is a DC signal, there is no difference in the output signal obtained at the output port. However, when the input signal is high frequency or even microwave signal, the situation changes qualitatively. For an easier comparison, the signal transmissions of Fig. 1.4a, b are put together as shown in Fig. 1.4c. It can be seen from this figure that when the corner is a right angle, the output port has no output signal, and a standing wave is formed on the input arm; in case of a chamfer at the corner, there is output signal at the output port.

1.2 Electromagnetic Power Flux

By learning the concept of microwave power flux, our readers can understand that: 1. Even in the case of DC, the energy transmitted by the voltage source to the electronic load can be transmitted through the free space; 2. the properties of devices like capacitors, inductors, and resisters are depend on its energy storage and energy consumption characteristics.

Electromagnetic field is a special form of material existence. Although the electromagnetic field has no static mass, it has the performance of energy and force. For example, solar energy is a kind of electromagnetic energy; the electrostatic force, the magnet's attraction to the ferromagnetic field, and the magnetic force generated around the current indicate that the electromagnetic field has force.

In this section, we will first study the possible transmission channels of electromagnetic energy from the perspective of electromagnetic power influx to further explain the coupling channels in EMC research. Then, the relationship between the properties of the resistors, inductors, capacitors, and their internal electromagnetic fields is studied from the perspective of electromagnetic fields to reveal the nature of the components: Resistors are energy-consuming components, and inductors and capacitors are energy storage components. This research is especially important when a system cannot be described by lumped variables like resistance, inductance, and capacitance.

1.2.1 *The Transmission of Electromagnetic Power Flux*

In this section, we explain that the electromagnetic fields are capable to carry energy, and wherever there is a field, there is electromagnetic energy.

From the electromagnetic field theory, we see that the electric field energy density for the electrostatic field problem is:

$$w_E = 1/2\varepsilon_0 \boldsymbol{E}(r) \cdot \boldsymbol{E}(r) \quad (\boldsymbol{J}/\boldsymbol{m}^3) \tag{1.4}$$

The total electric field energy distributed in space is:

$$W_E = \frac{1}{2}\int_V \varepsilon_0 \boldsymbol{E}(r) \cdot \boldsymbol{E}(r)\boldsymbol{d}V \quad (\boldsymbol{J}) \tag{1.5}$$

The integration area covers the whole space.

For a constant magnetic field problem, the magnetic field energy density is:

$$w_H(r) = \frac{1}{2}\mu_0 \boldsymbol{H}(r) \cdot \boldsymbol{H}(r)(\boldsymbol{J}/\boldsymbol{m}^3) \tag{1.6}$$

By integrating the full space, we can get the total magnetic field energy distributed in the whole space

$$W_H = \frac{1}{2}\int_V \boldsymbol{H}(r) \cdot \mu_0 \boldsymbol{H}(r)\boldsymbol{d}V \ (J) \tag{1.7}$$

The definition of $w_E(r)$ and $w_H(r)$ are also applicable to the time-varying field, i.e.,

$$w_E(r,t) = \frac{1}{2}\varepsilon_0 \boldsymbol{E}(r,t)\cdot\boldsymbol{E}(r,t)\quad \left(\boldsymbol{J/m^3}\right) \tag{1.8}$$

$$w_H(r,t) = \frac{1}{2}\mu_0 \boldsymbol{H}(r,t)\cdot\boldsymbol{H}(r,t)\quad \left(\boldsymbol{J/m^3}\right) \tag{1.9}$$

The total energy density of electromagnetic field is

$$w(r,t) = w_E(r,t) + w_H(r,t)\quad \left(\boldsymbol{J/m^3}\right) \tag{1.10}$$

By integrating the full space, we can get the total electromagnetic field energy distributed in the whole space

$$W = \frac{1}{2}\int_V (\varepsilon_0 \boldsymbol{E}(r,t)\cdot\boldsymbol{E}(r,t) + \mu_0 \boldsymbol{H}(r,t)\cdot\boldsymbol{H}(r,t))\boldsymbol{d}\,V \tag{1.11}$$

The analysis above shows that for static electric fields, electric field energy exists wherever $\boldsymbol{E} \neq 0$; for a constant magnetic field, the magnetic field energy exists wherever $\boldsymbol{H} \neq 0$; for electromagnetic waves, as long as the electromagnetic field is not equal to zero, electromagnetic energy exists.

The fact that wherever there is electromagnetic field there is electromagnetic energy indicates that electromagnetic energy can be propagated in space. Here, we use the practical application of the flashlight to explain that even in the case of DC, the energy output from the DC voltage source can be transmitted to the resistor in space between the voltage source and the load. The circuit model of the flashlight is consisted of a DC voltage source, a wire and a resistance. To facilitate an accurate, we modified the model in premise of retaining the working principle.

Now, we take Fig. 1.5 as an example to analyze how the energy supplied from the DC voltage source (battery) is transferred to the load (bulb). Here, we use column coordinate system (the variables are $\boldsymbol{r_c}$, φ and z, and the unit vectors are $\hat{i}_{r_c}$, $\hat{i}_\varphi$ and $\hat{i}_z$). Then, we build a analysis model as shown in Fig. 1.5c. There is a linear, uniform, cylindrical resistor rod (with electric conductivity σ,length d, and radius a). Its two ends are connected to two circular parallel plates with conductivity σ equals ∞,and radius $b > a$. At the position of $r_C = b$, the system is excited by a circularly symmetric voltage source and the potential difference between the plates is kept to be a constant V_0.

This is a system unrelated to φ. The electric field between the plates is a uniform field, i.e.,

$$\mathbf{E} = -\mathbf{i}_z \frac{V_0}{d}\quad (\mathrm{V/m}) \tag{1.12}$$

Under the excitation of an electric field, a current is generated in the resistance bar, and the current density is

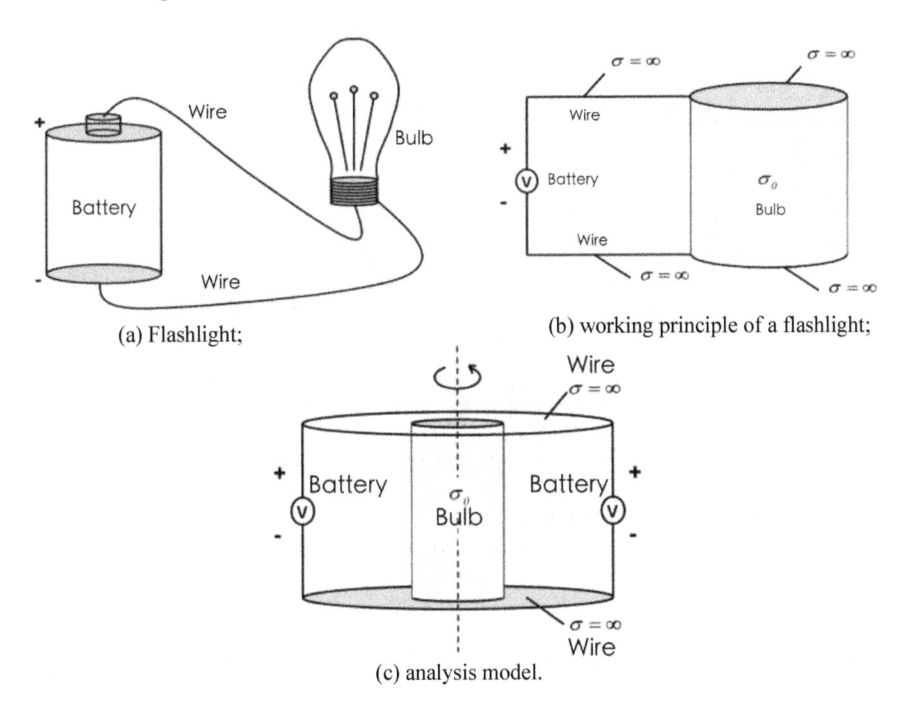

Fig. 1.5 Example of power flux analysis

$$\mathbf{J}(r) = \sigma \mathbf{E}(r) = -\mathbf{i}_z V_0 \sigma / d \quad (A/m^2)$$

The total current in the rod is $I_0 = \int_S \mathbf{J}(r) \cdot d\mathbf{a} = \sigma V_0 \pi a^2 / d \quad (A)$.
Using Ampere's loop law, the magnetic field of the system can be determined as:

$$\boldsymbol{H} = \begin{cases} -\boldsymbol{i}_\varphi \frac{I_0 r_C}{2\pi a^2} & (0 \leq r_C < a, 0 < z < d) \\ -\boldsymbol{i}_\varphi \frac{I_0}{2\pi r_C} & (a < r_C < b, 0 < z < d) \ (\boldsymbol{A/m}) \\ 0 & (r_C > b, 0 < z < d) \end{cases} \tag{1.13}$$

Poynting vector is $\mathbf{S}(r) = \mathbf{E}(r) \times \mathbf{H}(r)$. It is not difficult to see that in the area of $0 \leq r_C < a$, there is

$$\mathbf{S}(r) = \mathbf{E}(r) \times \mathbf{H}(r) = \left(-\mathbf{i}_z \frac{V_0}{d}\right) \times \left(-\mathbf{i}_\varphi \frac{I_0 r_C}{2\pi a^2}\right) = -\mathbf{i}_{r_C} \frac{V_0 I_0 r_C}{2\pi d a^2} \quad (\mathrm{W/m^2}) \tag{1.14}$$

In the area of $a < r_C < b$, there is

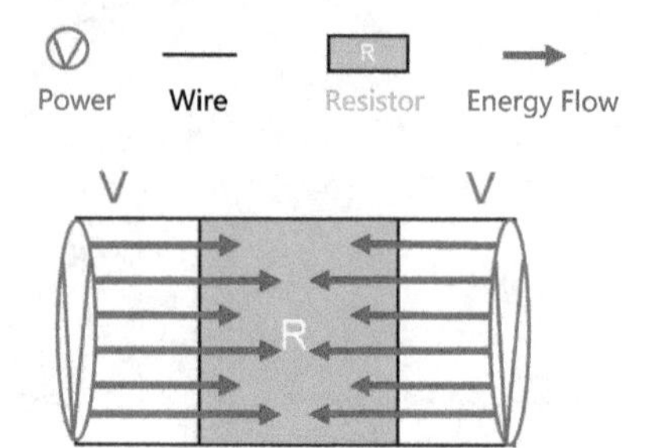

Fig. 1.6 Distribution of Poynting's vector $\mathbf{S}(r)$

$$\mathbf{S}(r) = \mathbf{E}(r) \times \mathbf{H}(r) = \left(-\mathbf{i}_z \frac{V_0}{d}\right) \times \left(-\mathbf{i}_\varphi \frac{I_0}{2\pi\, r_C}\right) = -\mathbf{i}_{r_C} \frac{V_0 I_0}{2\pi\, d\, r_C} \quad (\mathrm{W/m^2}) \tag{1.15}$$

Outside the system, i.e., in the area $r_C > b$, since $\mathbf{H}(r) = 0$, there is $S(r) = 0$.

Using Poynting's theorem, the distribution of electromagnetic power flux density can be obtained as:

$$\nabla \cdot \mathbf{S}(r) = \frac{1}{r_C}\frac{\partial}{\partial r_C}\left(r_C S_{r_C}\right) = \begin{cases} -\frac{V_0\, I_0}{\pi\, a^2 d} = -\sigma \frac{V_0^2}{d^2} & (0 \le r_C < a) \\ 0 & (r_C > a) \end{cases} \quad (\mathrm{W/m^3}) \tag{1.16}$$

It can be seen from formula (1.16), in the air domain of $a < r_C < b$, $\mathbf{S}(r)$ has no dispersion; that is, the field lines are continuous in this area. It indicates that only the electromagnetic power is transmitted in the air domain, and there is no loss of electromagnetic power.

However, inside the resistor rod, the divergence of $\mathbf{S}(r)$ is negative; thus, the field line should be terminated within the resistance bar, so the Poynting vector is absorbed inside the resistance bar, which indicates that there is a loss of electromagnetic power. It can be calculated that the electromagnetic power density of the loss in the resistance bar is

$$p_d = \sigma |\boldsymbol{E}(r)|^2 = \sigma V_0^2 / d^2 \quad (W/m^3)$$

Figure 1.6 is a schematic diagram of the distribution of the vector $\mathbf{S}(r)$. It can be clearly seen that the energy flux is transmitted from the DC voltage source to the loads through the space in between.

As it can be seen from Fig. 1.6, in an ideal conductor plate, there is no electromagnetic power flux and the flux only distributes in the air. It shows that the power applied to the resistor by the power supply does not flux through the conductor plate to the resistor, but is applied to the resistor through the air field around the resistor rod. This is different from the concept in circuit theory. According to the electromagnetic field theory, electromagnetic power is transmitted through space, and the conductor only serves to guide the electromagnetic field (i.e., boundary conditions).

1.2.2 Capacitors—Electrical Energy Storage

In circuit theory, when the power supply charges the capacitor to V_0, the energy provided is:

$$W = W_e = 1/2CV_0^2 \quad (J)$$

where C is the capacitance of the capacitor.

In the electromagnetic field theory, the electrical energy stored between the parallel plate capacitors is related not only with parallel plate capacitor plate area S and board spacing d, but also with the dielectric ε between the two plates, i.e.,

$$W_e = V_0^2 \varepsilon\, S/2d \quad \text{(J)}$$

From the above analysis, the capacitance of the parallel plate capacitor is

$$C = 2\, W_e/V_0^2 = \varepsilon\, S/d \quad \text{(F)}$$

It is consistent with the capacitance of the parallel plate capacitor in circuit theory.

This example shows that the power stored in the system can be used to find its capacitance. At the same voltage, the greater the power stored in the system, the greater the capacitance. If there is no electrical energy storage, the capacitance is zero. Thus, the capacitance of the system is a representation of the ability of the system to store electrical energy.

1.2.3 Inductor—Magnetic Energy Storage

In the circuit theory, the energy supplied by a current source to a single-turn coil is

$$W = W_m = 1/2L\, I_0^2 \quad \text{(J)}$$

where L is the amount of inductance.

In the electromagnetic field theory, the magnetic energy stored in single-turn coil is related to the circumference of the coil $l \times d$, width of the coil w, and the medium filled in the coil.

$$W_m = 1/2(\mu\, l\, d/w)I_0^2 \quad \text{(J)}$$

where μ is the magnetic conductivity.

From the analysis above, we can get the inductance of this coil as

$$L = 2W_m/I_0^2 = \mu\, l\, d/w \quad (\boldsymbol{H})$$

Table 1.1 Recommended maximum linear dimension of circuits and systems using DC or LF methods

f/kHz	10 GHz	1 GHz	100 MHz	10 MHz	1 MHz	100 kHz
λ/cm	3 cm	30 cm	3 m	30 m	300 m	3000 m
Maximum linear dimension (the electrical size is smaller than λ/20)	1.5 mm	1.5 cm	15 cm	1.5 m	15 m	150 m

which is consistent with the formula for the inductor inductance of a single-turn coil in circuit theory.

It can be seen from this example that the inductance of the system can be calculated from magnetic energy, which is a representation of the ability of the system to store magnetic energy.

1.2.4 Examples of Device Properties Analysis

Example 1.3 There is a capacitor with 10 nH self-inductance, and the capacitance measured at 2 MHz is 10 pF. What is the capacitance of the capacitor at 450 MHz?

Solution

$$
\begin{aligned}
f &= 0.45\,\text{GHz} \\
X_L &= 2\pi f_{GHz} L_{nH} = 28.3\,\Omega \\
X_C &= 1/(2\pi f_{Hz} C_F) = 159/(f_{GHz} C_{pF}) = 35.2\,\Omega \\
jX_T &= jX_L - jX_C = -j7\,\Omega = -j159/(f_{GHz} C_{pF}) \\
C'_{pF} &= -159/(f_{GHz} X_T) = 50\,\text{pF}
\end{aligned}
$$

where X_T is the total reactance.

When the frequency is less than the resonant frequency, the capacitor plays the major role. The closer to the resonant frequency, the more obvious the effect of self-inductance. At the resonant frequency, the capacitor is no longer a capacitor. When the frequency is greater than the resonant frequency, the capacitor is actually equivalent to an inductor.

Table 1.1 lists the wavelength corresponding to the frequency from 100 kHz to 10 GHz and the maximum linear dimension of circuits and systems that will not cause intolerable error when using DC or low frequency to analyze the characteristics of the circuit or system.

It can be seen from the above analysis and discussion that in an energy storage component, if the energy is stored in the form of electrical energy, the component will be capacitive to the outside; if the energy is stored in the form of magnetic energy, the component will exhibit inductivity to the outside. When the energy is stored as both electrical energy and magnetic energy, then when $W_e > W_m$, it is externally capacitive; when $W_e < W_m$, it is externally inductive. Usually, in an alternating field, the properties of the components are not single. Therefore, the influx of so-called lead inductance and distributed capacitance is often considered in circuit theory. Essentially, the nature of the component depends on the form in which it is stored.

1.3 The Reflection of Electromagnetic Wave

The electromagnetic shielding used in EMC usually utilizes the concept of electric wall in principle. The surface where the tangential component of electric field is zero is called the electric wall, which can shield the electric field.

1.3.1 Boundary Conditions of the Electromagnetic Field on the Ideal Conductor Surface

The boundary refers to the discontinuous interface in the electromagnetic field. If the electromagnetic properties of the two regions are different, there should be a change in the electromagnetic field in the crossed interface.

The boundary condition is the relationship of electromagnetic fields between the two sides of the interface in which such electromagnetic properties are discontinuous.

According to Faraday's law of electromagnetic induction from Maxwell's equations

$$\oint_C \boldsymbol{E} \cdot \mathrm{d}s = -\frac{\mathrm{d}}{\mathrm{d}t}\int_S \mu_0 \boldsymbol{H} \cdot \mathrm{d}\boldsymbol{a}$$

and the modified Ampere's circuital law

$$\oint_C \boldsymbol{H} \cdot \mathrm{d}s = \int_S \boldsymbol{J} \cdot \mathrm{d}\boldsymbol{a} + \frac{\mathrm{d}}{\mathrm{d}t}\int_S \varepsilon_0 \boldsymbol{E} \cdot \mathrm{d}\boldsymbol{a}$$

the tangential boundary conditions of the electric and magnetic fields can be concluded as

$$\boldsymbol{i}_n \times (\boldsymbol{E}_1 - \boldsymbol{E}_2) = 0, \boldsymbol{i}_n \times (\boldsymbol{H}_1 - \boldsymbol{H}_2) = \boldsymbol{K}$$

where $\boldsymbol{i}_n$ is the normal direction of the boundary; $\mathbf{E}_1$ and $\mathbf{E}_2$ represent the electric field intensity in area 1 and area 2; $\mathbf{H}_1$ and $\mathbf{H}_2$ represent the magnetic field intensity in region 1 and region 2, respectively.

It can be seen from $\boldsymbol{i}_n \times (\boldsymbol{E}_1 - \boldsymbol{E}_2) = 0$ that the tangential component of the electric field is continuous within the boundary. From $\boldsymbol{i}_n \times (\boldsymbol{H}_1 - \boldsymbol{H}_2) = \boldsymbol{K}$, we see that the boundary component of the magnetic field intensity on both sides of the tangential direction at the boundary is discontinuous when current exits on the boundary surface, and the mold of the difference is equal to the mold of the surface current density.

According to Gauss's law of electric field from Maxwell's equations

$$\oint_S \varepsilon_0 \boldsymbol{E} \cdot \mathrm{d}\boldsymbol{a} = \int_V \rho \mathrm{d}V = Q_{net}$$

and Gauss's law for magnetism

$$\oint_S \mu_0 \boldsymbol{H} \cdot \mathrm{d}\boldsymbol{a} = 0$$

the normal boundary conditions of electric and magnetic fields will be known as

$$\boldsymbol{i}_n \cdot (\boldsymbol{\varepsilon}_0 \boldsymbol{E}_1 - \boldsymbol{\varepsilon}_0 \boldsymbol{E}_2) = \eta, \boldsymbol{i}_n \cdot (\mu_0 \boldsymbol{H}_1 - \mu_0 \boldsymbol{H}_2) = 0$$

where $\boldsymbol{i}_n \cdot (\varepsilon_0 \boldsymbol{E}_1 - \varepsilon_0 \boldsymbol{E}_2) = \eta$ shows that the normal component of the electric flux vector on both sides of the boundary is discontinuous when there is a surface charge. $\boldsymbol{i}_n \cdot (\mu_0 \boldsymbol{H}_1 - \mu_0 \boldsymbol{H}_2) = 0$ shows that the normal component of the flux vector on both sides of the boundary is continuous.

We will explain the above principles with three examples: (1) A closed metal shell has a shielding effect from the electric field generated by a charged system with zero charge (Example 1.4); (2) a closed metal shell does not have shielding effect when the net charge is not equal to zero (Example 1.5); (3) a closed and grounded metal shell shields the electric field generated by a charged system whether the charge is equal to zero (Example 1.6).

Example 1.4 There is a hollow conductor spherical shell with inner radius a and outer radius b. An electric dipole is placed at the center of the sphere. What is the spatial potential distribution?

Solution The problem can be solved with the electrostatic field separation variable method.

First, divide the space into three areas: area I($0 < r_S < a$), area II($a < r_S < b$), and area III($r_S > b$).

The potentials in the three regions are $\Phi_1(r), \Phi_2(r), \Phi_3(r)$ respectively, which all satisfy the Laplace equation. The four boundary conditions are

$$r_S = a, \Phi_1(r) = \Phi_2(r) = C \text{ (Constant)}$$

$$r_S = b, \Phi_2(r) = \Phi_3(r) = C \text{ (Constant)}$$

$$r_S \to \infty, \Phi_3(r) \to 0$$

$$r_S \to 0, \Phi_1(r) \to \frac{p}{4\pi\varepsilon_0 r_S^2}\cos\theta$$

Then, the solved spatial distribution of the potential is [1]

$$\Phi(r) = \begin{cases} \frac{p}{4\pi\varepsilon_0}\left(-\frac{r_S}{a^3} + \frac{1}{r_S^2}\right)\cos\theta & (0 < r_S < a) \\ 0 & (a < r_S < b)\ (V) \\ 0 & (r_S > b) \end{cases} \tag{1.17}$$

It is easy to see that the potential outside the metal spherical shell is zero, which indicates that the metal spherical shell has shielded the electric field generated by the electric dipole.

Example 1.5 There is a hollow metal spherical shell with inner radius a and outer radius b, with a point charge placed at the center of the sphere. Calculate the spatial potential distribution.

Solution First, divide the space into three areas: area $I(0 < r_S < a)$, area $\mathrm{II}(a < r_S < b)$, and area $\mathrm{III}(r_S > b)$. The potentials in these three areas are $\Phi_1(r), \Phi_2(r), \Phi_3(r)$, respectively, all of which satisfy the Laplace equation. The four boundary conditions are

$$r_S = a, \Phi_1(r) = \Phi_2(r) = C\ (\text{Constant})$$

$$r_S = b, \Phi_2(r) = \Phi_3(r) = C\ (\text{Constant})$$

$$r_S \to \infty, \Phi_3(r) \to 0$$

$$r_S \to 0, \Phi_1(r) \to \frac{q}{4\pi\varepsilon_0 r_S}$$

Then, the solved spatial distribution of the potential is

$$\Phi(r) = \begin{cases} \frac{q}{4\pi\varepsilon_0}\left(\frac{a-b}{ab} + \frac{1}{r_S}\right) & (0 < r_S < a) \\ \frac{q}{4\pi\varepsilon_0 b} & (a < r_S < b)\ (V) \\ \frac{q}{4\pi\varepsilon_0 r_S} & (r_S > b) \end{cases} \tag{1.18}$$

It can be seen that there is still a field outside the spherical shell of the conductor in this case. That is to say, the spherical shell cannot shield the field of the point charge.

Example 1.6 There is a grounded hollow metal sphere with an inner radius a and an outer radius b, with a point charge placed at the center of the sphere. Calculate the spatial potential distribution.

Solution This problem can be solved with the electrostatic field separation variable method. First, divide the space into three areas: area $\boldsymbol{I}(0 < r_S < a)$, area $\text{II}(a < r_S < b)$, and area $\text{III}(r_S > b)$. The potentials in the three areas are $\Phi_1(r)$, $\Phi_2(r)$, $\Phi_3(r)$, respectively, all of which satisfy the Laplace equation. The four boundary conditions are

$$r_S = a, \Phi_1(r) = \Phi_2(r) = 0\,(\text{Constant})$$

$$r_S = b, \Phi_2(r) = \Phi_3(r) = 0\,(\text{Constant})$$

$$r_S \to \infty, \Phi_3(r) \to 0$$

$$r_S \to 0, \Phi_1(r) \to \frac{q}{4\pi\,\varepsilon_0 r_S}$$

Then, the solved spatial distribution of the potential is [1]

$$\Phi(r) = \begin{cases} \frac{q}{4\pi\varepsilon_0}\left(\frac{a-b}{ab} + \frac{1}{r_S}\right) & (0 < r_S < a) \\ 0 & (a < r_S < b)\ (\boldsymbol{V}) \\ 0 & (r_S > b) \end{cases} \tag{1.19}$$

It can be seen that there is no field outside the spherical shell of the conductor in this case. That is to say, the grounded metal spherical shell can shield the field of the point charge.

1.3.2 Air Electric Wall

In this section, we will introduce the reflection characteristics of the semi-infinite ideal metal plane electromagnetic wave and explain that the electric wall is not necessarily composed of an ideal metal conductor, but "air" can also be used for shielding.

An ideal metal conductor is defined as a metal conductor with an infinite conductivity; a semi-infinite ideal metal plane, an ideal metal plane, is infinite.

The reflection of a semi-infinite ideal metal plane electromagnetic wave refers to an ideal conductor plane in which electromagnetic waves are obliquely incident from a free space to a semi-infinite. There is no electromagnetic field in the ideal metal conductor, so it is only necessary to study the relationship between the reflected wave and the incident wave.

Suppose that the incident wave enters the interface between free space and a semi-infinite ideal metal plane at an angle of $\theta_i = \theta \neq 0°$. We take the vertical polarization as an example for discussion. In this case, the incident wave is a linearly polarized wave that electric field is perpendicular to the incident surface, as shown in Fig. 1.7. Take $z = 0$ as the interface and XOZ as the incident surface, because it is vertically polarized; $\tilde{\mathbf{E}}_i$ should be a complex vector in the y directions. Assume the electric field $\tilde{\mathbf{E}}_r$ of the reflected wave is still in the y direction, according to electromagnetic field boundary conditions and electric fields $\tilde{\mathbf{E}}$, magnetic field $\tilde{\mathbf{H}}$, and glass pavilion vector $\tilde{\mathbf{S}}$ in plane electromagnetic waves are vertical to each other; we obtain that the electromagnetic field of the synthetic wave in free space $z < 0$ is

$$\tilde{\boldsymbol{E}} = \tilde{\boldsymbol{E}}_i + \tilde{\boldsymbol{E}}_r = \boldsymbol{i}_y \dot{\boldsymbol{E}}_y \tag{1.20}$$

$$\tilde{\boldsymbol{H}} = \tilde{\boldsymbol{H}}_i + \tilde{\boldsymbol{H}}_r = \boldsymbol{i}_x \dot{\boldsymbol{H}}_x + \boldsymbol{i}_z \dot{\boldsymbol{H}}_z \tag{1.21}$$

where

$$\dot{\boldsymbol{E}}_y = -j2\dot{\boldsymbol{E}}_{i0} \sin \beta_z z e^{-j\beta_x x} \quad (\text{V/m}) \tag{1.22}$$

$$\dot{\boldsymbol{H}}_x = -j2\dot{\boldsymbol{E}}_{i0} \cos\theta \, \cos \beta_z z e^{-j\beta_x x}/\eta_0 \quad (\text{A/m}) \tag{1.23}$$

$$\dot{\boldsymbol{H}}_z = -j2\dot{\boldsymbol{E}}_{i0} \sin\theta \, \sin \beta_z z e^{-j\beta_x x}/\eta_0 \quad (\text{A/m}) \tag{1.24}$$

From the expression of the synthetic wave, we see that the electric field of the synthetic wave is a linear polarization field in y direction; but the magnetic field is an elliptically polarized field on XOZ plane. In z direction, the synthetic electromagnetic field exhibits the properties of standing waves. The traveling wave in x direction forms a guided wave propagating along the metal surface. Detailed derivation of the synthesis wave and the characteristics of the guided wave can be found from reference [1].

A typical application of the air electric wall technology is the AirMax VS connector, as shown in Fig. 1.8. The connector uses a virtual shield design with air as the high-efficiency dielectric, eliminating the need for staggered shielding, which significantly reduces the weight and price of the AC connector system. The connector achieves a rate of 2.5 GB/s and 6.25 GB/s and can be scaled to 12 GB/s high-speed computing and network system design.

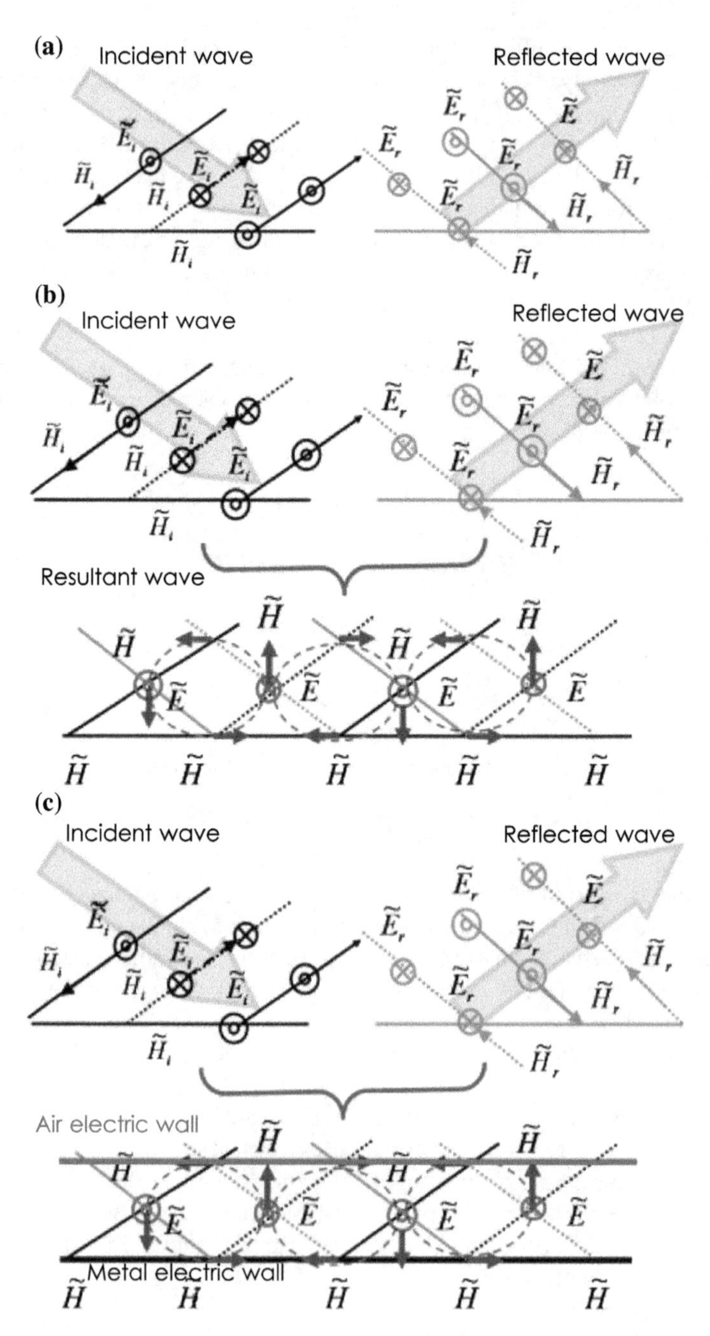

Fig. 1.7 Schematic diagram of the air electric wall: a schematic diagram of incident wave and reflected wave; b synthetic wave diagram; c metal electric wall and air electric wall

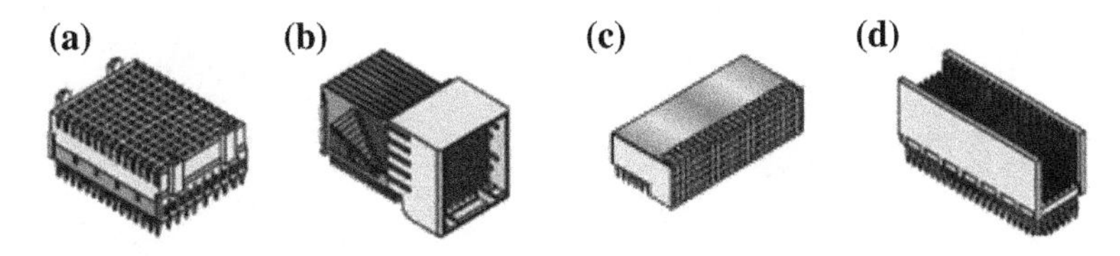

Fig. 1.8 Schematic diagram of the AirMax versus connectors

The formation of an air electric wall is the result of coherent electromagnetic waves in space. Using the coherence characteristics of electromagnetic waves is a useful technique for us to effectively control electromagnetic interference problems. For example, we can use the coherence characteristics of electromagnetic waves to find areas with weak synthetic fields and place susceptive devices and cables in these areas to reduce the mutual interference between devices.

Chapter 2
Microwave Technology

This chapter introduces the transmission line theory in microwave technology [2–4] which is closely related to EMC research. In particular, this chapter covers the concept of single-conductor transmission line and double-conductor transmission line in transmission line theory which is the basis to understand the mechanism and design method of electromagnetic shielding. Through this chapter, our readers will get knowledge of the effect of distributed variables and the cross talk problem in cable layout. We also make clear in this chapter that the characteristics of electronic circuit in DC are fundamentally different from that when the circuit's linearity is comparable to wavelength.

Microwave usually refers to the wave in the frequency band from 300 MHz to 3000 GHz, that is, from decimeter wave to submillimeter wave. Microwaves have special properties that low-frequency radio waves do not have. Since its frequency is several orders of magnitude higher than the low-frequency radio wave, some effects which are not obvious at low frequency are very significant in the microwave band. The most significant effect is that it takes time for electromagnetic waves to travel from one end of the circuit to the other end, because the propagation velocity of the electromagnetic waves is finite (the velocity of electromagnetic waves in free space is the speed of light). This effect is called the delay effect, which makes each point in the circuit exhibit a different phase. The wavelength corresponding to the microwave is very short, so that the size of a general object is much larger than or comparable to the wavelength. Therefore, the general object exhibits a very strong distributed parameter effect, which means that the electromagnetic energy is dispersed over the entire object. This distributed parameter effect is fundamentally different from the lumped parameter effect of traditional low-frequency circuits (the energy is concentrated in the lumped elements). In addition, the skin effect and radiation effect of the high-frequency current in the microwave frequency band are more obvious; the quantum effect also appears in the high frequency range of the microwave frequency band [3].

The basic theory of microwave technology is the classical electromagnetic theory, based on Maxwell's equation [5]. The basic research method is the "field solution" method, which involves the solution of partial differential equations. Since it is often

© National Defense Industry Press and Springer Nature Singapore Pte Ltd. 2019

D. Su et al., *Theory and Methods of Quantification Design on System-Level Electromagnetic Compatibility*, https://doi.org/10.1007/978-981-13-3690-4_2

difficult to directly solve the electromagnetic field, the "microwave equivalent circuit theory," using the thought called "changing field into circuit," has gradually formed, with inspiration from the equivalent circuit method in the low-frequency circuit.

2.1 The Theory of Microwave Transmission Line

In EMC research, "case shielding" seems to be a perennial problem. The effective design of cases for the required shielding performance involves concepts of "single-conductor transmission line" and "double-conductor transmission line" in microwave technology.

In Fig. 2.1, the screen, vents, and slits have single-conductor effect, while the power lines, data lines, buttons, and switches have double-conductor effect.

2.1.1 Overview of Microwave Transmission Line

In the low-frequency case, it is sufficient to transfer energy from the power source to the load with two wires, and there are no special requirements for the form of the wires. However, when the wave frequency is very high, the distance between the two wires is comparable to the wavelength such that the energy is radiated into the space through the wire. In other words, the two wires act as antennas to a certain extent. To avoid the radiation loss, the transmission wires should be packed in a closed form like coaxial lines. As the frequency of the waves further increases, new problems arise: The optimal working state of a coaxial line is single-mode operation, and the outer diameter of the coaxial line is limited by the single-mode working condition

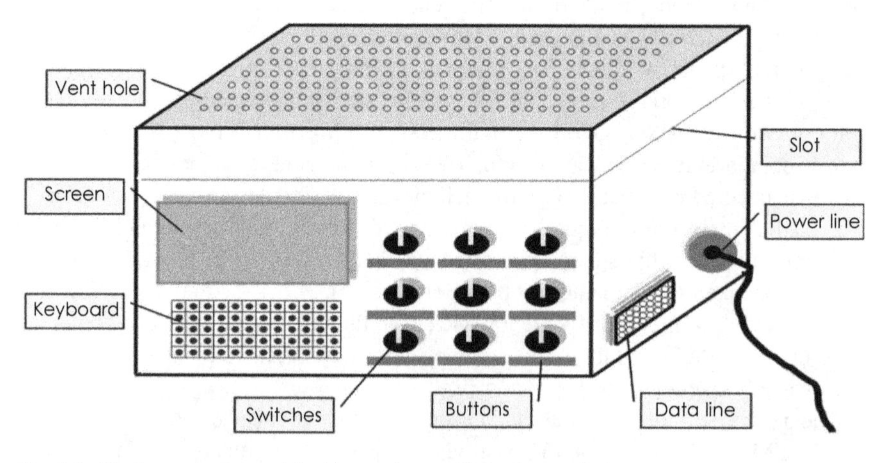

Fig. 2.1 Single-conductor and double-conductor effects in a typical case

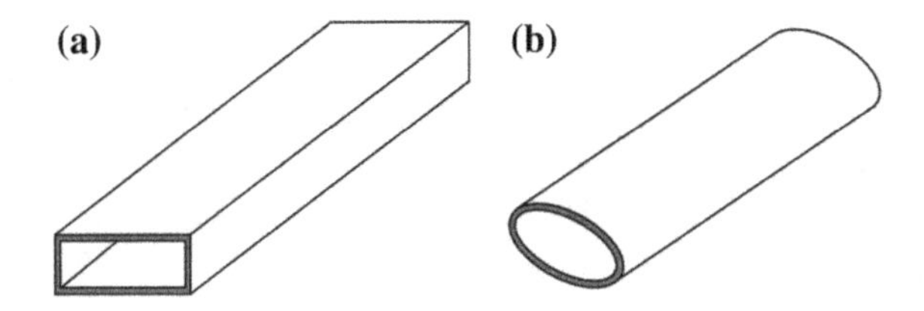

Fig. 2.2 Typical single-conductor transmission line. **a** Rectangle waveguide and **b** circular waveguide

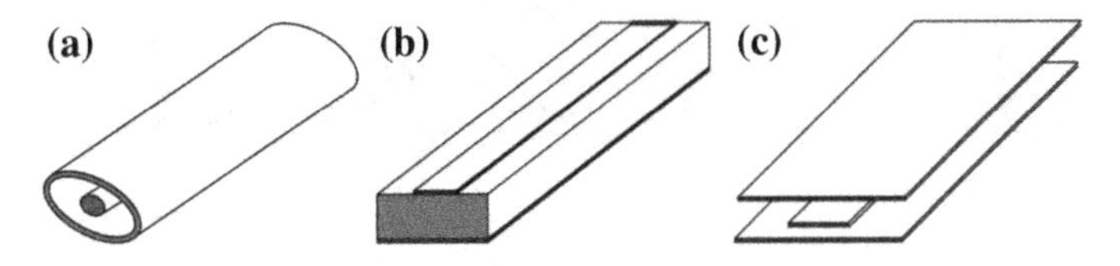

Fig. 2.3 Typical double-/multi-conductor transmission lines. **a** Coaxial line; **b** microstrip line; and **c** stripline

(there is an optimal ratio between the inner and outer diameters of the coaxial line). When the frequency is high, the cross-sectional dimension of the coaxial line must be reduced. As a result, the field strength between the inner and the outer conductors increases under the same voltage and may easily cause breakdown, which limits the transmission power of the coaxial lines. Therefore, the operating frequency and transmission power of the coaxial line are quite limited.

Theoretical research shows that waveguides are characterized by low loss, large power capacity, and high operating frequency, but its working frequency band is much narrower than that of the coaxial lines.

The waveguide transmission line is constituted of a single conductor, so we call it as single-conductor transmission line. The coaxial line is composed of two conductors inside and outside, so it is called a double-conductor transmission line. The commonly used single-conductor microwave transmission line is rectangular waveguide and circular waveguide, as shown in Fig. 2.2. Double-conductor transmission lines include coaxial lines and microstrip lines, as shown in Fig. 2.3a, b. The stripline in Fig. 2.3c is a multi-conductor transmission line.

2.1.2 Transmission State and Cutoff State in the Microwave Transmission Line

Figure 2.4 shows a columnar uniform transform system with a cross section of arbitrary shape. A uniform transmission line means that the cross section of the transmission line system is equal everywhere, and arbitrary shape means that the cross section can be in any shape. It can be seen from Fig. 2.4 that this case may be

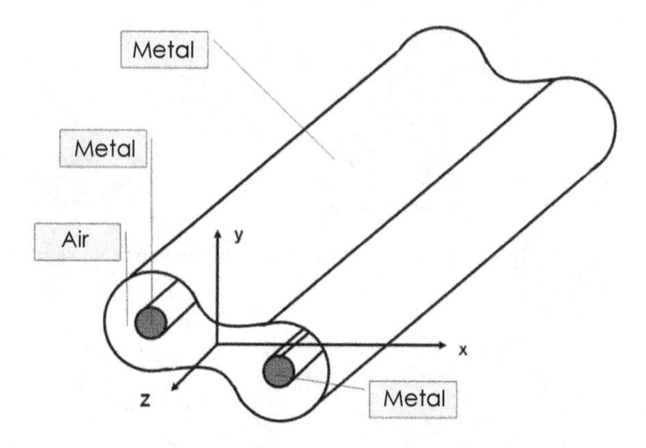

Fig. 2.4 Columnar uniform transmission line system with arbitrary shape

constituted by a plurality of transmission line conductors. For the sake of simplicity, it is assumed that metal under vacuum is an ideal conductor.

We will solve the electromagnetic wave existing in the transmission line system.

The expression for the electromagnetic field of the cross section in the transmission line system can be determined as

$$\begin{aligned} \nabla_{\mathrm{T}}^{2}\mathbf{E}(x,y)+k_{c}^{2}\mathbf{E}(x,y)=0 \\ \nabla_{\mathrm{T}}^{2}\mathbf{H}(x,y)+k_{c}^{2}\mathbf{H}(x,y)=0 \end{aligned} \tag{2.1}$$

where k_c is the cutoff wave number, $k_c^2 = k^2 + \gamma^2$, k is the phase constant in free space, and γ is the propagation constant. The relationship between k, the wavelength λ, and the frequency f is:

$$k = 2\pi/\lambda = 2\pi f/c$$

k_c is an important special parameter, and λ_c corresponding to k_c is called the cutoff wavelength, $\lambda_c = 2\pi/k_c$; the corresponding frequency is defined as cutoff frequency, and $f_c = c/\lambda_c = ck_c/2\pi$.

The propagation constant can be expressed as

$$\gamma=\alpha+\mathrm{j}\beta=\sqrt{k_c^2-k^2}=k\sqrt{\left(\frac{k_c}{k}\right)^2-1}=\frac{2\pi}{\lambda}\sqrt{\left(\frac{\lambda}{\lambda_c}\right)^2-1}=\frac{2\pi f}{c}\sqrt{\left(\frac{f_c}{f}\right)^2-1} \tag{2.2}$$

where α is the attenuation constant and β is the phase constant.

When k_c is a positive real number, there are two distinct cases in the transmission line system shown in Fig. 2.4, namely the transmission state and the off state.

1. Transmission state

When $\lambda < \lambda_c$ or $f > f_c$, $\lambda = \mathrm{j}\beta$ is a pure imaginary number, and the solution of Maxwell's equations in the transmission line system is

$$\mathbf{E}(x, y, z, t) = \mathbf{E}(x, y)e^{-j(\omega t \mp \beta z)} \tag{2.3}$$

This solution indicates that there is a fluctuation process, and $(\omega t \mp \beta z)$ indicates that there are electromagnetic waves transmitted in the +z- and -z-directions in the transmission line. The phase constant is

$$\beta = \frac{2\pi}{\lambda}\sqrt{\left(\frac{\lambda}{\lambda_c}\right)^2 - 1} = \frac{2\pi f}{c}\sqrt{\left(\frac{f_c}{f}\right)^2 - 1} \tag{2.4}$$

We usually take $\lambda < \lambda_c$ or $f > f_c$ as the propagation conditions of the wave.

2. Cutoff state

When $\lambda > \lambda_c$ or $f < f_c$, $\lambda = \alpha$ is a pure real number, and the solution of Maxwell's equation in the transmission line system is

$$\boldsymbol{E}(x, y, z, t) = \boldsymbol{E}(x, y)\boldsymbol{e}^{\mp\alpha z}\boldsymbol{e}^{-j\omega t} \tag{2.5}$$

It shows an in situ vibration process with only an axial attenuation field in the transmission line and no wave propagation. The attenuation constant is

$$\alpha = \frac{2\pi}{\lambda}\sqrt{\left(\frac{\lambda}{\lambda_c}\right)^2 - 1} = \frac{2\pi f}{c}\sqrt{\left(\frac{f_c}{f}\right)^2 - 1} \tag{2.6}$$

We usually take $\lambda > \lambda_c$ or $f < f_c$ as the cutoff conditions of the wave, i.e., the cutoff wavelength and the cutoff frequency.

2.1.3 The Concept of TEM Mode, TE Mode, and TM Mode in Microwave Transmission Line

In Fig. 2.4, z-direction is the longitudinal direction of the microwave transmission line system, i.e., the propagation direction of energy transmission line; x–y-direction is the transverse direction of the microwave transmission line, i.e., perpendicular to the direction of the energy transmission of the transmission line. Since the traveling wave in the microwave transmission line system satisfies Maxwell's equation, in general, the lateral component and the longitudinal component of the distribution function in the transmission line are not independent of each other.

From the expression of the distribution function, the guided waves can generally be divided into the following three categories:

(1) Transverse electric wave (TE wave): The characteristic is that the longitudinal component of the electric field is zero, and the longitudinal component of the magnetic field is not zero, i.e., $\boldsymbol{E}_z = 0$ and $\boldsymbol{H}_z \neq 0$.

(2) Transverse magnetic wave (TM wave): The characteristic is that the longitudinal component of the electric field is not zero, and the longitudinal component of the magnetic field is zero, i.e., $\boldsymbol{E}_z \neq 0$ and $\boldsymbol{H}_z = 0$.
(3) Transverse electromagnetic wave (TEM wave): It is characterized that the longitudinal component of the electric field and the longitudinal component of the magnetic field are both zero, i.e., $\boldsymbol{E}_z = 0$ and $\boldsymbol{H}_z = 0$.

Characteristic difference of TEM wave from the TE wave and the TM wave: When the longitudinal components of the electric field and the magnetic field are both zero, in order to obtain a nonzero solution, there must be $k_c = 0$, which means $f_c = 0$. It shows that in the case of TEM waves, the cutoff frequency is zero, so the TEM electromagnetic waves of all frequencies are in a propagating state.

Since $\boldsymbol{E}_z = 0$ and $\boldsymbol{H}_z = 0$, the distribution function becomes a two-dimensional vector of only x and y.

Note: TEM waves cannot be transmitted by any system. For example, in a hollow waveguide, since the TEM wave does not satisfy the Maxwell equation, there will not be any TEM wave in the hollow waveguide. However, in a double-conductor or multi-conductor transmission line system, TEM waves can exist.

In circuit theory, the ratio of voltage to current has an impedance dimension, which represents the transfer state of energy. In microwave technology, due to the distribution characteristics of the system, we use the ratio of electric field to magnetic field to describe the state of energy transfer, and the ratio also has an impedance dimension.

For TE waves, the wave impedance is

$$\eta_{\mathrm{TE}} = \frac{\omega\mu_0}{\beta} = \left(\sqrt{\frac{\mu_0}{\varepsilon_0}}\right)\left(\frac{1}{\sqrt{1-\left(\frac{\lambda}{\lambda_c}\right)^2}}\right) = 120\,\pi\frac{1}{\sqrt{1-\left(\frac{\lambda}{\lambda_c}\right)^2}}\,(\Omega) \tag{2.7}$$

For TM waves, the wave impedance is

$$\eta_{\mathrm{TM}} = \frac{\beta}{\omega\varepsilon_0} = \sqrt{\left(\frac{\mu_0}{\varepsilon_0}\right)}\sqrt{1-\left(\frac{\lambda}{\lambda_c}\right)^2} = 120\pi\sqrt{1-\left(\frac{\lambda}{\lambda_c}\right)^2}\,(\Omega) \tag{2.8}$$

For TEM waves, the wave impedance is

$$\eta_{\mathrm{TEM}} = \frac{\beta}{\omega\varepsilon_0} = \sqrt{\left(\frac{\mu_0}{\varepsilon_0}\right)} = 120\,\pi\,(\Omega) \tag{2.9}$$

2.1.4 Main Characteristics of the Coaxial Line [4]

The working mode of the coaxial line is TEM wave. In the traveling wave state (electromagnetic wave propagates forward in coaxial line), the average power that can pass through the coaxial line can be expressed as

$$P = Z_c I^2/2 \tag{2.10}$$

In order to avoid the breakdown in the coaxial line, it is required that the maximum power allowed in the traveling wave state in the coaxial line should meet

$$P\text{max} = \frac{b^2}{120}\mathbf{E}_b^2 \ln\left(\frac{D}{d}\right) \tag{2.11}$$

where $\boldsymbol{E}_b$ is the magnitude of the electric field intensity at the inner conductor of the coaxial line at the time of breakdown, b is the radius of the inner conductor, d is the diameter of the inner conductor, and D is the diameter of the outer conductor.

If there is loss from the inner and outer conductors of the coaxial line, the surface resistance of the inner conductor is R_{sd}, and the surface resistance of the outer conductor is R_{sD}, then the attenuation coefficient of the coaxial line can be expressed as

$$\alpha = \frac{1}{120\,\pi \ln \frac{D}{d}}\left(\frac{R_{sD}}{D} + \frac{R_{sd}}{d}\right) \tag{2.12}$$

If the medium is filled between the inner and outer conductors of the coaxial line, the loss of the medium should also be considered. The best working state of the coaxial line is when filled with nondispersion medium and working in TEM wave mode (dispersion medium is a medium in which the propagation speed of electromagnetic waves is a function of frequency. Any medium with loss is dispersion medium.). However, when the operating frequency of the electromagnetic wave is sufficiently high, a high-order mode occurs in the coaxial line. Therefore, in order to ensure that the coaxial line is in a single-mode transmission state, there are certain restrictions toward the inner and outer diameter dimensions and operating frequency of the coaxial line.

The inner and outer dimensions of the coaxial line should meet the requirement

$$\frac{D}{d} < 4 \text{ and } \lambda > 1.73(D + d)$$

Based on the calculation, the size of the coaxial line can be considered from two aspects.

(1) The maximum passing power: In the case of fixed diameter of the outer conductor of the coaxial line, the maximum corresponding power capacity of the coaxial

line is: The inner–outer radius ratio is 1.65, and the characteristic impedance is 30 Ω.

(2) The minimum attenuation coefficient: In the case of fixed diameter of the outer conductor of the coaxial line, the minimum attenuation condition of the coaxial line is: The inner–outer radius ratio is 3.6, and the characteristic impedance is 77 Ω.

At present, the coaxial line with 75 Ω or 50 Ω is usually used according to past experiences, the former has minimum attenuation, and the latter has large passing power and small attenuation coefficient.

2.1.5 Main Characteristics of the Waveguide Transmission Line

The most common waveguide transmission lines are rectangular waveguides and circular waveguides. The boundary conditions of the two are different due to the different geometrical structures. Here, the main characteristics of the waveguide transmission line will be introduced from both a general and the EMC design perspective.

There is no TEM wave in a rectangular waveguide, but there can be TE waves and TM waves. The characteristics of the TE wave and the TM wave in the rectangular waveguide are analyzed below.

As mentioned in Sect. 2.1.2, k_c is an important parameter, and it gives the propagation conditions of different modes.

As shown in Fig. 2.5, in a rectangular waveguide, according to the size of the waveguide (a is the wide side, and b is the narrow side), k_c can be expressed as

$$(k_c)_{mn}^2 = \left(\frac{m\pi}{a}\right)^2 + \left(\frac{n\pi}{b}\right)^2, \quad (m, n = 0, 1, 2, \ldots) \tag{2.13}$$

where m and n represent the various modes that can exist in the waveguide. m and n can be any positive integer or zero (they cannot be zero at the same time; otherwise, a nonzero solution cannot be obtained).

Fig. 2.5 Rectangular waveguide transmission lines

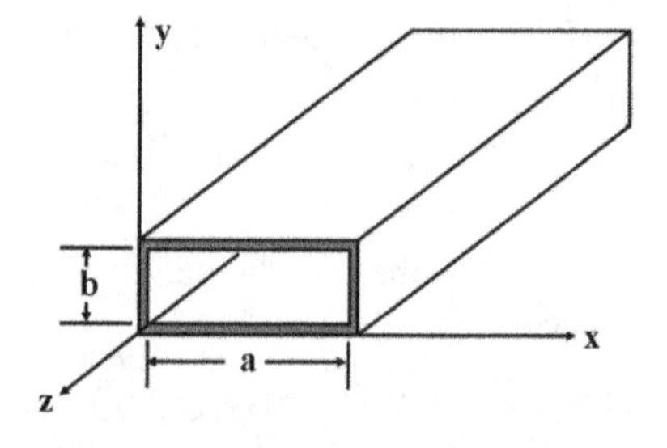

Different m and n correspond to different TE waves and TM waves, denoted as TE_{mn} and TM_{mn}, or we can use $\boldsymbol{H}_{mn}$ instead of TE_{mn} and $\boldsymbol{E}_{mn}$ instead of TM_{mn}. TE_{mn} can be explained as: T means transverse, so TE refers to transverse electric wave. $\boldsymbol{H}_{mn}$ can be explained as: There is a magnetic field (H field) in the direction of propagation. Therefore, the two are only expressed differently.

TE_{mn} and TM_{mn} are also known as the mode (waveform) in the waveguide transmission line, and m and n represent the mode number. For different modes, there are different cutoff wavelengths:

$$(\lambda_c)_{mn} = \frac{2\pi}{(k_c)_{mn}} = \frac{2}{\sqrt{\left(\frac{m}{a}\right)^2 + \left(\frac{n}{b}\right)^2}}, \quad (m, n = 0, 1, 2, \ldots) \tag{2.14}$$

Only the modes that meet the propagation requirement $\lambda < (\lambda_c)_{mn}$ (refer to the transmission state and cutoff state in microwave transmission lines in Sect. 2.1.2) can be propagated in a rectangular waveguide.

For any given a, b, the larger m and n are, the larger the corresponding cutoff frequency f_c is. In all possible modes, the cutoff wavelength λ_c corresponding to the TE_{10} mode is the maximum, and the corresponding cutoff frequency f_c is the minimum. Therefore, TE_{10} is called the lowest mode in a rectangular waveguide (or a fundamental mode), and other modes are called higher-order modes.

When TE_{10} is in the traveling wave state, the theoretical value of the maximum power that can pass is

$$P_{\max} = \frac{ab}{480\pi} E_b^2 \sqrt{1 - \left(\frac{\lambda}{2a}\right)^2} \tag{2.15}$$

Considering some practical factors results in the drop of the maximum power, and the maximum power allowed by waveguide in engineering is 25–30% of the theoretical value.

The attenuation coefficient of TE_{10} is

$$\alpha = \frac{R_s}{120\pi b\sqrt{1 - \left(\frac{\lambda}{2a}\right)^2}} \left[1 + 2 \times \frac{b}{a}\left(\frac{\lambda}{2a}\right)^2\right] \tag{2.16}$$

Obviously, the larger the cross-sectional size of the waveguide, the smaller the attenuation. However, an increase in the cross-sectional size leads to a multi-mode phenomenon. Therefore, the size of the rectangular waveguide needs to be balanced.

For the rectangular waveguide with a single transmission mode TE_{10}, the size selection principles are: (1) Single-mode transmission should be guaranteed; (2) loss and attenuation should be as small as possible; (3) the power capacity should be as large as possible; (4) the dispersion should be as small as possible.

The waveguide of which $b = a/2$ can be taken as the standard waveguide form, which has the characteristics of the maximum passing power with a guaranteed bandwidth.

It is worth pointing out that the above explanations emphasize TE_{10} in the traveling wave state. An important condition for ensuring the traveling wave state is that the electromagnetic wave is far from the electromagnetic discontinuity; that is, the cross-sectional dimension of the waveguide is much smaller than the longitudinal length.

When the longitudinal length is not very long, the high-order mode will not be completely in the off state, which is called the withering mode (having a certain propagation ability, but the amplitude is gradually attenuated). The degree of attenuation of the withering mode is related to the longitudinal length. The longer the longitudinal length, the greater the attenuation of the withering mode. It is a very important concept in the structural design of shielding.

The field solution of a circular waveguide is different from that of a rectangular waveguide, but its basic principle and analysis method are similar. Circular waveguides also have single-mode transmission problems, high-order mode problems, cut-off wavelengths, etc. We will not discuss the details here. Our readers can refer to reference [3] for more explanation.

2.1.6 *The Distributed Parameter Effect of Microwave Transmission Line*

The distributed parameter effect is produced when the length of the microwave transmission line is much longer than that of the electromagnetic wave. It is also called the long-term theory in some contexts. It is important to note that the length of the transmission line does not depend on its geometric length, but depends on its electric length, which is defined as the ratio of geometric length to wavelength. The distributed parameter effect is called "electric length" long when the phase of the electric length is not equal everywhere. For example, a 50 km transmission line is not considered to be very long because it operates at 50 Hz and has a wavelength of 6,000 km. But a 10 cm transmission line that works at 10 GHz is considered to be very long, because the wavelength of 10 GHz is only 3 cm.

This book will use the transmission line theory to analyze the distributed parameter effect problem. Figure 2.6 shows a uniform transmission line and circuit model. In this figure, a finite length is cut off on the coming line (upper line) and the returning line (lower line) and is divided into several differential segments, each of which has its own unit length resistance, self-inductance, and mutual inductance.

Taking two sets of independent equations in integral form which are derived from Maxwell's equations as basic laws, the equation of uniform transmission line with mutual inductance can be derived [6]

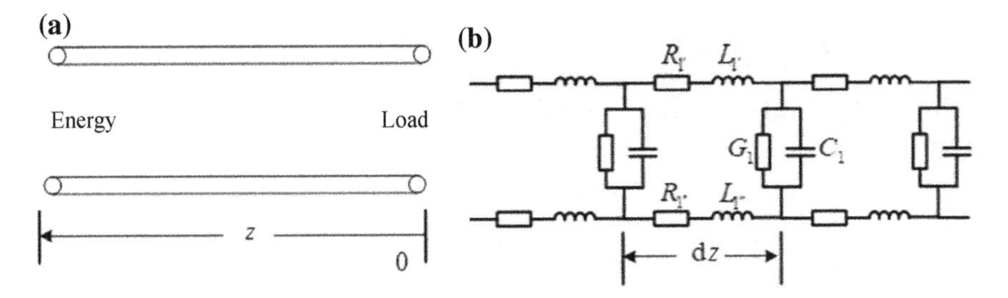

Fig. 2.6 Transmission lines and their equivalent circuits. **a** Transmission lines and **b** equivalent circuit

$$\begin{aligned} -d\dot{V}/dz &= (R_{1'} + R_{1''})\dot{I} + j\omega(L_{1'} + L_{1''} + M_{1'} + M_{1''})\dot{I} \\ -d\dot{I}/dz &= G_1\dot{V} + j\omega C_1\dot{V} \end{aligned} \tag{2.17}$$

where $R_{1'}$, $R_{1''}$, $L_{1'}$, $L_{1''}$, $M_{1'}$, and $M_{1''}$, respectively, represent the resistances, self-inductance coefficients, and mutual inductance coefficients of unit length of the incoming line and the return line.

When the mutual inductance of the transmission line is ignored, and there is $R_1 = R_{1'} + R_{1''}$ and $L_1 = L_{1'} + L_{1''}$, then Eq. (2.17) can be simplified as:

$$\begin{aligned} -d\dot{V}/dz &= Z_1\dot{I} \\ -d\dot{I}/dz &= Y_1\dot{V} \end{aligned} \tag{2.18}$$

where Z_1 is the series impedance of unit length, $Z_1 = R_1 + j\omega L_1$; Y_1 is the parallel admittance of unit length, $Y_1 = G_1 + j\omega C_1$.

The equation of voltage and current on the line is also called the telegraph equation. The above equations show that the voltage change on the transmission line is caused by the reduction of the series impedance. The change in current is caused by the shunt action of the shunt admittance.

By integrating the telegraph Eq. (2.18), we can get

$$d^2\dot{V}/dz^2 = Z_1Y_1\dot{V} = \gamma^2\dot{V} \tag{2.19}$$

The general solution is

$$\dot{V} = \dot{V}_1e^{\gamma z} + \dot{V}_2e^{-\gamma z} \text{ and } \dot{I} = \left(\dot{V}_1e^{\gamma z} - \dot{V}_2e^{-\gamma z}\right)/Z_c \tag{2.20}$$

where $\dot{V}_1$ and $\dot{V}_2$ are the undetermined complex constant, γ is the propagation constant, and Z_c is the characteristic impedance. The formula of γ and Z_c is

$$\gamma = \sqrt{Z_1Y_1} = \sqrt{(R_1 + j\omega L_1)(G_1 + j\omega C_1)}$$

$$Z_c = Z_1 / \gamma = \sqrt{Z_1 / Y_1} = \sqrt{(R_1 + j\omega L_1) / (G_1 + j\omega C_1)} \tag{2.21}$$

The voltage and current of the incident wave are:

$$\dot{V}_+ = \dot{V}_1 e^{\gamma z}, \quad \dot{I}_+ = \dot{V}_1 e^{\gamma z} / Z_c \tag{2.22a}$$

The voltage and current of the reflected wave are:

$$\dot{V}_- = \dot{V}_2 e^{-\gamma z}, \quad \dot{I}_- = -\dot{V}_2 e^{-\gamma z} / Z_c \tag{2.22b}$$

The relationship between the voltage and current of the incident wave and characteristic impedance, and the relationship between voltage and current of the reflected wave and characteristic impedance are as follows

$$Z_c = \dot{V}_+ / \dot{I}_+ = -\dot{V}_- / \dot{I}_- \tag{2.23}$$

However, the ratio of the total voltage to the total current in a transmission line in general is not equal to the characteristic impedance.

The total voltage in the transmission line is

$$\dot{V} = \dot{V}_+ + \dot{V}_- \tag{2.24a}$$

The total current in the transmission line is

$$\dot{I} = \dot{I}_+ + \dot{I}_- = (\dot{V}_+ - \dot{V}_-) / Z_c \tag{2.24b}$$

If the loss of the microwave transmission line is assumed to be small enough, we have $R_1 << \omega L_1, G_1 << \omega C_1$. When the loss of the transmission line can be ignored, the characteristic parameters of the nondestructive transmission line can be obtained.

The propagation constant is

$$\gamma = \mathrm{j}\omega\sqrt{L_1 C_1} = \mathrm{j}\beta \tag{2.25}$$

The phase constant is

$$\beta = \omega\sqrt{L_1 C_1} \tag{2.26}$$

The characteristic impedance is

$$Z_c = \sqrt{L_1 / C_1} \tag{2.27}$$

The general solution of the nondestructive transmission line is:

$$\dot{V} = \dot{V}_1 e^{j\beta z} + \dot{V}_2 e^{-j\beta z}, \quad \dot{I} = \left(\dot{V}_1 e^{j\beta z} - \dot{V}_2 e^{-j\beta z}\right) / Z_c \tag{2.28}$$

Assume the initial phase angle of $\dot{V}_1$ is φ_1, and the initial phase angle of $\dot{V}_2$ is φ_2; the relationship between voltage and current in Fig. 2.6 can be obtained. Since the z-axis in Fig. 2.6 is from the load to the source, the incoming voltage wave $\dot{V}_+(\dot{V}_1)$ propagates in –z-direction with the factor $e^{j\beta z}$; the reflection wave $\dot{V}_-(\dot{V}_2)$ propagates in +z-direction with the factor $e^{-j\beta z}$ (let the time factor be $e^{j\omega t}$).

1. Reflection coefficient
The reflection coefficient is defined as the ratio of the reflected voltage to the incident voltage, that is

$$\Gamma = |\Gamma| e^{j\theta} = \dot{V}_- / \dot{V}_+ = V_2 e^{j(\varphi_2 - \varphi_1 - 2\beta z)} / V_1 \tag{2.29}$$

The reflection coefficient is a function of the reference position. If the reference position is set at z = 0, the reflection coefficient of the terminal load is

$$\Gamma_0 = |\Gamma_0| e^{j\theta_0} = \dot{V}_- / \dot{V}_+ = V_2 e^{j(\varphi_2 - \varphi_1)} / V_1 \tag{2.30}$$

where

$$|\Gamma_0| = V_2 / V_1, \quad \theta_0 = (\varphi_2 - \varphi_1)$$

There is the following relationship between the reflection coefficient Γ at any position z and the reflection coefficient Γ_0 at the reference point

$$\Gamma = \Gamma_0 e^{j-2\beta z} = |\Gamma| e^{j\theta}.$$

Apparently, there is

$$|\Gamma| = |\Gamma_0|, \quad \theta = \theta_0 - 2\beta z \tag{2.31}$$

Equation (2.31) provides the transformation formula of the reflection coefficient when the reference surface moves: When the reference surface moves, the modulus of the reflection coefficient on the nondestructive transmission line remains unchanged, and the phase is linearly related to the moving distance.

2. Input impedance
The ratio of the total voltage to the total current on the reference surface is defined as the input impedance, and its reciprocal is the input admittance.

$$Z_i = 1 / Y_i = \dot{V} / \dot{I} \tag{2.32}$$

Input impedance can also be expressed with reflection coefficient

$$Z_i = \frac{1}{Y_i} = \frac{\dot{V}}{\dot{I}} = Z_c \frac{\dot{V}_+ + \dot{V}_-}{\dot{V}_+ - \dot{V}_-} = Z_c \frac{1 + \Gamma}{1 - \Gamma} \tag{2.33}$$

and then we have

$$\Gamma = \frac{Z_i - Z_c}{Z_i + Z_c} = \frac{Y_c - Y_i}{Y_c + Y_i} \tag{2.34}$$

3. Terminal equation

Using the terminal condition at z = 0, constants $\dot{V}_1$ and $\dot{V}_2$ can be determined.

Since z = 0, we can get:

$$\dot{V}_0 = \left(\dot{V}_1 e^{j\beta z} + \dot{V}_2 e^{-j\beta z}\right)\Big|_{z=0} = \dot{V}_1 + \dot{V}_2 \tag{2.35a}$$

$$\dot{I}_0 = \left(\left(\dot{V}_1 e^{j\beta z} - \dot{V}_2 e^{-j\beta z}\right) / Z_c\right)\Big|_{z=0} = \left(\dot{V}_1 - \dot{V}_2\right) / Z_c \tag{2.35b}$$

where $\dot{V}_0$ and $\dot{I}_0$ are the voltage and current at $z = 0$.

Then,

$$\begin{aligned} \dot{V}_1 &= \left(\dot{V}_0 + \dot{I}_0 Z_c\right)/2 \\ \dot{V}_2 &= \left(\dot{V}_0 - \dot{I}_0 Z_c\right)/2 \end{aligned} \tag{2.36}$$

The solution to the terminal equation can be obtained by putting the terminal Eq. (2.36) into the terminal Eqs. (2.35a, 2.35b)

$$\begin{aligned} \dot{V} &= \dot{V}_0 \cos \beta z + j \dot{I}_0 Z_c \sin \beta z \\ \dot{I} &= \dot{I}_0 \cos \beta z + j \frac{\dot{V}_0}{Z_c} \sin \beta z \end{aligned} \tag{2.37}$$

The terminal Eq. (2.37) relates the voltage $\dot{V}$ and the current $\dot{I}$ on the reference surface from the terminal load z to the terminal load voltage $\dot{V}_0$ and terminal load current $\dot{I}_0$.

Using the terminal Eq. (2.37), the relationship between the input impedance Z_i and the input admittance Y_i, with the terminal load impedance Z_0 and the terminal load admittance Y_0, can be obtained as

$$\begin{aligned} Z_i &= \frac{\dot{V}}{\dot{I}} = Z_c \frac{Z_0 \cos \beta z + j Z_c \sin \beta z}{Z_c \cos \beta z + j Z_0 \sin \beta z} = Z_c \frac{Z_0 + j Z_c \mathrm{tg}\, \beta z}{Z_c + j Z_0 \mathrm{tg}\, \beta z} \\ Y_i &= \frac{\dot{I}}{\dot{V}} = Y_c \frac{Y_0 \cos \beta z + j Y_c \sin \beta z}{Y_c \cos \beta z + j Y_0 \sin \beta z} = Y_c \frac{Y_0 + j Y_c \mathrm{tg}\, \beta z}{Y_c + j Y_0 \mathrm{tg}\, \beta z} \end{aligned} \tag{2.38}$$

where $Z_0 = 1/Y_0 = \dot{V}_0/\dot{I}_0$.

4. The state of traveling wave, pure standing wave, and traveling–standing wave in transmission line

Because the load in the transmission line is different, the state in the transmission line is also different. Generally, there are three typical states: traveling wave state, pure standing wave state, and traveling–standing wave state.

(1) Traveling wave state

If the load absorbs all the incident energy and there is no reflection, it is only a one-way transmission from the source to the load in the transmission line. This state is called a traveling wave state.

Because the reflected wave is zero, i.e., $\dot{V}_{-} = 0$, we can get

$$\dot{V} = \dot{V}_1 e^{j\beta z} \tag{2.39}$$

$$\dot{I} = \dot{V} e^{j\beta z} / Z_c \tag{2.40}$$

Both sides of Eqs. (2.39) and (2.40) are multiplied with $e^{j\omega t}$, and the instantaneous voltage and current expression can be obtained

$$v(z, t) = V_1 \sin(\omega t - \beta z + \varphi_1) \tag{2.41a}$$

$$i(z, t) = \frac{V_1}{Z_c} \sin(\omega t - \beta z + \varphi_1) \tag{2.41b}$$

(2) Pure standing wave state

If the active power is not absorbed by the load, the power of the incident wave will be reflected completely. This state is called the pure standing wave state. When the load is fully reflected, the modulus of the reflection coefficient must be 1, i.e.,

$$|\Gamma_0| = \left|(Z_0 - Z_c) / (Z_0 + Z_c)\right| = 1$$

$$|Z_0 - Z_c| = |Z_0 + Z_c|$$

Z_c of the transmission line without consumption is a real number, and the above formula is only valid when $Z_0 = 0$, $Z_0 = \infty$, or $Z_0 = jX_0$, that is, in the situation of short circuit, open circuit, or pure reactance load, respectively.

The voltage and current on the transmission line are

$$\dot{V} = j\dot{I}_0 Z_c \sin\beta z = j\dot{I}_0 Z_c \sin(2\pi z / \lambda) \tag{2.42a}$$

$$\dot{I} = \dot{I}_0 \cos\beta z = \dot{I}_0 \cos(2\pi z / \lambda) \tag{2.42b}$$

The instantaneous value is

$$v(z,t) = I_0 Z_c \sin(2\pi z/\lambda)\cos(\omega t + \varphi_0) \tag{2.43a}$$

$$i(z,t) = I_0 \cos(2\pi z/\lambda)\sin(\omega t + \varphi_0) \tag{2.43b}$$

By comparing the pure standing wave with the traveling wave, we see the following differences:

(1) In the pure standing wave state, the voltage and the current along the line no longer have the characteristics of forward propagation, but they simply oscillate in place; in λ/2, the magnitude of each point is different, but the phase is exactly the same; but in the other λ/2, the phase is completely opposite.
(2) There is a 90° phase difference between the voltage and the current of the pure standing wave, and the power in this situation is reactive power.
(3) There is a 90° phase difference between the voltage and current in space.

(3) Traveling–standing wave state

If the load absorbs some of the energy of the incident wave, the rest is reflected back. This state between the traveling wave and the pure standing wave is called the traveling–standing wave state. The load corresponding to the standing wave state is neither characteristic impedance nor open circuit, short circuit, or pure reactance.

The expression of traveling–standing wave voltage is

$$\dot{V} = \dot{V}_1 e^{j\beta z} + \dot{V}_2 e^{-j\beta z} = 2\dot{V}_2 \cos\beta z + (\dot{V}_1 - \dot{V}_2)e^{j\beta z}, \quad (|\dot{V}_1| > |\dot{V}_2|) \tag{2.44}$$

where the first term is pure standing wave and the second term is traveling wave.

It can be seen from Eq. (2.44) that the traveling–standing wave is actually the superposition of the traveling wave and the pure standing wave.

5. Voltage standing wave ratio

The voltage standing wave ratio ρ is the ratio of the maximum voltage to the minimum voltage, i.e.,

$$\rho = V_{\max}/V_{\min} \tag{2.45}$$

The maximum voltage occurs when the incident wave voltage is in phase with the reflected wave voltage, i.e., $V_{\max} = V_+ + V_-$. The voltage minimum occurs when the incident wave voltage and the reflected wave voltage are opposite, i.e., $V_{\min} = V_+ - V_-$. So, the voltage standing wave ratio can be expressed as

$$\rho = \frac{V_+ + V_-}{V_+ - V_-} = \frac{1+|\Gamma|}{1-|\Gamma|} \tag{2.46}$$

After proper transformation of (2.46), we have

$$\Gamma = \frac{\rho - 1}{\rho + 1} \tag{2.47}$$

Voltage standing wave ratio and reflection coefficient are used to represent the above three states:

(1) Traveling wave state: $|\Gamma| = 0$, $\rho = 1$;
(2) Pure standing wave state: $|\Gamma| = 1$, $\rho = \infty$;
(3) Traveling–standing wave state: $0 < |\Gamma| < 1$, $1 < \rho < \infty$.

2.2 Application of Transmission Line Theories in EMC Research

The basic ideas and methods of the transmission line theory have a wide range of applications in EMC research. Now, we use the discussion of case shielding effectiveness in Fig. 2.1 as an example, to further explain how the microwave transmission line theory can be applied to EMC problems.

2.2.1 Application of the Single-Conductor Transmission Lines in EMC Research

The cutoff frequency characteristic of a single-conductor transmission line has an important guiding significance to improve the shielding effectiveness of the ventilation holes and slits of the case.

1. Shielding design for the ventilation holes
The size of the aperture should be calculated according to the shielding requirements (including the frequency band to be shielded and the shielding requirement) and the wavelength corresponding to the highest frequency to be shielded.

The specific calculation formula is: Suppose the highest frequency that needs to be shielded is f_c, the corresponding wavelength is λ_c, and the diameter of the aperture cannot be greater than $\lambda_c / 10$. For example, if the maximum frequency to be shielded is 10 GHz (the wavelength corresponding to 10 GHz is 3 cm), the diameter of the aperture should not be greater than 3 mm.

2. Shielding design of slots
The slot generally has a characteristic that it is narrow in one direction and long in the perpendicular direction. The problem of decreasing cutoff frequency caused by the long slots must be considered. The specific solution is to set "glitches" that meet the cutoff requirements along the longest direction according to the highest frequency that needs to be cut off, so that the long slots are cut into short slits as shown in Fig. 2.7.

In the shielding design of the aperture and slot, in addition to the relationship between the aperture size and the cutoff wavelength, it is also necessary to consider

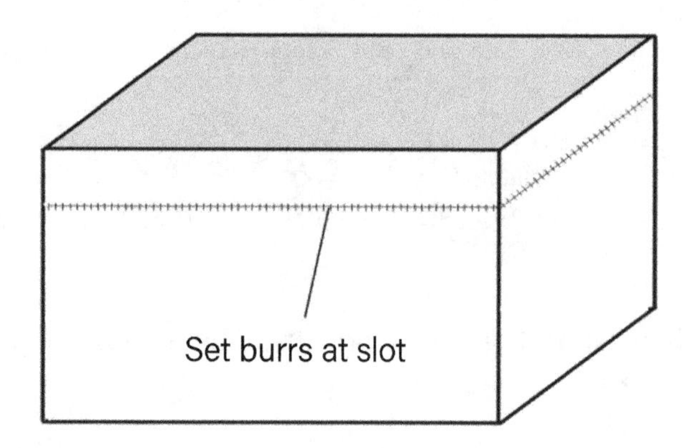

Fig. 2.7 Set a glitch at the long slot

the attenuation path length of the withering mode, that is, the wall thickness of the aperture and the slot.

When the shielding requirement is high, the following measures can be taken to improve the shielding effect:

① Concave–convex design at the joint between the case and the cover to increase the cutoff length of the withering mode as shown in Fig. 2.8a;
② Reed design at the joint between the case and the cover to convert the long slot into short slots as shown in Fig. 2.8b;
③ Flange design at the joint between the case and the cover to increase the cutoff length of the withering mode as shown in Fig. 2.8c;
④ The cable passes through box which can be equivalent to the double-conductor transmission line effect, as shown in Fig. 2.8d.

2.2.2 Application of Multi-conductor Transmission Lines in EMC

The shield design of the keyboard, buttons, switches, power lines, and data lines of the case in Fig. 2.1 needs to refer to the characteristics of the double-conductor and multi-conductor transmission line.

Shielding design of the above units exploits the basic principles of the double-conductor or multi-conductor transmission line as shown in Fig. 2.8d. The factors that affect the shielding effectiveness include:

(1) The aperture size of the cable passing through the case. We should try to enable the closed case with shielding capacity, so the aperture of the through-wall should be as small as possible.

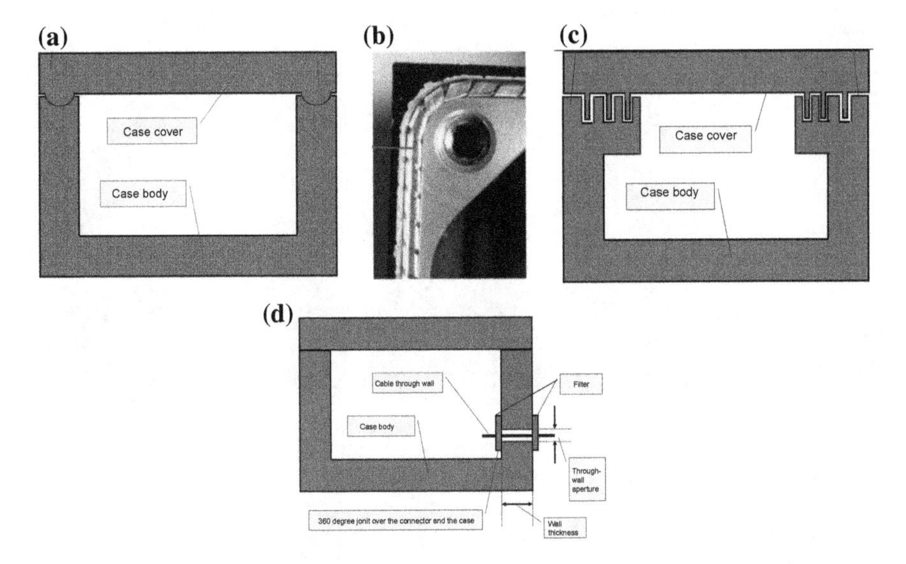

Fig. 2.8 Design to improve the shielding effectiveness of the case. **a** Concave–convex design at the joint between the case and the cover; **b** reed design at the joint between the case and the cover; **c** flange design at the joint between the case and the cover; and **d** the cable passes through the box

(2) The thickness of the case wall. Appropriately increase the thickness of the case wall where the cable passes through, so that the cavity may have some contribution to the shielding.

(3) The through-wall connector which electrically connects to the case in 360°. The through-wall cable makes it possible that the electromagnetic energy from the DC may overflow the wall. Therefore, the connector and the case connecting in 360° can effectively ensure the overall electrical continuity of the case.

(4) A connector with filtering capability. The double-conductor effect makes it possible for all frequency signals starting from DC to leak. Therefore, only the useful frequency signal carried by the through-wall signal line is allowed to pass, and the signal amplitude of other frequencies is effectively attenuated by the filter to improve the shielding effectiveness of the case.

Chapter 3
Antenna Theory and Engineering

This chapter explains the field generated by the alternating electric dipole, which is closely related to EMC research in the antenna theory and engineering [1, 7, 8]. Through this chapter, our readers will learn the basic characteristics of the near field, far field, and transition zone and understand that the EMC problems usually happens with the coexistence in the near field, far field, and transition zone. This chapter also explains the radiation characteristics of the antenna. Our readers will get the knowledge of some basic concepts of the antenna, including the antenna pattern and gain. In the last section, an example is provided to illustrate that the out-of-band characteristics of the antenna have become an important factor in the EMC design and antenna layout analysis of the whole aircraft.

3.1 Field of Alternating Electric Dipole

The field generated by the alternating electric dipole is closely related to EMC problems due to the leakage of fields from slots, apertures, and cables.

An alternating electric dipole refers to two alternating point charges $+q(t)$ and $-q(t)$, with the same value but opposite signs. The distance between the two point charges is $dl(dl \to 0)$, generally represented by $\dot{q}dl$. From the view of the law of charge conservation, alternating electric dipole $\dot{q}dl$ can also be regarded as current element $\dot{I}dl$

Let the time change factor be $e^{-j\omega t}$, and then the electromagnetic field generated by the alternating dipole is [1]:

$$\tilde{\boldsymbol{H}}(\boldsymbol{r}) = \boldsymbol{i}_{\varphi}\frac{\dot{I}dl\sin\theta}{4\pi}\left(\frac{j\beta}{r_S} + \frac{1}{r_S^2}\right)e^{j\beta r_S}(\mathrm{A/m}) \tag{3.1a}$$

$$\tilde{\boldsymbol{E}}(\mathrm{r}) = \boldsymbol{i}_{r_S}\dot{\boldsymbol{E}}_{r_S} + \boldsymbol{i}_{\theta}\dot{\boldsymbol{E}}_{\theta}(\boldsymbol{r}) \tag{3.1b}$$

© National Defense Industry Press and Springer Nature Singapore Pte Ltd. 2019
D. Su et al., *Theory and Methods of Quantification Design on System-Level Electromagnetic Compatibility*, https://doi.org/10.1007/978-981-13-3690-4_3

where

$$\begin{cases} \dot{\boldsymbol{E}}_{r_S}(\boldsymbol{r}) = \frac{\dot{I}dl\cos\theta}{2\pi\omega\varepsilon}\left(\frac{\beta}{r_S^2} - \frac{j}{r_S^3}\right)e^{-j\beta r_S}(\mathrm{V/m}) \\ \dot{\boldsymbol{E}}_{\theta}(\boldsymbol{r}) = \frac{\dot{I}dl\sin\theta}{4\pi\omega\varepsilon}\left(\frac{j\beta^2}{r_S} + \frac{\beta}{r_S^2} - \frac{1}{r_S^3}\right)e^{-j\beta r_S}(\mathrm{V/m}) \end{cases} \tag{3.1c}$$

For the sake of convenience, formulas (3.1a) and (3.1c) can be rewritten as

$$\dot{\boldsymbol{H}}_{\varphi}(\boldsymbol{r}) = \frac{\dot{I}dl\beta^2\sin\theta}{4\pi}\left[\frac{1}{j\beta r_S} + \frac{1}{(j\beta r_S)^2}\right]e^{-j\beta r_S}(\mathrm{A/m}) \tag{3.2a}$$

$$\dot{\boldsymbol{E}}_{r_S}(\boldsymbol{r}) = \frac{\dot{I}dl\beta^2\eta 2\cos\theta}{4\pi}\left[\frac{1}{(j\beta r_S)^2} + \frac{1}{(j\beta r_S)^3}\right]e^{-j\beta r_S}(\mathrm{V/m}) \tag{3.2b}$$

$$\dot{\boldsymbol{E}}_{\theta}(\boldsymbol{r}) = \frac{\dot{I}dl\beta^2\eta\sin\theta}{4\pi}\left[\frac{1}{j\beta r_S} + \frac{1}{(j\beta r_S)^2} + \frac{1}{(j\beta r_S)^3}\right]e^{-j\beta r_S}(\mathrm{V/m}) \tag{3.2c}$$

where $\eta = \sqrt{\mu/\varepsilon}$ is the wave impedance of the medium.

The electromagnetic field excited by the alternating current elements represented by (3.1a)–(3.1c) is very important in the discussion of radiation problems. Next, we will analyze these three equations, which will lead to some basic concepts of radiation problems.

3.1.1 *Near Field*

The region that satisfies $\beta r_S << 1$ is called the near-field region of the electric dipole. In the near-field region, there is

$$(\beta r_S)^{-3} >> (\beta r_S)^{-2} >> (\beta r_S)^{-1}$$

$$e^{-j\beta r_S} \approx 1$$

Therefore, the magnetic field in the near-field region can be expressed as

$$\tilde{\boldsymbol{H}}(\boldsymbol{r}) \approx \boldsymbol{i}_{\varphi}\dot{I}dl\frac{\sin\theta}{4\pi r_S^2}(\mathrm{A/m})$$

The electric field in the near-field region can be expressed as:

$$\tilde{\boldsymbol{E}}(\boldsymbol{r}) \approx \left(\boldsymbol{i}_{r_S}2\cos\theta + \boldsymbol{i}_{\theta}\sin\theta\right)\frac{\dot{p}}{4\pi\varepsilon r_S^3}(\mathrm{V/m})$$

It should be noted that the near-field magnetic field is exactly the same as the static magnetic field and the near-field electric field is identical with respect to the field excited by static electric dipole. Therefore, the near-field region is called as "quasi-static field." From the Poynting vector of electromagnetic field, it can be concluded that the complex Poynting vector of the static field is a pure imaginary number with zero active power density. Therefore, the electromagnetic energy of the static field is not radiated, as if this part of the energy was trapped in the near zone. Therefore, the static field is usually called the "bounded field." Based on this fact, we can say that the main component of the near-field region generated by the alternating electric dipole is the bounded field.

3.1.2 Far Field

The region which satisfies $\beta r_S >> 1$ is the far-field region of the electric dipole. In the far field, since $\beta r_S >> 1$, there is

$$(\beta r_S)^{-1} >> (\beta r_S)^{-2} >> (\beta r_S)^{-3}$$

Therefore, the magnetic field in the far-field region can be expressed as:

$$\tilde{\boldsymbol{H}}(\boldsymbol{r}) \approx \boldsymbol{i}_\phi \mathrm{j} \dot{I} dl \frac{\sin\theta e^{-\mathrm{j}\beta r_S}}{2\lambda r_S} (\mathrm{A/m})$$

The electric field in the far-field region can be expressed as:

$$\tilde{\boldsymbol{E}}(\boldsymbol{r}) \approx \boldsymbol{i}_\theta \mathrm{j} \eta \dot{I} dl \frac{\sin\theta e^{-\mathrm{j}\beta r_S}}{2\lambda r_S} (\mathrm{V/m})$$

It is not difficult to find that the main components of $\dot{H}_\phi(\boldsymbol{r})$ and $\dot{E}_\theta(\boldsymbol{r})$ in the far-field region have the same phase, and their Poynting vector shown below is real.

$$\tilde{\mathbf{S}}(\boldsymbol{r}) = \tilde{\boldsymbol{E}}(\boldsymbol{r}) \times \tilde{\boldsymbol{H}} * (\boldsymbol{r}) \big/ 2 = \boldsymbol{i}_{r_S} \eta \left| \dot{H}_\phi \right|^2 \big/ 2 \ (\mathrm{W/m^2})$$

It indicates that there is an active power flow in the $\boldsymbol{i}_{r_S}$ direction, and the electromagnetic energy is propagating outward. In addition, from the expression of the electromagnetic field in the far field, the main part of the far field is inversely proportional to r_S. Thus, their Poynting vector should be inversely proportional to r_S^2. Then, we can make a flux integral on the Poynting vector on the surface A of a sphere with the radius r_S, and we will get a constant unrelated with r_S, i.e.,

$$\oint_A \tilde{\mathbf{S}}(\boldsymbol{r}) \cdot \mathrm{d}\boldsymbol{a} = \frac{\pi \eta I^2 dl^2}{3\lambda^2} (\mathrm{W})$$

Therefore, the electromagnetic power carried by the main part of far-field region will be fully radiated. The amount of field in the far field that is inversely proportional to r_S is called the radiation field

In the electromagnetic field, apart from the bounded field and the radiation field, there is another item proportional to $1/r_S^2$, which is generally referred to as the inducted field.

3.2 Basic Antenna Concepts

This section introduces several basic concepts of the antenna [1], including directivity function, antenna pattern, radiation power, radiation resistance, main lobe, half-power beamwidth (HPBW), antenna side lobe, antenna gain, and antenna feed system.

3.2.1 Directivity Function and Pattern

In the radiation field of the alternating electric dipole, the electromagnetic field amplitude changes in accordance with $\sin\theta$. When $\theta = 0$ and $\theta = \pi$, the radiation field of alternating electric dipole is zero, and it is the strongest when $\theta = \pi/2$. It shows that the electromagnetic radiation generated by the alternating electric dipole is directional.

The directivity function is often used to indicate the directivity of the radiator. The directivity function $F(\theta, \phi)$ is defined as

$$F(\theta, \phi) = |E(\theta, \phi)| / |E_{\max}|$$

where $|E(\theta, \phi)|$ is the amplitude of the radiation electric field in the direction (θ, ϕ), and $|E_{\max}|$ is the amplitude of the radiated electric field in the direction of the maximum radiation field.

The directivity function of the alternating electric dipole is $F(\theta, \phi) = |\sin\theta|$.

The representation of the directivity function in space is the antenna pattern. For alternating electric dipoles, the curve of $F(\theta, \phi)$ is quite like the number "8" on the plane where ϕ is constant. Since it only contains the electric power line of the radiation field on this plane, it is often called the E-plane. On the plane where $F(\theta, \phi)$ is a circle, only the magnetic field lines of the radiation field are contained. Therefore, the plane is often called the H-plane.

3.2.2 Radiation Power

The electromagnetic power radiated by the alternating electric dipole can be calculated from its field. With the alternating electric dipole as the center, using r_S as a radius to make a closed sphere A, then the electromagnetic power through the surface A is

$$\dot{P} = \oint_A \tilde{\mathbf{S}} \cdot \mathrm{d}\boldsymbol{a}$$

where $\tilde{\mathbf{S}}$ is the complex Poynting vector.

The radiated power of the alternating electric dipole can be calculated from the far field

$$\dot{P}_\Sigma = \frac{\pi \eta I^2 \mathrm{d}l^2}{3\lambda^2} (\mathrm{W})$$

For alternating electric dipoles in free space, the wave impedance is $\eta_0 = 120\pi\,(\Omega)$. Therefore, the radiation power is

$$\dot{P}_\Sigma = 40\pi^2 I^2 (\mathrm{d}l/\lambda_0)^2 (\mathrm{W})$$

Obviously, the total radiation power is proportional to $(\mathrm{d}l/\lambda_0)^2$. $\mathrm{d}l/\lambda_0$ is often called the electrical size of an alternating electric dipole. For alternating electric dipoles, the electrical size is always small. Therefore, its radiation capacity is quite limited.

3.2.3 Radiation Resistance

Radiation resistance measures the radiation capacity of a radiator. Here, the term "resistance" in circuit theory is used to reflect the radiation of the radiator; that is, the power radiated by the radiator is equivalent to the power absorbed by a resistor, and this equivalent resistance value is defined as the radiation resistance of the radiator. Thus, when the current fed to the antenna by the transmitter is fixed, the bigger the radiation resistance of the antenna is, the stronger the radiation capability is. Referring to the form of the circuit formula, the radiation power P_Σ is

$$P_\Sigma = I^2 R_\Sigma / 2 (\mathrm{W})$$

The radiation resistance of the alternating electric dipole in free space is

$$R_\Sigma = 80\pi^2 (\mathrm{d}l/\lambda_0)^2 (\Omega)$$

Since the electrical size of the alternating electric dipole is small, its radiation resistance is quite small, which indicates that its radiation capability is not strong. For example, in free space, when $dl/\lambda_0 = 0.1$, it is already a very large electrical size for the alternating electric dipole, but the radiation resistance R_Σ is only 7.89 Ω.

3.2.4 Antenna Beamwidth and Gain

This section will introduce main lobe, half-power beamwidth (HPBW), side lobe, antenna gain, and so on closely related to EMC design. When analyzing the difference between the characteristics of the bare antenna and the characteristics of the installed antenna, we need to pay special attention to the large influence of the fuselage on the antenna pattern after the antenna is installed in the EMC calculation. Then, we can further explain the change of the antenna pattern caused by the fuselage and analyze the influence of the fuselage on the working distance of the installed antenna system.

1. Antenna pattern

The radiation of the transmitting antenna is not uniform in all directions of the space, and the energy received by the receiving antenna in all directions of the space is not uniform either. The direction selectivity of the antenna can be described by the antenna pattern [9].

The main lobe and its beamwidths, side lobes, and back lobes of the antenna can all be obtained from the antenna pattern.

(1) Main lobe is defined as the lobe with the highest radiation intensity in the pattern.
(2) The main lobe beamwidth is defined as the angular separation between the points at which the main lobe power drops to half the peak on the pattern, i.e., the angle between the 3 dB points.
(3) Side lobes are defined as other lobes than the main lobe.
(4) Back lobes are defined as lobes in the opposite direction of the main lobe.

2. Gain

The power gain is a variable describing the ability of the antenna to concentrate energy in a certain direction.

The power gain of an antenna is usually defined as the value of the gain at peak direction of the antenna's main lobe. When the input power of antenna A is $P_{in}(\text{A})$, its main peak value is $P_{out}(\text{A})$. When using nondirectional reference antenna B with peaks $P_{out}(\text{B}) = P_{out}(\text{A})$, the input power of B is $P_{in}(\text{B})$. Then, the power gain of antenna A is

$$D(\theta, \phi) = \frac{P_{in}(\text{B})}{P_{in}(\text{A})} \tag{3.3}$$

The bare antenna characteristics are defined as antenna characteristics before the antenna is installed (on another platform).

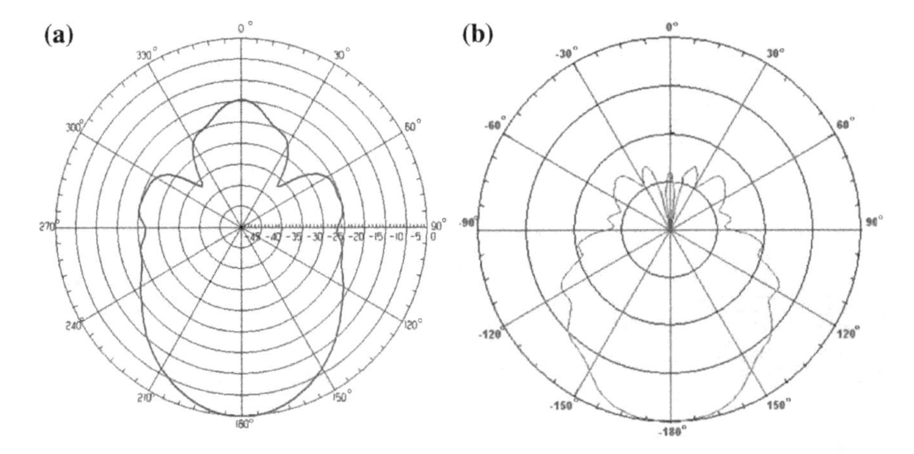

Fig. 3.1 Change of the antenna pattern before and after the installation of an antenna. **a** The antenna pattern before installation; **b** the antenna pattern after installation

The characteristics of the installed antenna are defined as the integrated antenna characteristics after the secondary radiation caused by the new electromagnetic boundary generated on the surface of the fuselage, after the bare antenna is installed.

Generally, the boundary of the fuselage makes the performance of the antenna characteristics of the bare antenna decrease. This drop is manifested in the radiation characteristics (directionality, gain, etc.) and impedance characteristics of the antenna. Figure 3.1 shows the change of the antenna pattern before and after the installation of an antenna.

The fuselage often makes the pattern of the airborne antenna become somehow "fat," which means that the antenna's working distance decreases, and the energy outside the main direction increases; as a result, the antenna isolation between the antenna and other antennas on the same aircraft decreases, and the mutual interference increases. Therefore, in the design and evaluation of EMC of the whole aircraft, it is necessary to consider the change of the antenna characteristics after the bare antenna is installed.

The following example illustrates the effect of the secondary radiation from the fuselage on the characteristics of the bare antenna [10].

Figure 3.2 shows the effect of the gun bay of an aircraft on the radiation characteristics of the antenna. Since the fuselage and the gun bay are made of metal materials, the influence of the antenna toward the antenna pattern is relatively large. Figure 3.2b analyzes the influence of the distance between the antenna and the belly on the main lobe and the HPBW when the antenna is moved down in the vertical direction. Figure 3.2c analyzes the influence of the distance between the antenna and the gun bay on the main lobe and the HPBW when the antenna is moved in the backward direction. It can be seen that under certain conditions, the secondary radiation of the fuselage has a significant influence on the radiation characteristics of the airborne antenna.

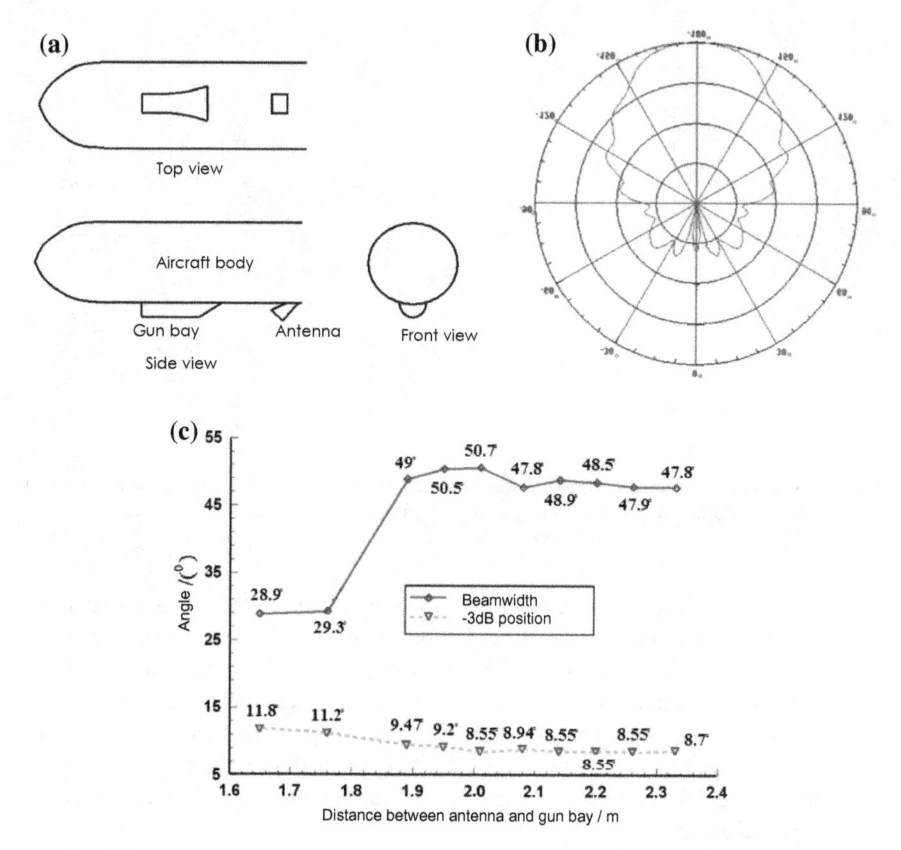

Fig. 3.2 Influence of a machine gun bay in the belly of an aircraft on the radiation characteristics of the antenna. **a** Aircraft fuselage; **b** the change of the antenna lobe in vertical direction; **c** the change of the antenna lobe in horizontal direction

3.2.5 *The Impact of Antenna Out-of-Band Characteristics Toward Aircraft EMC*

In the past, during the design of the airborne antenna and the analysis of the antenna layout of the whole aircraft, the influence of the out-of-band characteristics of the antenna has been rarely considered. However, with the increasing number of airborne antennas and the widening operation frequency band, the influence of the out-of-band characteristics of the airborne antenna (including the directional characteristics and the impedance characteristics) to the whole aircraft's EMC cannot be ignored anymore. Figure 3.3a shows the pattern of an airborne antenna at the center operating frequency, at an out-of-band frequency point 1 lower than the center frequency, and at an out-of-band frequency point 2 higher than then center frequency. Figure 3.3b–d illustrates the change of the main lobe and side lobe of the airborne antenna at the

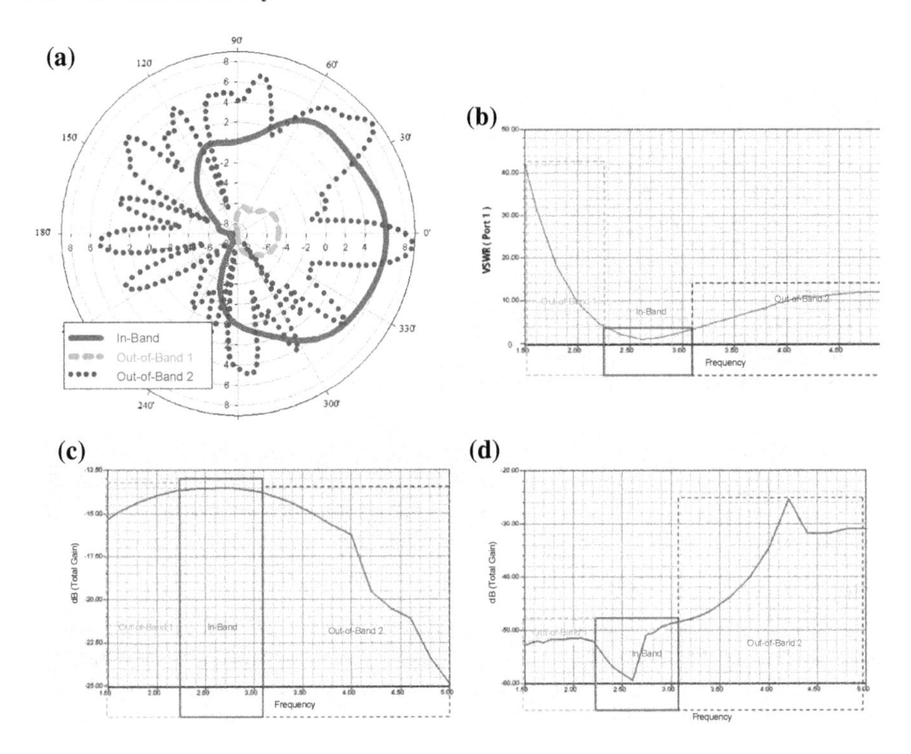

Fig. 3.3 Airborne antenna characteristics with changes in frequency. **a** In-band and out-of-band pattern characteristics of an airborne antenna; **b** in-band and out-of-band SWR of an airborne antenna; **c** in-band and out-of-band main lobe characteristics of an airborne antenna; **d** in-band and out-of-band side lobe characteristics of an airborne antenna

center operating frequency, at an out-of-band frequency point 1 lower than the center frequency, and at an out-of-band frequency point 2 higher than then center frequency.

Figure 3.3a indicates that the change of pattern characteristics of the antenna inside and outside the operation band is very intense. Figure 3.3b–d shows that, if the out-of-band SWR does not increase rapidly (as Fig. 3.3b), the main lobe gain does not decrease rapidly (as Fig. 3.3c), and the side lobe gain increases rapidly (as Fig. 3.3d); thus, the effect of the out-of-band characteristics of the antenna on the whole antenna isolation cannot be ignored. Taking Fig. 3.3 for example, the ratio of the main and side lobes of the antenna within the operating band is about 25 dB. However, at the frequency point 2–3 times of the center frequency, the ratio of the main and side lobes has reduced to 2.5 dB and the SWR is 10. The antenna is therefore suffering from EMC problems.

A large number of practical problems have shown that when the EMC design accuracy requirements are high, the influence of the out-of-band characteristics of the antenna on EMC needs to be considered.

3.2.6 *Antenna Feed System*

This section will introduce the concept of the antenna feed system. Through this section, our readers will understand that antenna isolation is only one of the many important tasks in analyzing system-level EMC problems. To design the EMC of the whole aircraft, it is also necessary to grasp the RF front-end characteristics of the antenna feed system and analyze the isolation between the equipment besides antenna isolation.

The antenna system of a typical radio application system consists of a feed network and an antenna. To improve the receiving sensitivity, a low-noise amplifier (LNA) is sometimes used to connect to the receiving antenna. Since antennas are both the transmitting and receiving ports of the electromagnetic energy in communication electronic equipment, they are most likely to form mutual radiation interference. In other words, electronic systems are highly possible to couple to each other through antennas. Therefore, how to achieve high isolation between antennas becomes the key to achieve compatibility between electronic equipment.

1. Input impedance and antenna matching

The input impedance of antenna is the impedance presented by the input of the antenna, that is, the ratio of the input voltage to the current at the antenna feed end. The input impedance of the antenna is affected by other antennas and adjacent objects. It consists of real and imaginary parts as

$$Z_{in} = R_{in} + j X_{in} \tag{3.4}$$

where the input resistance (R_{in}) represents power loss. The power can be consumed in two ways, namely radiated out in the form of electromagnetic waves and heat loss from the antenna. Input reactance X_{in} indicates the storage of the power in the near field of the antenna.

The best scenario connection between the antenna and the feed is that the input impedance of the antenna is pure resistance and is equal to the characteristic impedance of the feed. Under this condition, there is no power reflection at the feed port, and there is no standing wave on the feed. The input impedance of the antenna changes gently with frequency. The feeding network is used to complete the matching work of the antenna, eliminate the reactance component in the input impedance, and make the resistance component as close as possible to the characteristic impedance of the feed. The input impedance of the general communication and radar antenna is 50 Ω.

The quality of antenna matching can be measured by parameters such as reflection coefficient, traveling wave ratio, standing wave ratio (SWR) and return loss. There is a certain numerical relationship between these four parameters. In general, SWR and return loss are used frequently.

2. Return loss

The return loss is the reciprocal of the absolute value of the reflection coefficient and it is expressed in decibels. The value of the return loss is between 0 dB and infinity. The smaller the return loss, the worse the match and vice versa. 0 dB means total reflection, and on the contrary, infinity means complete match.

3. Voltage standing wave ratio (VSWR)

The impedance of the antenna feed should be exactly matched with the input impedance of the antenna. Otherwise, there will be reflected waves at the antenna port, and there will be electromagnetic waves flowing to the signal source on the feeder. The electromagnetic wave formed by the combination of the reflected wave and the incident wave is called a standing wave, and the ratio of the maximum value to the minimum value of the amplitude of a standing wave signal is called the voltage standing wave ratio (VSWR). It is the reciprocal of the traveling wave coefficient and its value is between 1 and infinity. If the SWR equals to 1, it indicates a perfect match. And if the SWR equals to infinity, it indicates the total reflection, which means a complete mismatch. In the communication system, the SWR is generally required to be less than 2–3. However, the VSWR should be less than 2 in practical applications, because excessive VSWR will decrease the efficiency of the transmitter power and cause increased interference within the system, and further affects system's performance.

$$\text{Return Loss} = 20\lg\frac{VSWR + 1}{VSWR - 1} \tag{3.5}$$

4. Antenna isolation

The antenna isolation is defined as the ratio of the transmitting power (P_t) of the transmitting antenna to the receiving power (P_r) received by the receiving antenna:

$$L = \frac{P_t}{P_r} \tag{3.6}$$

Typically, in engineering applications, it is expressed in dB:

$$L = 10\lg\frac{P_t}{P_r}(\text{dB}) \tag{3.7}$$

When both antennas are in a far field from each other, their energy coupling is mainly from the radiation field. When the polarization between the transmitting and receiving antennas is not completely matched, the loss due to polarization mismatch also needs to be taken into account. The mismatch loss from circular polarization is about 3–4 dB at both vertical and horizontal direction, and the mismatch loss between vertical polarization and horizontal polarization is about 20–35 dB.

If the antennas cannot be located in their far-field region at the same time, then the mutual interference between the two antennas is not mainly from the radiation field, but is caused by the near-field bound field or the near-field induction field. Since the concept of antenna power gain is established in the far field, Eq. (3.6) is not applicable to near-field antenna isolation analysis. In the near-field case, the system composed of the transmitting and receiving antennas can generally be regarded as a two-port network and the mutual interference performance can be solved by analyzing the scattering matrix of the two-port network.

Part II
Methods and Applications of Quantitative System-Level EMC Design

Electromagnetic compatibility (EMC) is an important ability of electronic information systems. It is the basis and prerequisite for electromagnetic environment adaptability. It covers the whole life cycle from demonstration and development to application of electronic information systems. It is the bottleneck technology for the informatization of electronic information systems. It is a general support technology which ensures that the electronic information systems adapt to the complex electromagnetic environment. It is also the technological basis to improve the survivability, combat effectiveness, and stability of electronic information systems. In summary, it is a major support technology for electronic information systems and it will keep developing with the advancement of electronic information technology.

In this part, we will discuss the quantitative design of system-level EMC of electronic information systems.

Electronic information systems refer to systems consist of electronic, electrical, electromechanical, and information subsystems/equipment, such as aircrafts, satellites, ships, and vehicles. The electronic information subsystem/equipment is an abbreviation for electronic, electrical, electromechanical, and information subsystems/equipment, including communication, radar, navigation, identification, control, power, and lighting. The electronic information subsystems/equipment can be further classified into subsystems/equipment subordinate to the platform and subsystems/equipment subordinate to the tasks, as shown in Fig. II.1. In this book, the authors use aircraft as an example, but the theory and methods can be adopted by other systems.

The quantitative design of system-level EMC is an important issue related to the development, performance, and security of electronic information systems. The electronic information systems usually have extremely complicated EMC problems due to the system characteristics, such as a wide operating bandwidth, overlapping frequency of transmitting and receiving equipment, large transmitting power of transmitting equipment, high sensitivity of receiving equipment, a large number of antennas, and limited geometric space. In real world, it is also common that the actual electronic information systems have EMC problems caused by the

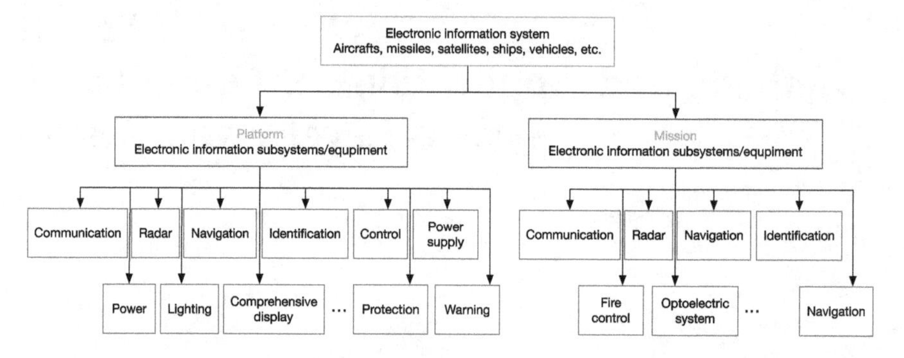

Fig. II.1 Classification of electronic information subsystems/equipment

coexistence of new and old electronic information devices, or the coexistence of platform-level and mission-level electronic information devices. Moreover, electronic information systems that contain full-band, high-power transmitting equipment have potential safety issues triggered by high-power transmitting equipment. In addition, most electronic information systems operate in complex and changeable electromagnetic environment. Therefore, EMC is not only an important ability for achieving self-compatibility of electronic information systems, but also an important basis for electronic information systems to obtain the electromagnetic environment adaptability. The compatibility of electronic information systems can be achieved through the quantitative design of system-level EMC.

System-level EMC quantitative design has three key elements: system level, design, and quantification. Specifically, "system level" means that we should work from the perspective of the whole aircraft; "design" should be carried throughout each phase of scheme design—principle prototype design—and engineering prototype design and emphasize each state of conceptual design, digital design, semi-physical design, and physical design; "quantitative" is relative to qualitative and it refers to the quantification of indicators.

Taking aircraft as an example, the difficulties of quantitative design of EMC are mainly as follows:

(1) The aircraft is a electrically large structure (The geometric size of aircrafts can reach several tens of meters. It means that for operating frequencies of 18 GHz or 40 GHz, the aircraft will have a size of several thousand wavelengths or more).

(2) There may be tens of electronic information equipment and antennas installed on the aircraft. The electromagnetic interferences and electronic jamming signals have many different types, which makes the interference correlation relationship of the whole system complex and changeable.

(3) The greatest challenge in quantification during the demonstration, prediction, and design phases is the lack of data. In most cases, only performance data, such as the receiver's frequency, sensitivity, power, antenna gain, and beam width, can be obtained. The EMC demonstration and design, however, require not only performance data, but also a large amount of nonfunctional data (e.g., the parasitic effect),

as well as the correlation between the functional and nonfunctional data. Therefore, the quantification of the whole aircraft's performance and the EMC performance indicators is usually carried out under the condition that the subsystems/equipment are almost *gray boxes*.

This book summarizes the authors' work for the past 20 years in aircraft EMC analysis, engineering design, and troubleshooting. Taking the aircraft platform, which is a typical electronic information system, as an example, the authors present the issues encountered in the quantitative design of system-level EMC and introduce the method of aircraft EMC design. From the aspects of system-level EMC theory, top-down EMC quantitative design method, and the electromagnetic design of the whole aircraft, the authors describe the quantitative design technology on system-level EMC and the quality control technology in implementation of system-level EMC design, discuss the principal, methodology, and technical measures, and provide the framework of EMC demonstration and design indicator system for the whole aircraft. The authors also present in detail the three-key technologies of quantitative design on system-level EMC, namely the whole digital aircraft system with EMC, behavioral modeling and simulation technology of EMC and field–circuit coupling co-analysis technology. Then, the book introduces the EMC performance control of the whole aircraft and provides design case studies. The methodologies discussed in this book have been proved in applications in various information systems. We hope this book will help our readers in prediction, design, and troubleshooting of the whole lifecycle EMC for electronic information systems.

Chapter 4
Basic Concepts of Quantitative System-Level EMC Design

The definition of EMC, the three aspects, and the terminologies of EMC can be found in related regulations [11, 12]. In this chapter, we will further explain these concepts from the perspective of quantitative design on system-level EMC [12].

4.1 Basic Definitions of EMC

This section presents the concepts of electromagnetic interference (EMI), electromagnetic compatibility (EMC), electromagnetic vulnerability (EMV), electromagnetic environment (EME), electromagnetic environment adaptability, spectrum management, spectrum certification, spectrum supportability, etc.

4.1.1 Electromagnetic Interference

Any electrical and electronic system will produce electronic signals or electromagnetic emissions (including intentional and unintentional) while operating. When these electronic signals or electromagnetic emissions cause undesirable or unacceptable responses, malfunctions, performance degradation, earlier-than-expected discovery, positioning, identification by the enemy, the electronic signals or electromagnetic emissions are defined as electromagnetic interference (EMI).

It should be noted that EMI includes two aspects, namely emission and susceptibility.

In order to better explain the definition, several terminologies need to be defined first:

(1) Intentional emission: It is the electromagnetic emissions generated by the system antenna in the operating frequency band.

© National Defense Industry Press and Springer Nature Singapore Pte Ltd. 2019

D. Su et al., *Theory and Methods of Quantification Design on System-Level Electromagnetic Compatibility*, https://doi.org/10.1007/978-981-13-3690-4_4

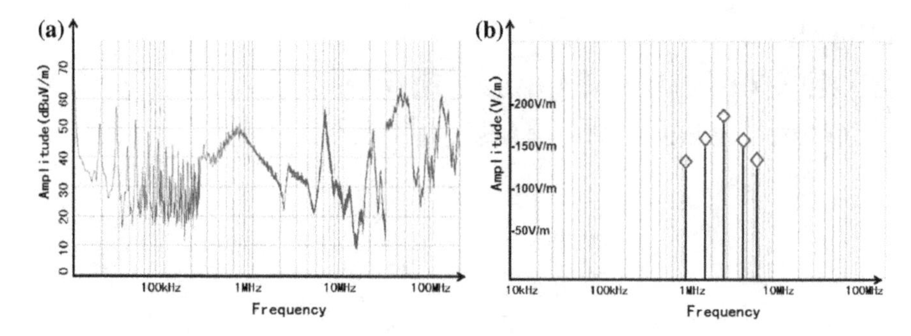

Fig. 4.1 Result of electromagnetic interference test of a system. **a** Electromagnetic emission; **b** electromagnetic susceptibility

(2) Unintentional emission: It is the electromagnetic emissions generated from the part that should not generate emissions (e.g., cables, apertures, slots, keyboards), or the electromagnetic emissions generated by the antenna outside the operating frequency band.
(3) Responses, malfunctions, performance degradation: The influence of the three on the electrical and electronic systems is progressive. Response has the lightest influence that can be perceived but will not affect the system operation; malfunction means that part of the system is affected but the overall system performance remains normal; performance degradation is the most serious, which means that the overall system performance has been affected and dropped.
(4) Discovery, positioning, identification: The three terms are progressive in degree of being discovered by the enemy. Discovery refers to the discovery of suspicious objects; positioning refers to locating of the suspicious objects; identification refers to the determination of whether the object is enemy or not and other attributes of the object.

The electromagnetic emission and electromagnetic susceptibility characteristics are inherent attributes of the electrical and electronic systems which are determined by design principles, manufacturing, inherent structures, etc. The main purpose of EMI research is to objectively describe and accurately grasp the electromagnetic emission and electromagnetic susceptibility characteristics of the electrical and electronic systems through simulation analysis and physical testing.

Figure 4.1 shows the results of the EMI test of the electrical and electronic systems. Since the main purpose of the EMI test is to objectively understand the electromagnetic emission and electromagnetic susceptibility of the tested product, there is no "limit requirement" of electromagnetic emission and electromagnetic susceptibility in EMI detection.

Electromagnetic susceptibility usually attracts more attention than electromagnetic emission (because electromagnetic susceptibility exposes problems of the system itself, while electromagnetic emission affects others). However, we think the latter is equally important, because the path of electromagnetic emission is often

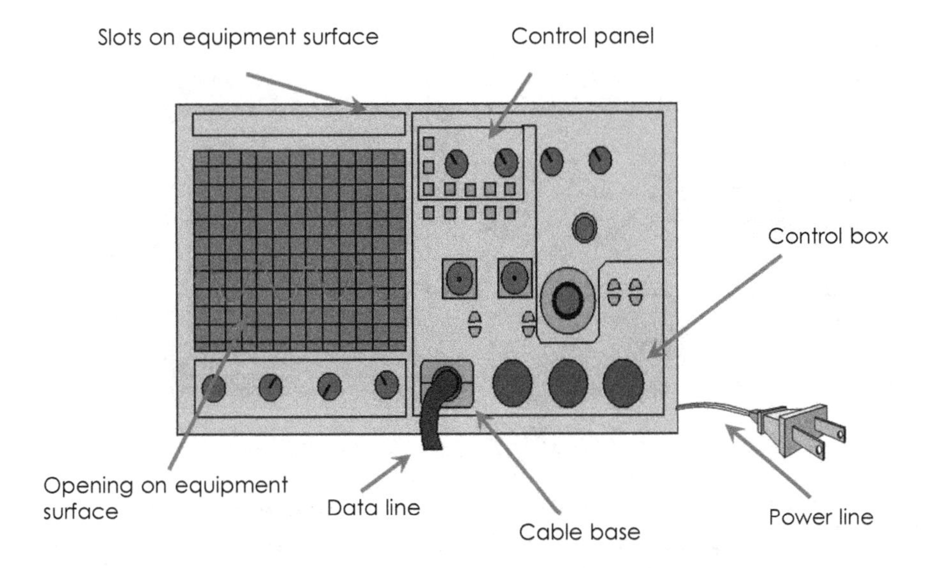

Fig. 4.2 Schematic diagram of EMI

shared by the interference signals of electromagnetic susceptibility due to the reciprocity. In Fig. 4.2, the intentional and unintentional radio frequency energy can be picked up by the system or emitted from the system equally through apertures or slots on the surface of the equipment, power lines, data lines, cable bases, control boxes, control panels, etc.

In addition, we need to distinguish between *interference* and *jamming*: The former is used in EMC field; the latter is used in the field of electronic warfare.

4.1.2 Electromagnetic Compatibility

First, we need to distinguish between electromagnetic compatible and electromagnetic compatibility (EMC).

Electromagnetic compatible is a state where all systems, equipment, and devices that work with the electromagnetic spectrum can perform their own functions in common electromagnetic environment.

Electromagnetic compatibility is an ability that enables all systems, equipment, and devices to work with the electromagnetic spectrum. This ability ensures that they will not cause unacceptable or unexpected performance degradation due to electromagnetic emissions or responses under predetermined operating conditions.

The definition of EMC is explained below. There are three points in the definition of EMC that the reader should pay attention to:

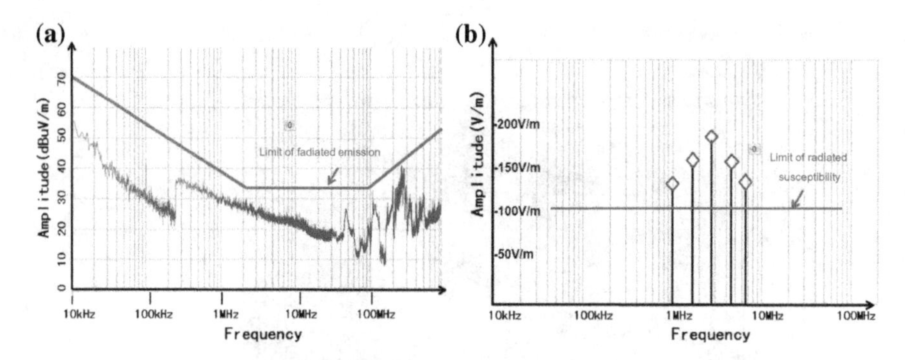

Fig. 4.3 EMC test result of a system. **a** Limit of electromagnetic emission; **b** limit of electromagnetic susceptibility

(1) **An ability**. It emphasizes that EMC, like other functional indicators, is an indicator that electronic information systems must satisfy. That is, if the EMC of an electronic information system fails, the electronic information system is unqualified.
(2) **Predetermined operating conditions**. It generally includes the external electromagnetic environment where the electronic information system platform operates to perform the intended tasks (mainly affecting the system EMC during normal use) and the electromagnetic environment within the electronic information system platform itself (mainly affecting the electromagnetic compatibility within the system). Therefore, the *predetermined operating conditions* should be taken as an input for the EMC verification and design. It should also be included in the overall demonstration and design of the electronic information system, which requires to add the EMC requirements analysis, EMC brief design and detailed design into the design of electronic information systems.
(3) **Unexpected performance degradation**. It emphasizes that there must be a margin in EMC design (*margin* is a unique requirement of EMC), which is due to the inherent electromagnetic susceptibility and electromagnetic vulnerability (EMV) of electronic information systems.

Based on EMI, EMC sets limits to the electromagnetic emission and electromagnetic susceptibility. Electromagnetic emission has an upper limit, while electromagnetic susceptibility has a lower limit, as shown in Fig. 4.3.

Our readers can get a better understanding of the limit required by regulations for the electromagnetic emission and electromagnetic susceptibility by comparing Fig. 4.1 with Fig. 4.3. These are basic EMC requirements of electronic information systems.

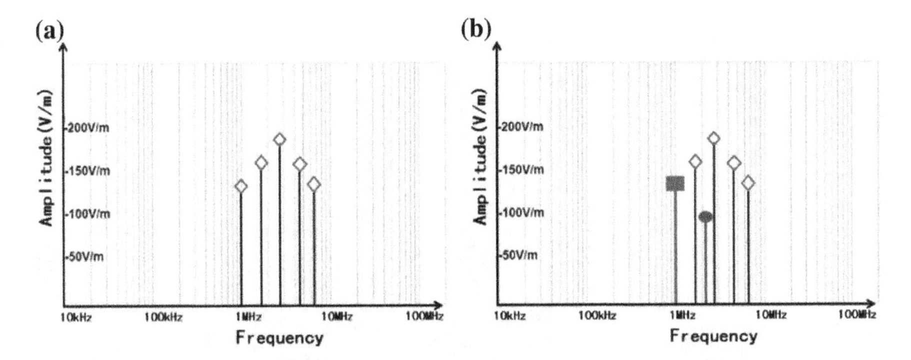

Fig. 4.4 Relationship between electromagnetic susceptibility and electromagnetic vulnerability. **a** Electromagnetic susceptible frequency points; **b** electromagnetic vulnerable points

4.1.3 Electromagnetic Vulnerability

Electromagnetic vulnerability (EMV) refers to the electromagnetic susceptibility that causes serious failures, degraded performance, and safety threats to personnel or equipment. Typically, it refers to more serious problem that might affect personnel or system safety or might be exploitable by enemy forces. A system is said to be vulnerable if its performance is degraded below a satisfactory level as a result of exposure to the stress of an operational electromagnetic field or transient. EMV can be considered a special subset of the susceptibility side of EMI which is defined in Sect. 4.1.1, and it is also an inherent attribute of the system. The root cause of EMV is the oversusceptibility of the system.

Our readers can better understand the relationship between EMV and electromagnetic susceptibility (EMS) by comparing Figs. 4.1 and 4.4. EMS frequency points that cause system and personnel safety problems are referred to as EMV frequency points (the frequency points corresponding to ■ and ◊ in Fig. 4.4b); thus, EMV is a special subset of electromagnetic susceptibility. However, it should be noted that not all EMV frequency points are marked in the EMS test report (the frequency point corresponding to ■ in Fig. 4.4b). In practice, it is possible that certain EMV frequency point occurs outside the susceptible frequency marked in the electromagnetic susceptibility test report (the frequency point corresponding to ◊ in Fig. 4.4b). The main reason is that the frequency points of EMS test are not dense enough.

EMV is directly related to the safety of the system and personnel, so it is necessary to add a sufficient safety margin to the tested product at the vulnerable frequency, as shown in Fig. 4.5. The current military standard has a safety margin requirement of 6 dB for non-pyrotechnics and a safety margin requirement of 16.5 dB for pyrotechnics.

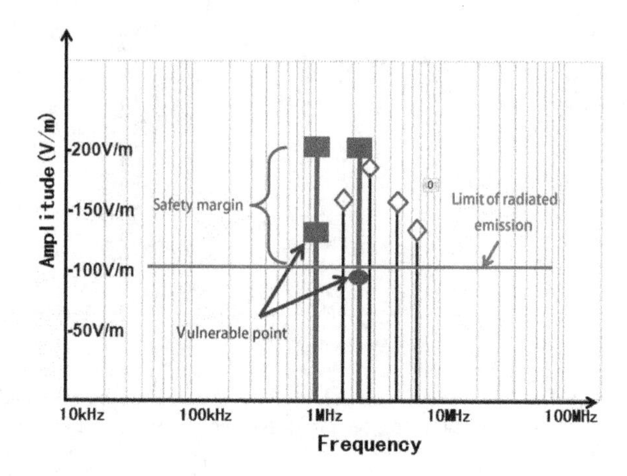

Fig. 4.5 Safety margin

4.1.4 Electromagnetic Environment

Electromagnetic environment (EME): The distribution of electromagnetic emission energy in a range of frequency changes over time, i.e., the various conduction and radiation emissions that may be encountered by an electronic information system when performing its assigned mission in its predetermined operational environment.

According to different electromagnetic emission characteristics, the EME can be divided into two kinds: electromagnetic environment involved in EMC research (which is called background environment) and electromagnetic environment involved in electronic warfare research (which is called threat environment).

The EME in this book mainly refers to the background environment, since this book focuses on EMC research.

From the definition of the electromagnetic environment, we can see that the electromagnetic environment function is a function of frequency, time, and electromagnetic emission power, i.e.,

$$\text{EME} = \text{function}(f, t, P)$$

where f represents frequency, t represents time, and P is electromagnetic emission power.

Complexity can be used to describe how complex electromagnetic environment is. This book will give a description of EME complexity from the perspective of the EME's impact on the electronic information system (i.e., from the perspective of EMC).

For a certain equipment, if it is susceptible due to a certain kind of electromagnetic emission, the emission is called effective EMI. And we can say that the EME containing such electromagnetic emission signal has a certain degree of complexity. On the other hand, regardless of the EME situation, it is not complex as long as it does not produce significant EMI toward susceptible equipment. Therefore, the EME complexity is determined by the number of elements of electromagnetic emission contained in the environment and the probability of them constituting effective electromagnetic interference, as well as the number of electromagnetic susceptible elements. In other words, the EME complexity is a function of the number of electromagnetic emission elements, the probability of effective electromagnetic interference, and the number of electromagnetic susceptible elements

$$\text{CEME} = \text{function}\big(\text{element}_{\text{Electromagnetic-emission}},\ \text{element}_{\text{Electromagnetic-susceptibility}}\ \text{P}_{\text{Availble-electromagnetic-interference}}\big)$$

The electromagnetic emission element is defined as a basic unit of electromagnetic emission with certain independence (or orthogonality) and representativeness (or completeness).

The electromagnetic emission element that causes the susceptibility is defined as effective EMI. Since EMC has statistical characteristics, the degree of EMI is described using the probability of effective EMI.

Electromagnetic emission elements and electromagnetic susceptibility elements are functions of the inherent attributes of the electronic information systems. These inherent attributes reflect the basic components of a general electronic information system, such as the power supplies, crystal oscillators, and frequency converter, and their signal characteristics can be expressed as wideband, harmonics, and modulated signals.

The overall electromagnetic emission and electromagnetic susceptibility characteristics of an electronic information system shown externally are determined by its working principle, manufacturing process, and structure and can generally be represented by a linear combination of electromagnetic emission elements and electromagnetic susceptibility elements.

4.1.5 Electromagnetic Environment Effect

Electromagnetic environment effect (E3) [13] refers to the impact of the EME upon the operational capability of electronic information systems, including electromagnetic interference (EMI), electromagnetic compatibility (EMC), electromagnetic vulnerability (EMV), electromagnetic pulse (EMP), electronic protection (EP),[1] hazards of electromagnetic radiation (to personnel (HERP), ordnance (HERO), and volatile materials such as fuel (HERF)), and natural phenomena effects of lightning and precipitation static (p-static).

Sections 4.1.1–4.1.3 have described the related concepts of EMI, EMC, and EMV. The EMP, EP, hazards of electromagnetic radiation (to personnel (HERP), ordnance

[1] Electronic Protection is described MIL-HDBK-237C [12]. It is worth noticing that in MIL-HDBK-237D (the revised version of MIL-HDBK-237), EP has been removed, but the authors think EP is an important concept in EMC and E3, so EP is retained in this book.

(HERO), and volatile materials such as fuel (HERF)), lightning, and precipitation static (p-static) are explained below.

(1) **Electromagnetic pulse (EMP)**: Electromagnetic radiation from a nuclear explosion caused by Compton recoil electrons and photoelectrons from photons scattered. Once the radiation is picked up by the electronic system or equipment, sudden voltage and current changes will occur, which will cause temporary or permanent damage to the electronic system or equipment. Since the damage range of EMP can be hundreds of kilometers, EMP weapons are a method of electromagnetic attack.
(2) **Electronic protection (EP)**: Used to be called electronic protection means (EPM) or electronic counter-countermeasure (ECCM). It refers to measures for protecting people, equipment, and facilities from electromagnetic spectrum produced by an enemy or one's own that reduces, suppresses, and destroys one's the combat effectiveness. EP includes various technologies and measures such as antielectron interference, anti-reconnaissance, and anti-destruction.
(3) **Electromagnetic radiation hazard (RADHAZ)**: When humans, equipment, ordnance, or fuel are exposed to dangerous electromagnetic radiation environment, the electromagnetic energy density will be sufficient to cause ignition, combustion of volatile materials, harmful biological effects of human, false triggering of electrically initiated devices (EIDs), failure or gradual degradation of critical safety circuits. These phenomena are usually referred to as electromagnetic radiation hazards on people (HERP), electromagnetic radiation hazards on ordnance (HERO), and electromagnetic radiation hazards on fuels (HERF).
(4) **Lightning**: The impact of lightning on electrical and electronic systems has become an important issue for systems with composite materials such as aircraft and ships. On the one hand, lightning strikes directly on aircraft and ships can instantly produce extremely high electromagnetic energy, causing catastrophic disasters to the system; on the other hand, currents generated by lightning on the surface of the system will have secondary effects, which will cause destruction to the critical equipment in the system.
(5) **Precipitation static (P-static)**: It is an EM disturbance caused by a random ESD buildup as a result of the flow of air, moisture, or airborne particles over the structure or components of a vehicle moving in the atmosphere, such as an aircraft or spacecraft [13].

4.1.6 Electromagnetic Environment Adaptability

Electromagnetic environment adaptability (EEA) refers to the system's ability to achieve its intended function, performance, and/or malfunction in the expected EME, and its ability to not produce unacceptable electromagnetic emissions to its environment.

It is worth pointing out that EEA and the response under EME are two completely different concepts. The former requires that the electronic information system must adapt to the EME, and the latter only needs the performance of the electronic information system under the EME.

EEA test is composed of two types: the assessment test and the performance evaluation test. Assessment test needs to be quantifiable, repeatable, and physical; i.e., the generated excitation environment, the EME acting on the equipment under test (EUT), and the electromagnetic energy induced by the EUT should be quantifiable; the test conditions, test data, and test results should be repeatable. The assessment items and indicators should physically reflect the characteristics of the EUT. Usually, the equipment used in the assessment test is general-purpose or special-purpose instrument that has been calibrated. The site used is usually a dedicated test site that can be calibrated. The state of the EUT is mostly static or detachable dynamic. The performance evaluation test, on the other hand, emphasizes that the test site and the excitation source should simulate the actual state of use as much as possible on the basis of assessment test. The performance evaluation test should be a comprehensive dynamic test with conditions close to the actual use.

4.1.7 Spectrum Management

Spectrum management (SM) [13] is the planning, coordinating, and managing the joint use of the electromagnetic spectrum through operational, engineering, and administrative procedures, with the objective of enabling electronic systems to perform their functions in the intended EME without causing or suffering unacceptable EMI. The main content of SM is spectrum allocation and spectrum certification.

Electromagnetic spectrum management is a method of achieving electromagnetic compatibility between systems through management when the electromagnetic compatibility of electronic information systems has been decided. Therefore, if the electromagnetic compatibility of each electronic information system is good enough, there can be more spectrum allocation solutions for spectrum management in actual use; otherwise, the difficulty of spectrum coordination will be increased.

4.1.8 Spectrum Certification

Spectrum certification (SC) [13]: Regular checks of the conformity between the spectrum in use and the spectrum allowed in the development and usage of the electrical and electronic systems are necessary. In the development and use of electrical and electronic systems, deviations may occur between the spectrum in use and the spectrum allowed due to design, manufacturing, structure, degraded performance of electronic devices, and structural changes during maintenance and repair. Such deviations often result in excessive occupancy of the limited spectrum resources which will further cause mutual interference problems. Therefore, the purpose of SC is to regularly test and confirm the legitimacy of spectrum usage in electronic information systems.

However, SC often faces many technical problems. For example, to achieve SC for large-scale RF towers, stereoscopic detection methods must be exploited.

4.1.9 Spectrum Supportability

Spectrum supportability (SS) [14] refers to the grade to which the spectrum and bandwidth of the military system can be effectively used. SS must ensure that the military system has the ability to work together with other systems in the same EME and achieve full performance. The evaluation of the SS of the system or equipment is based on two points: (1) equipment spectrum certification (ESC), which guarantees the effectiveness of frequency of use; (2) spectrum certification of sovereign nations and related information of electromagnetic compatibility.

The relationship among spectrum supportability, electromagnetic environment adaptability, spectrum certification, electromagnetic compatibility, and electromagnetic interference is shown in Fig. 4.6:

(1) Equipment must pass the spectrum certification. The equipment spectrum certification is closely related to its electromagnetic compatibility and electromagnetic interference. Only when the equipment has good electromagnetic compatibility and free of electromagnetic interference, can the equipment pass spectrum certification.
(2) The system must satisfy the requirement of electromagnetic environment adaptability and spectrum supportability. These qualities can be determined through test and evaluation (T&E). Specially, leftover engineering issues can be detected and solved through the pre-compliance test/evaluation and dynamic test/evaluation.
(3) When multiple systems are used together or in coordination, problems in electromagnetic environment adaptability and spectrum conflicts of ordnance may

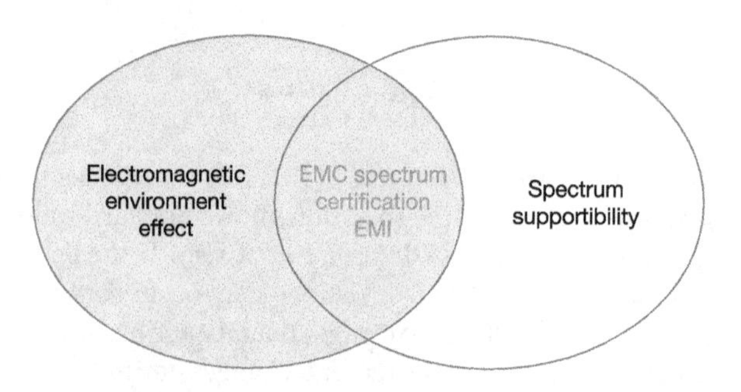

Fig. 4.6 Relationship between EEA and SS

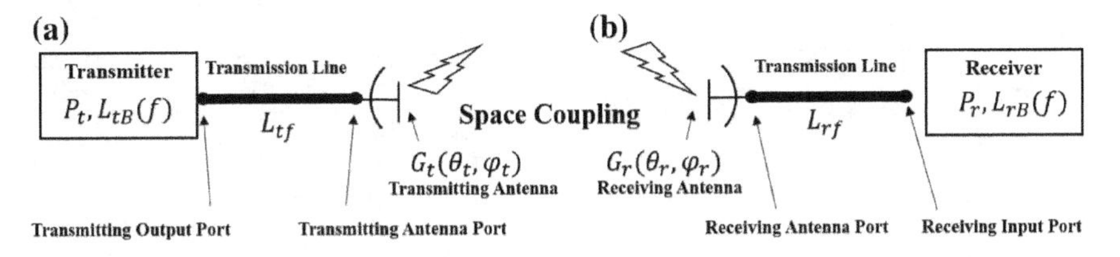

Fig. 4.7 Paths of the receiver of the aircraft that receives EMI

occur, and it may affect the normal task of the system or limit the certain functions.

(4) The electromagnetic environment adaptability and spectrum supportability should be evaluated to minimize the limitations and vulnerability of systems, subsystems, and equipment. The evaluation should be performed through the life cycle which includes the stages of design, development, operation, etc.

4.2 Essences of Quantitative EMC Design

In order help our readers better understand the quantitative system-level EMC design, we will introduce and discuss the elements of electromagnetic coupling, the development phases of EMC, and system-level EMC in this section.

4.2.1 Identification of Electromagnetic Coupling Elements

Interference sources, coupling paths, and susceptive equipment are the three aspects (or elements) of EMC.

In practice, it is most difficult to identify the coupling path, because both the conductor and the air can be used as the transmission channel for the electromagnetic energy. Moreover, the electromagnetic fields involved in EMC problems often include the coexistence of near/far/inductive fields.

The following four examples illustrate the difficulty of coupling path identification.

Example 4.1 As shown in Fig. 4.7, the electromagnetic emission received by the receiver on the aircraft may come from three paths: circuit–circuit coupling of the receiver, field–circuit coupling between the receiving antenna and receiver cable, and the field coupling of receiving antenna.

The interference received by the circuit inside the receiver under conduction mode is referred to as the interference received through the circuit–circuit path. The emission received from field by the cable of the receiver is referred to as the interference

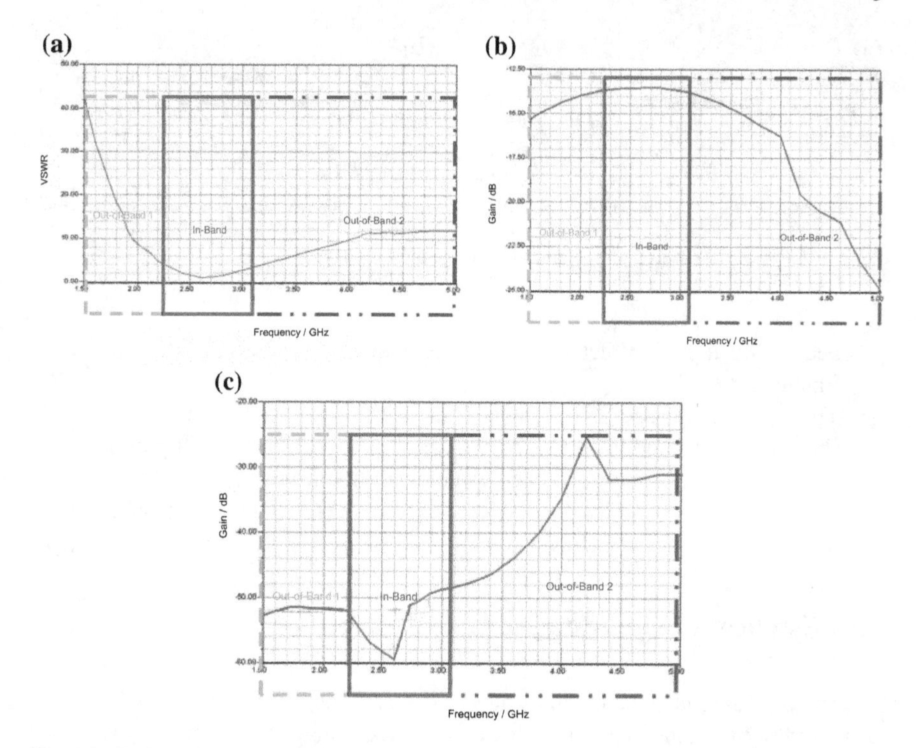

Fig. 4.8 Standing wave, main beam, and side lobe with changing frequency of an airborne receiving antenna. **a** Standing wave frequency characteristic; **b** main lobe frequency characteristic; **c** side lobe frequency characteristic

received through field–circuit coupling path. The emission received by the antenna, which is generated by other transmitting antennas or other EME signals, is referred to as interference received through the field–field coupling channel. Specifically, field coupling channel interference can be received by the main beam of the receiving antenna as shown in Fig. 4.8b, which is called direct coupling. The interference can also be received by the side lobe of the receiving antenna, as shown at the out-of-band part in Fig. 4.8c, which is called indirect coupling.

Example 4.2 Figure 4.9 shows the four coupling paths of the emission from the source to the receiver:

Path 1—The emission radiates directly from the source to the receiver.
Path 2—The emission radiates directly from the source and is then picked up by a power cable or signal/control cable connected to the receiver. The emission is thus conducted to the receiver.
Path 3—The emission radiates from the power, signal, or control cables of the source.
Path 4—The emission is conducted directly from the source through the utility power cable or through a common signal cable/control cable to the receiver.

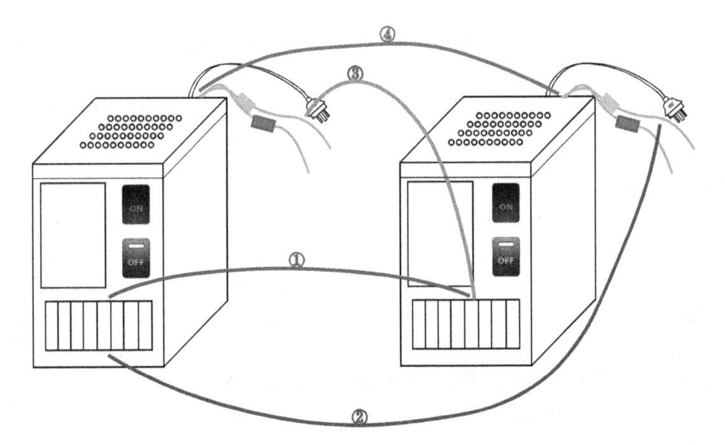

Fig. 4.9 Schematic diagram of the four coupling paths of EMI from source to receiver

Based on the four coupling paths described above, it is obvious that the main mechanism of emission coupling from the source to the receiver is conduction and radiation.

Cables often appear in bundles in real-world projects. Thus, the emission carried by various power/signal/control cables connected to the source can be easily coupled to the power/signal/control cables connected to the receiver.

It is worth noticing that the emission caused by the cable is diffusive (whether it is sending out interference or picking up interference), so it is very difficult to solve the interference problem caused by the cable. Especially when the interference picked up by the cable is at the same frequency with the functional signal transmitted in the cable, the interference can hardly be filtered out. Therefore, it is fundamentally important to strictly control the coupling path caused by cables.

Example 4.3 Possible "antenna" coupling effects on radio frequency circuit boards (referred to as circuit boards in this section). An engineer tested a circuit board during the debug phase and found it was working properly. Then, the engineer made via holes at the four corners of the board to fix it into a case. However, after installation, the performance of the circuit board was abnormal. The board was then removed from the case, and the engineer found that the performance of the circuit board was inconsistent with the result of the previous test. The performance before punching could not be reproduced after several tests.

After careful investigation by an EMC specialist, it was found that the reason for this phenomenon was the mounting holes that the engineer made at four corners of the circuit board. The holes were located on the surface of the circuit board where electromagnetic currents were distributed. The holes actually cut the electromagnetic current and formed an "antenna" effect, which caused the system to operate abnormally. Since the cutting was irreversible, the performance of the circuit board could not be restored to the state before.

If the engineer could have analyzed the electromagnetic current on the board surface using the analysis software or the electromagnetic field scanning instrument before mounting the holes, we believe the above problem can be avoided.

Example 4.4 A metal partition was installed between two sections of a system. Later, the metal partition was removed in order to reduce the weight during a refit process. Then, it was found that an EMI problem occurred in the equipment located in the compartment.

The reason for this problem is that the metal partition has an electric shielding effect. The isolation between the front and rear cabin sections was reduced; thus, EMI problem occurred.

The four examples illustrate that there are many kinds of electromagnetic coupling paths. Only by understanding the coupling mechanism can we effectively avoid the emergence of new EMC problems.

4.2.2 Three Stages of EMC Technology Development

EMC studies have been through three stages of development: problem solving, standard design, and system design. The corresponding EMC research methods are problem-solving method, standard design method, and system design method.

(1) Problem-solving stage: In the design process of electrical and electronic systems, EMC is not fully considered. When there is an EMI problem later, engineers start to analyze the cause and find solutions.
    Problem-solving method: The core concept of this method is "solving the problem only after it occurs." This method is limited by many factors, and it is often difficult to implement the solution. Moreover, in terms of complex systems, it is even more difficult to solve the EMC problem.
(2) Standard design stage: Standardize EMC work and implement EMC design according to the standard requirements. At present, there is relatively complete EMC design specifications in the field of circuit boards and components design. Some effective qualitative or quantitative specification requirements have also been formed in grounding, bonding, shielding, cable layout, and static electricity prevention. EMC issues can be effectively reduced by designing in strict accordance with the specifications.
    The commercial circuit design software that uses the design specifications plays an important role in reducing the interference of the electrical and electronic systems. As an example, the software may specify the minimum spacing limit for two parallel laid RF printed lines in order to avoid signal cross talk caused by too small spacing.
    However, for complex systems, the significance of the standard specification is often "in theory" due to the variations and complexity of the system.
(3) System design stage: EMC performance of systems, subsystems, equipment, circuit boards, and components is analyzed and predicted, indicators are reason-

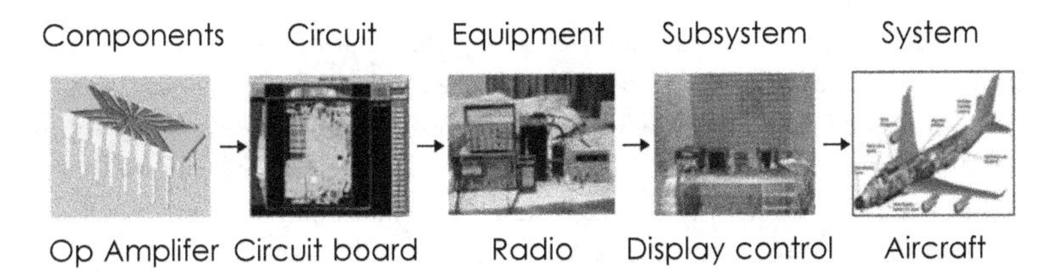

Fig. 4.10 Concept of levels in EMC

ably allocated, and the system performance can be optimized by continuously iterating during the entire system design phase.

The "system design method" emphasizes top-level design, process control, and continuous iteration, and therefore, it requires technical methods and physical environment, which includes analysis, prediction, design, evaluation, and other methods; models, databases, and other resources; testing, verification, and other test facilities. The system design method embodies the iterative and spiral-up optimization ideas. A large number of applications indicate that system design method plays an important role in effective solving of system-level EMC problems.

With the advancement of technology, the system design method is moving toward the direction of all digital design, digital physical collaborative design, and multi-site collaborative design.

4.2.3 System-Level EMC

Correct understanding of the system-level concept in EMC research is very important for quantitative system-level EMC design. The following section describes the system-level concept in the EMC field.

According to different functions, forms, and methods of problem solving, EMC problems can be classified into five levels: device level, circuit board level, equipment level, subsystem level, and system level, as shown in Fig. 4.10.

It should be pointed out that the form of system level has been changed significantly with advancement in technology, from single-entity systems (such as aircraft, ships, satellites, vehicles) to multi-entity systems (such as drone systems). Therefore, the concept of system-level EMC has also been broadened.

The concept of system-level EMC also includes intra-system EMC issues, inter-system EMC issues, and issues between the system and the environment.

(1) **Intra-system EMC issues**. The whole unit, which is used to accomplish the intended task and can achieve mutual EMC through design synchronization during the development phase, is called the "system," e.g., single-entity systems such as aircraft and ships, multi-entity systems such as drone systems, and even aircraft carrier systems that contain aircraft carrier and carrier-based aircraft.

When the whole system is intended to accomplish one task, and in the development phase, mutual EMC can be realized through design synchronization (if any unit in the whole is not electromagnetically compatible with other units, the overall intended task cannot be completed), and the EMC problem that occurs within the entire system should be the intra-system EMC issue.

(2) **Inter-system EMC issues**. The unit that is used to accomplish the tasks but unable to achieve mutual EMC through design synchronization during the development phase is called "multi-system." If aircraft fleets developed at different times are used, these aircraft cannot achieve mutual EMC through design synchronization at the development phase. Even though the development of aircraft in the later stage has fully considered the EMC with "old aircraft," its design often needs to reconcile with the performance of the "old aircraft" which can hardly be changed.
Ship fleets developed at different times and new carrier-based aircraft used for existing aircraft carrier systems often have such problems.

(3) **EMC issues between the system and the environment**. The system needs to complete its task in a new environment (such as the unexpected natural or man-made electromagnetic environments), which has not been considered during the system development phase.
The reason that we make distinction among intra-system EMC issues, inter-system EMC issues, and EMC issues between the system and the environment is that there are different approaches to achieve system EMC for different types of issues. EMC of all three types can be achieved through system-level EMC design, control, and evaluation in the development phase. The latter two can also be achieved through comprehensive technical methods such as spectrum management.

4.2.4 Characteristics of System-Level EMC

EMC is usually concealed and is closely related to the inherent attributes of electronic information systems, so that EMC indicators, EMC models, EMC design, EMC detection, EMC evaluation, and EMC testing all have the following specialties:

(1) Different from other electrical indicators, the EMC indicators have probabilistic statistical characteristics. Since the EMC of the whole aircraft is the result of the comprehensive action of all the electronic information systems of the aircraft, both the electromagnetic emission characteristics and the electromagnetic susceptibility characteristics of the whole aircraft have probabilistic statistical characteristics. Therefore, the EMC data sample of the aircraft, the data acquisition method, and statistical method are crucial for the determination of the EMC indicators of the aircraft.

(2) Different from other electrical design, EMC design is a combination of functional design and non-functional design. Functional design, also known as nor-

mal signal design, is the complete electrical design based on functional specifications; e.g., radio stations are designed based on operating frequency, RF bandwidth, working mode, receiver sensitivity, demodulation mode, out-of-band spurious and harmonics suppression, transmit power, modulation method, and parameters. Non-functional design is to predict and minimize the effect of non-functional signals with the presence of both functional and non-functional signals.

The EMC design consists of functional design and non-functional design. The design process includes design for normal signal and the design with the combination input of normal signal and abnormal signal. Specifically, non-functional signals refer to the interference caused by the aircraft on the airborne radio and the sum of airborne electromagnetic environment signals of the airborne radio. The interference signal caused by the aircraft mainly refers to the interference due to unsatisfactory loading conditions of onboard power supply and impedance, and the interference of other onboard electronic information devices to radio stations. The electromagnetic environment signal mainly refers to the EMI to the aircraft during operation, such as the impact of civil communications on the aircraft.

The design using normal signal and non-functional signal together as a design input means that the normal design signal and the interference signal from the internal and external environment of the aircraft are used as the input of the radio station to evaluate its impact. Only when the airborne radio station has a corresponding protective design for the interference signal, can it have the capacity to be installed and be compatible with other equipment on the aircraft.

(3) Different from other tests, EMC tests are used for diagnosis through external inspection. The system-level EMC test is performed on the whole aircraft, and the data generated is comprehensive from the whole system in operation. Therefore, the identification of the aircraft's EMC problem is similar to "diagnosis with traditional Chinese doctors," which means detecting the internal problems through observing external appearance. One major task of the system-level EMC test is to detect new EMC problems due to interconnects and couplings in subsystems who has passed EMC tests, respectively. This diagnostic kind of test makes the entire EMC test complex and important.

(4) Different from equipment- and subsystem-level test, system-level EMC tests belong to large-scale system tests. According to the system scale, EMC problems can be categorized into different levels, including device level, circuit board level, equipment level, subsystem level, and system level. Since the aircraft assembles a large number of electronic information and control equipment which are distributed on a large scale, the test antenna can only cover parts of the aircraft and some equipment in electromagnetic emission test and electromagnetic susceptibility test. Therefore, the system-level EMC test is essentially different from the tests at device level, circuit board level, equipment level, and subsystem level both in terms of indicators and methods.

Table 4.1 Difference between EMI in EMC and jamming in electronic warfare

EMI	Jamming
Intentional	Unintentional
Emission	Transmission
Susceptive	Damage

Table 4.2 Technical domain attributions of common interference phenomena

Type of disturbance source	Type of disturbance	Classification of subjects
Self-interference in our system	Unintentional interference	EMC
Inter-interference in our system	Unintentional interference	EMC/E3
Interference in our civil environment	Unintentional interference	EMC/E3
Interference from our equipment in the battlefield	Unintentional interference	EMC/E3
Interference from enemy equipment in the battlefield	Unintentional interference	EMC/E3
Intentional interference from enemy equipment in the battlefield	Intentional interference	Electronic warfare

4.2.5 Interpretations of the EMI in Different Fields

There are different interpretations of EMI in different fields. In the field of EMC, interference refers to the phenomenon of unintentional signal generation. In the field of electronic warfare, interference generally corresponds to intentional interference. Due to the different technologies adopted in the two cases, to avoid confusion, the interference in the field of EMC is still called EMI, and the interference in electronic warfare is called jamming, as listed in Table 4.1.

Table 4.2 lists the technical domain attributions of common interference phenomena (attribution is not unique and is subject to change).

4.3 Basic Concept of EMC Quantitative Design

4.3.1 Interference Correlation Relationship

The mutual interference among the subsystems and equipment in the system is called the interference relationship. It is usually obtained based on the analysis of the energy coupling relationships among the equipment in the system, which requires to con-

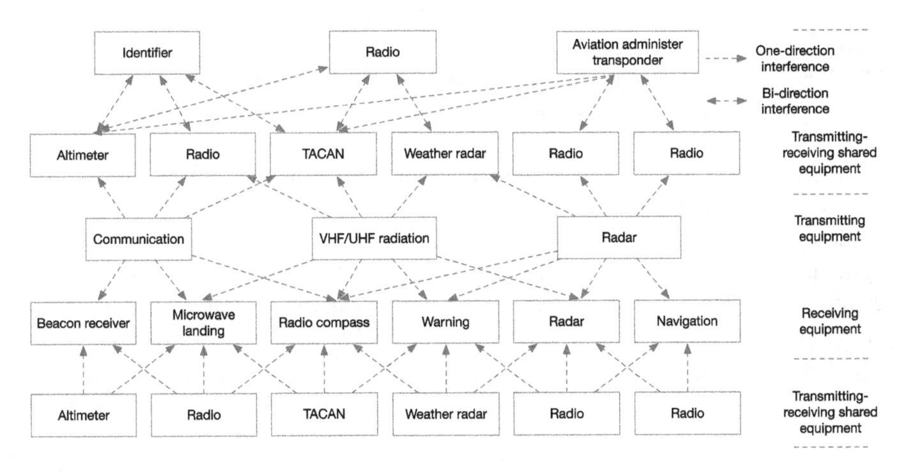

Fig. 4.11 Interference relationships among some airborne equipment

sider not only the energy coupling generated by the antenna of the transmitter and receiver, but also the energy coupling caused by the cases, cables, connectors, and even the power supply and grounding. Since the electromagnetic energy coupling in large-scale systems is usually complex, when analyzing interference relationships, a hierarchical analysis approach should be adopted; i.e., the energy coupling generated by the antenna of the transceiver system is considered firstly, then the coupling generated by cables and cases is considered, and finally, grounding and other couplings are considered. Figure 4.11 shows the interference relationships between some of the aircraft's airborne equipment. The interference relationships can also be expressed in tabular form.

In general, an aircraft includes transmitting equipment that radiates through antenna ports and other radiation equipment, as well as susceptive equipment that receivers coupled through antenna ports and other susceptive equipment. The interference correlation relationship within the entire system is complex, and the number of equipment is large. Figure 4.11 only describes the partial interference correlation relationship of the onboard equipment including communications, radar, altimeter, TACAN, and weather radar through the antenna port. In fact, when analyzing and evaluating EMC of the whole system, the number of radiation source ports and the number of susceptive ports are even bigger, and the coupling is more complicated.

4.3.2 *Interference Correlation Matrix*

The mathematical model to describe the interference correlation relationship quantitatively is called the interference correlation matrix. The matrix can reflect the mutual interference relationship of the whole aircraft system in a comprehensive, clear, and accurate manner [13]. It provides an important technical method for analyzing, evaluating, and optimizing the EMC of the whole aircraft and can be used to allocate the EMC indicators of the whole aircraft to subsystems/equipment.

The interference correlation matrix includes multiple forms, such as the coupling interference correlation matrix, isolation interference correlation matrix, and impedance interference correlation matrix. It depends on the specific analysis requirements to decide which type of matrix to use.

Equation (4.1) describes the coupling interference matrix (**A**) of a steady-state system. There are M interference ports and N susceptive ports in the system. The coupling function between the j-th susceptive port and the i-th interference port is $H_{i,j}(t, f)$:

$$\mathbf{A} = \begin{bmatrix} H_{1,1}(t, f) & \cdots & H_{1,j}(t, f) & \cdots & H_{1,N}(t, f) \\ \vdots & & \vdots & & \vdots \\ H_{i,1}(t, f) & \cdots & H_{i,j}(t, f) & \cdots & H_{i,N}(t, f) \\ \vdots & & \vdots & & \vdots \\ H_{M,1}(t, f) & \cdots & H_{M,j}(t, f) & \cdots & H_{M,N}(t, f) \end{bmatrix} \tag{4.1}$$

4.3.3 System-Level EMC Requirements and Indicators

1. System-level EMC requirements

System-level EMC requirements include three aspects: the overall EMC requirements, the overall EMC technical requirements, and the EMC management requirements.

The overall EMC requirements are usually clearly stated in the general requirements for product development and are defined in terms of use. When making the overall EMC requirements for a product, comprehensive consideration must be given to the natural or man-made EME where the product operates, the electromagnetic emission occurs when the product is in operation, the electromagnetic susceptibility that affects the function of the product, the EMC problems happened to the product before, and the requirements and methods for EMC evaluation of the product, etc.

The overall EMC technical requirements are based on the technical aspects of the overall requirements and are generally specified in the overall development plan. When making the overall EMC technical requirements, natural and man-made EME where the products operates, the electromagnetic emission generated by the product while in operation, the electromagnetic susceptibility affecting the function of the product, etc., need to be provided quantitatively. For example, the electric field intensity and magnetic field intensity corresponding to the EME, the time-domain feature of the transient field, the spectrum occupancy of the frequency-domain signal, and the characteristics of the pulse signal need to be provided. The overall technical EMC requirements should be based on the hierarchical levels of the system, subsystems, and equipment. The limits of the product's EMC test should also be specified.

EMC management puts forward requirements from the perspective of quality and standardization. It is usually specified in the overall development requirements. EMC management commonly includes the operating mechanism, management content, and control measures (milestones, control content, evaluation elements) of the product.

2. System-level EMC indicators

System-level EMC indicators consist of three items: demonstration indicator, design indicator, and test indicator. The demonstration indicator is used to control the overall EMC performance of the product. The design indicator is derived from the demonstration indicator. It is not only the input indicator of the system's EMC design, but can also be developed into the system's EMC test indicator. The test indicator specifies the EMC limit requirement of the system. It is used to test whether the EMC design of the system is effective and whether the actual EMC performance of the system meets the requirements of the EMC demonstration indicator. The relationship among the three indicators is shown in Fig. 4.12.

(1) Demonstration indicators

Finding indicators that objectively reflect the overall EMC of a product is a topic that needs continuous research. The EMC research team of Beihang University has investigated this issue for more than 10 years and has obtained preliminary research results—the EMC condition (EMC(s)) that characterize the EMC of the whole aircraft at system level [12]. EMC(s) quantitatively describes EMC of the system, subsystem, or component as a whole. It represents the extent to which the object being analyzed satisfies EMC requirement considering the four major factors including electromagnetic emission, electromagnetic susceptibility, isolation, and safety margin.

When the complete system of the aircraft has been integrated: If all susceptive equipment can work normally according to design specifications, and fuel and personnel safety also meet the requirement, then EMC(s) equals 1. On the other hand, if all susceptive equipment is affected by interference, fuel and personnel are exposed to radiation that exceeds the standard, then EMC(s) equals 0. In engineering practice, generally $0 < \text{EMC}(s) \leq 1$. The purpose of establishing the indicator for EMC condition in the demonstration phase is to control the overall EMC level of the aircraft. The purpose of EMC design and optimization is to make EMC(s) close to or equal to 1.

(2) Design indicators

The EMC design indicator has a close relationship with the system design indicator. It specifies system requirements from the perspective of EMC. EMC design indicators have eight categories, i.e., frequency resource planning, EME, transmitter performance (transmitting frequency, transmitting power, harmonic suppression, out-of-band suppression), receiver performance (receiving frequency, receiving sensitivity, adjacent channel suppression, out-of-band suppression), antenna layout performance (antenna pattern, polarization, antenna isolation, suppression capabilities), equipment and cable layout, interconnection characteristics (frequency band coupling, shielding effectiveness, power and grounding), and damage threshold.

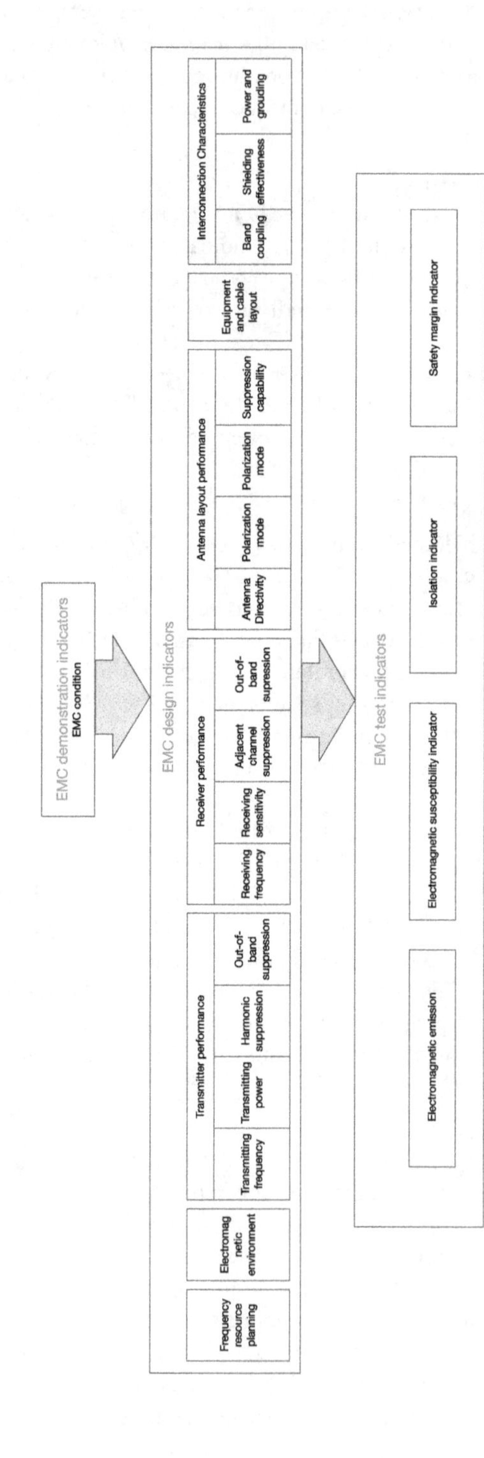

Fig. 4.12 Relationship among EMC demonstration indicators, design indicators, and test indicators

The frequency resource planning, EME, antenna layout, equipment and cable layout, interconnection characteristics, and damage threshold are indicators directly associated with the system-level design and are generally designed and controlled by the project manager. The corresponding control parameters are isolation and safety margin. Transmitter and receiver performance is based on the overall system isolation and safety margin requirements, which need to be determined in close consultation with the transmitting equipment, receiving equipment, and subsystems. The corresponding control parameters are directly related to the performance of the transmitting/receiving equipment.

Among the eight categories of design indicators we just introduced, EME and transmitter performance are the "sources" of the EMI profile. Receiver performance is an important factor in determining susceptibility profiles. Frequency resource planning, antenna layout performance, equipment and cable layout, and interconnection characteristics are important ways to avoid the susceptibility profile and the interference profile reaching each other. And the damage threshold is an important basis to determine the safety margin.

For large-scale systems with dozens or even hundreds of equipment and subsystems such as airplanes, in order to formulate the design indicators for the system, subsystem, and equipment, it usually requires EMC models and EMC simulation methods.

(3) Test indicators

EMC test requirements come in system level and equipment/subsystem level. GJB 1389 sets out detailed requirements for system-level EMC, while GJB 151 sets out requirements for equipment/subsystems and provides test methods in detail.

According to the three aspects of EMC, the EMC evaluation test requirements include conduction emission (CE), conduction susceptibility (CS), radiation emission (RE), and radiation susceptibility (RS). The requirements apply to both the system-level test and the subsystem-/equipment-level test. Conduction emission and radiation emission are two types of tests to evaluate the product's external electromagnetic emission characteristics. In these tests, the product is subject to the investigation as the "interference source." On the other hand, the conduction and radiation susceptive tests evaluate the product's characteristics of being interfered by external electromagnetic signals. In these tests, the product is treated as a "susceptive object." Sometimes, the same product can be both the "interference source" and the "susceptive object." The product's own antenna, cable, case, and reference ground can serve as the "interference sources" to other products. At the same time, the product can also be interfered by others and become the "susceptive object."

System-level EMC test requirements include safety margin, intra-system EMC, external RF EME, lightning, electromagnetic pulse (EMP), subsystems and equipment EMI, electrostatic charge control, electromagnetic radiation hazards (EMRADHAZ), and E3 hardness in the life cycle, electrical bonding, external grounds, TEMPEST, emission control (EMCON), and EM spectrum compatibility.

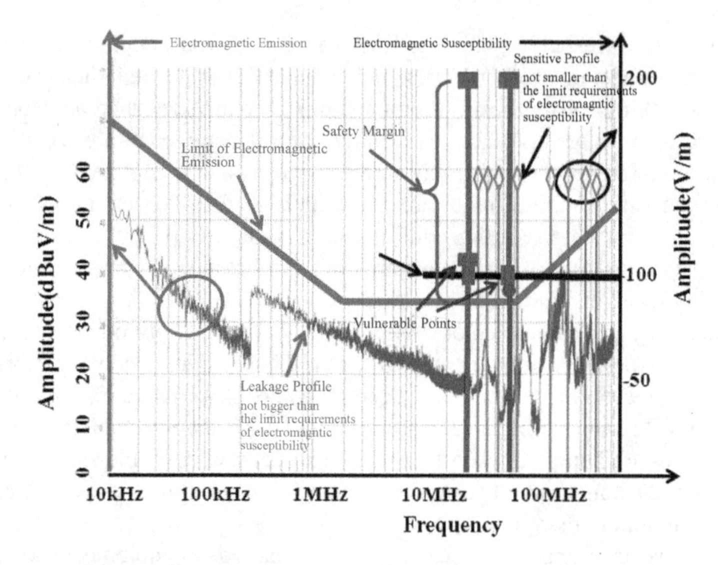

Fig. 4.13 Relationships among electromagnetic emission profile, electromagnetic susceptibility profile, vulnerable points, and safety margin

4.3.4 Electromagnetic Emission Profile and Electromagnetic Susceptibility Profile

The amplitude–frequency characteristic of the product's electromagnetic emission is called the electromagnetic emission profile. The product's amplitude–frequency characteristic of electromagnetic susceptibility is called the electromagnetically susceptibility profile, as shown in Fig. 4.13.

In Fig. 4.13, the solid line using the left axis as the vertical axis is the electromagnetic emission profile. The dotted line with ◊ using the right axis as the vertical axis is the electromagnetic susceptibility profile. Points marked with ■ are the vulnerable points. Points marked with ◊ are electromagnetic susceptive points and their corresponding threshold. The thick line between ◊ and ■ is the magnitude corresponding to the electromagnetic safety margin. And the difference between electromagnetic susceptibility profile that satisfies the limit requirements (needs to consider the safety margin at the vulnerable points) and the electromagnetic emission profile is the degree of isolation.

Product isolation requirement can be derived based on the analysis of product emission profile and susceptibility profile. Safety margin requirement can be further determined for the vulnerable points. Electromagnetic emission profile, electromagnetic susceptibility profile, isolation, and safety margins reflect the interrelationship of the three aspects of EMC: Electromagnetic emission profile is an objective reflection of unintentional emission of the product and describes the electromagnetic emission generated by the emission sources; the electromagnetic susceptibility

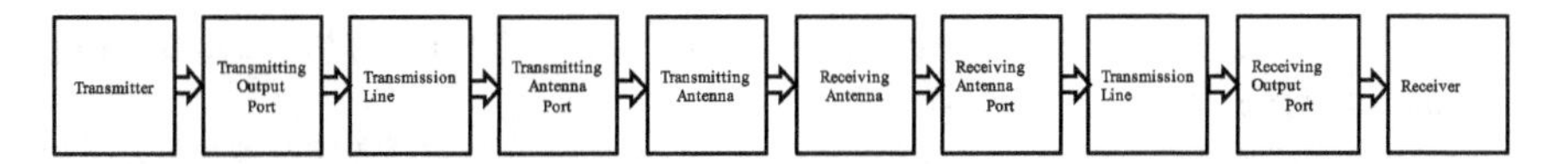

Fig. 4.14 Factors affecting the isolation between transmitter and receiver

profile reflects the product's response to electromagnetic emission and describes the susceptibility of the susceptive object. The isolation is to ensure that the interference profile of the interference source does not intersect with the susceptibility profile of the susceptive object, and it has a certain segmentation degree of isolation required by the margin; the safety margin is to prevent the vulnerability of the product due to its susceptibility, and it sets safety margin requirement to the vulnerable points and its surroundings.

4.3.5 Equipment Isolation

Equipment isolation is proposed relative to antenna isolation. Antenna isolation calculation is one of the most important topics in the system-level EMC analysis and prediction. The antenna isolation requirement is directly related to the installation position of the airborne antenna and may even affect the geometric layout and performance design of the transceiver system.

In practical applications, however, it is not rare that airborne transceiver systems that have met the design requirements for antenna isolation still suffer from mutual interference problems. After long-term investigations, we found that one of the main reasons leading to this problem is that the antenna isolation only serves as an intermediate parameter in the EMC design of the transceiver system, and it does not fully reflect the EMC of the transmitter and receiver; i.e., the antenna isolation only isolates between transmitting antenna—spatial channels—and receiving antenna. However, the factors affecting the EMC between transmitter and receiver also include the receiver's receiving characteristics, the transmitter's transmission characteristics, and the connectors and cables connecting the receiver/transmitter to the receiving/transmitting antenna. Moreover, the nonlinearity of the RF front end, out-of-band characteristics, bonding characteristics of the cable and connector, and shielding characteristics all contribute to EMC issues between transmitter and receiver.

In order to describe the EMC of the transceiver system accurately and improve the effectiveness of the prediction and design, it is necessary to investigate the isolation of the equipment. The factors affecting the isolation between the transmitter and the receiver are shown in Fig. 4.14.

Spatial isolation between the transmitting and receiving antennas needs to be calculated first, no matter if we are solving antenna isolation or equipment isolation.

Spatial isolation, antenna isolation, and equipment isolation, and the relationship among the three are defined as follows.

(1) **Spatial isolation**: Assuming that both the receiving antenna and the transmitting antenna are omnidirectional point source antennas, the isolation between the two antennas is spatial isolation, which reflects the contribution of various boundary conditions of isolation points.
(2) **Antenna isolation**: Receiving gain of the receiving antenna and the transmitting gain of the transmitting antenna after installation are included on the basis of spatial isolation. The influence of the aircraft fuselage needs to be taken into consideration when we calculate the receiving gain of the receiving antenna after installation. Similarly, the transmitting gain of the transmitting antenna after installation is also affected by the aircraft fuselage. In the antenna isolation, the transmitting and receiving antennas are usually directional, and the directionality is affected by the aircraft fuselage where the antennas are installed.
(3) **Equipment isolation**: Based on the isolation of the antenna, the characteristics of the RF front end of the receiver, the characteristics of the connector and the connecting cable between the receiver and the receiving antenna, the characteristics of the transmitter RF front end, and the characteristics of the connector and connection cable between the transmitter and the transmitting antenna are taken into account.

4.3.6 Quantitative Allocation of Indicators

Quantitative allocation of indicators refers to the implementation of the overall technical requirements of the system into equipment/subsystems.

Quantitative allocation of indicators is usually based on the principle of the three aspects of EMC, namely electromagnetic emission, electromagnetic susceptibility, and isolation.

For instance, the equipment isolation between the transmitter and the receiver is 180 dB. After subtraction of the isolation between the transmitting antenna and the receiving antenna (the isolation caused by the pattern, polarization and antenna position of receiving antenna and the transmitting antenna), the isolation between the EMI source and the susceptive equipment should be no less than 110 dB (the isolation requirement on the left side of Fig. 4.15a). Based on the actual capabilities of EMI sources and susceptive equipment, the indicators can be allocated through the following five ways.

(1) The out-of-band emission attenuation of the interference shall not be less than 85 dB, and the suppression capability of susceptive equipment to the out-of-band emission of the interference source shall not be less than 25 dB, as shown in Fig. 4.15a.
(2) The out-of-band emission attenuation of the interference source should not be less than 85 dB, but the suppression of out-of-band emissions from the

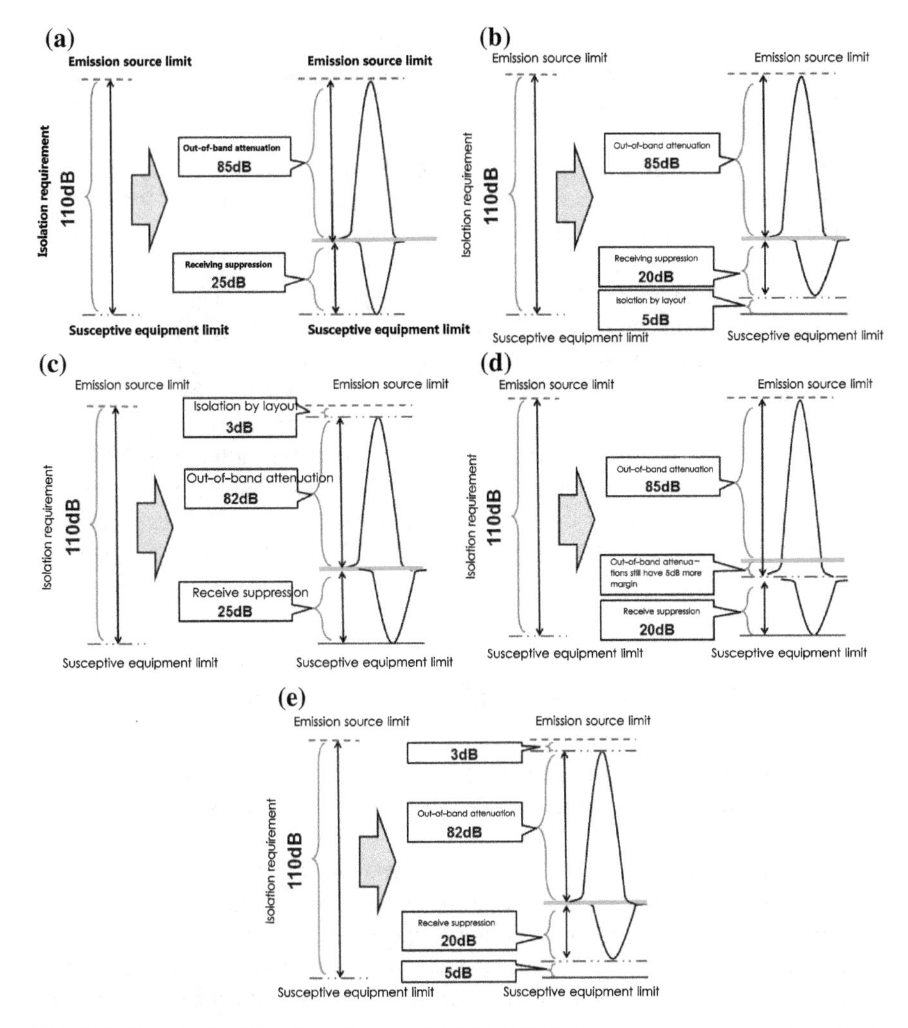

Fig. 4.15 Quantitative allocation of system-level EMC indicators

susceptive equipment to the interference source can only reach 20 dB. Therefore, the relative layout of the interference source and susceptive equipment needs to be adjusted to provide an additional 5 dB of spatial isolation; i.e., the isolation requirement between the EMI source and the susceptive equipment needs to be modified to "no less than 105 dB," as shown in Fig. 4.15b.

(3) The out-of-band emission attenuation of the interference source can reach 82 dB but cannot reach 85 dB. Meanwhile, the maximum suppression capability of susceptive equipment to the out-of-band emission of the interference source can only reach 25 dB. Then, the system needs to provide an extra 3 dB spatial isolation by adjusting the relative layout of interference source and susceptive

equipment; i.e., the isolation requirement between the EMI source and the susceptive equipment needs to be revised to "no less than 107 dB," as shown in Fig. 4.15c.

(4) The suppression capability of the susceptive equipment to the out-of-band emission of interference sources can only reach 20 dB, but the out-of-band emission attenuation of the interference source has the potential to achieve 90 dB, which is equivalent to the transmission equipment sharing 5 dB indicator for susceptive equipment, as shown in Fig. 4.15d. Similarly, if the transmitting equipment has insufficient capacity but the susceptive equipment has potentials, the susceptive equipment should share the indicator for the transmitting equipment.

(5) The out-of-band emission attenuation of the emission source can reach 82 dB but cannot reach 85 dB; the suppression capability of susceptive equipment to the out-of-band emission of the interference source can reach 20 dB but cannot reach 25 dB. Under this condition, the solution is either to allocate the indicators that the transmitting equipment and susceptive equipment cannot achieve to the system or to reduce the isolation requirements. It is worth noticing that decreasing isolation requirements means reducing performance, as shown in Fig. 4.15e.

4.3.7 The Construction of EMC Behavioral Model

The EMC behavioral model is proposed by the EMC Research Team of Beihang University to solve the top-level quantification, demonstration, and design of the system.

Equipment parameters and circuit design layout are often required when using simulation analysis software to do performance simulation analysis of the system, equipment, circuits, etc. However, it often happens that when the system is demonstrated at the top level, the subsystems and equipment in the system only have major functional indicators, but does not come with detailed design scheme, component parameters, or circuit design layout.

Based on the analysis and research of various state-of-the-art simulation technologies, this book proposes the concepts of system-level behavioral simulation model, subsystem/equipment behavioral simulation model, etc. [12].

The behavioral modeling of EMC targets to model the external characteristics (behavior), rather than the physical or internal structure of the system. This method focuses on the behavioral trends of the system and establishes homomorphism model through induction method. In other words, the informal description method is used to determine the main observation variables of the system. The observed data is then processed through induction and informal description. New data is then generated through extrapolation. The model data is thereby generated, and the system behavioral model is established. This method is called behavioral modeling.

The definition of behavioral modeling can be summarized as: modeling based only on the electromagnetic emission to the outside and the susceptive response

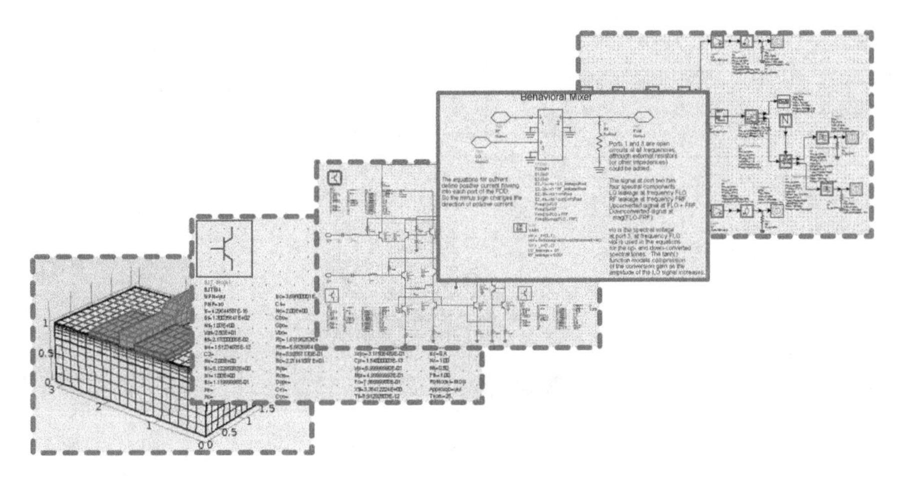

Fig. 4.16 Hierarchy of electronic system simulation model

to the external electromagnetic signal. With this method, the circuits and equipment within system, subsystems, and equipment are considered as a black box and internal characteristics are unnecessary to be extracted.

Figure 4.16 illustrates the electronic system simulation model hierarchy. The bottom layer is the physical model of the semiconductor structure. After the port parameters are extracted, device circuit models such as the transistor model are established, which are used to further build integrated circuit models such as amplifiers and mixers; the top layer is a complex system digital simulation model. The behavioral model is the bridge between the circuit model and the system simulation model. If the behavioral model is used to describe the subcircuit, the complexity of the system calculation problem caused by the existence of a large number of components can be avoided. This kind of behavior description will greatly improve the simulation speed and efficiency [16].

EMC behavioral simulation is the application of behavioral models to system-level EMC simulation. An electronic information system usually contains multiple subsystems and complex coupling relationships. By classifying and layering, the system can be divided into many subsystems with different functional characteristics, which are required for EMC analysis. Then, the behavioral model for the subsystems can be built one by one. The EMC behavioral modeling and simulation process are shown in Fig. 4.17. After the classification, the performance parameters of the subsystem are extracted, and a specific mathematical model is selected for description. During the modeling process, the model is continuously revised according to the performance verification, and the reliability of the model is thus improved. Finally, the model is used for system-level EMC simulation.

The EMC behavioral model has certain similarities with the behavioral model of analog electronic circuit devices, e.g., the large signal model and small signal model of semiconductor triodes. It is worth noticing that in the simulation of system EMC,

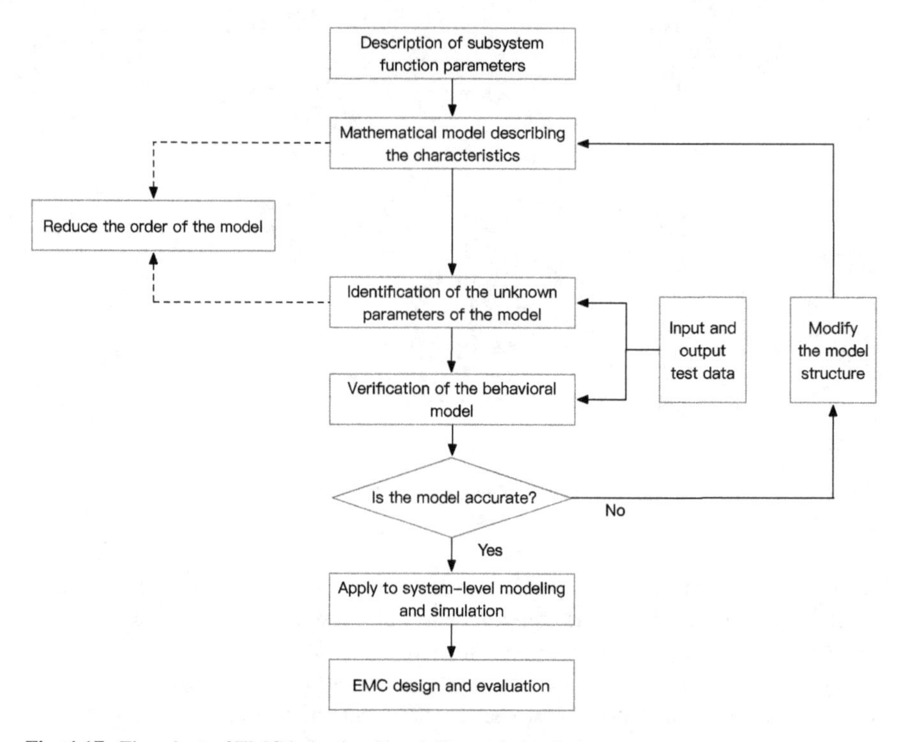

Fig. 4.17 Flowchart of EMC behavioral modeling and simulation

there are not only circuit-level problems, but also a large number of electromagnetic field coupling problems. Therefore, in the behavioral model, the influence of the electromagnetic field must be considered. In other words, to build an accurate behavioral model, we need to reform the field information (including near-field, far-field, and transition zone), the field–circuit coupling, and the distribution effect of the circuit–circuit parameter to the interference source at the circuit level.

Based on the EMC behavioral model, we can establish an EMC model of the whole aircraft that can be used in the simulation. The model is referred to as an aircraft EMC digital model. Based on the digital model of aircraft EMC, we can further carry out electromagnetic emission characteristics prediction and control, electromagnetic susceptibility characteristics prediction and protection design, and electromagnetic vulnerability prediction, etc.

4.3.8 The Behavior Simulation of EMC

EMC behavioral simulation technology is proposed as a key to solve the EMC quantitative demonstration and design from the top level of the system.

The behavioral model makes it possible to quantitatively demonstrate and design EMC of the system. It provides the necessary technical foundation for EMC from conceptual design to detailed design and lays the foundation for EMC behavioral simulation. Behavioral simulation method refers to the analysis of system characteristics and the external responses based on the behavioral model of the system.

EMC behavioral simulation: According to the EME where the system operates, the EMI signal and the functional signal are commonly equivalent to the input signal of the behavioral simulation, and based on the EMC behavioral model of the system and subsystem/equipment, the system performance can be simulated.

Figure 4.18 shows the process of behavioral modeling, model verification, and behavioral simulation analysis. It includes the following three stages.

(1) Stage of concept description: In the initial stage of system design, the concept and requirements of the system are proposed, a conceptual model is established, and the key attributes of the system are analyzed.
(2) Stage of behavioral modeling: An executable behavioral model of system functions is constructed. Then, we quantify the conceptual model established in the first stage and use a formal modeling language and modeling tools to create an executable abstract model. As a result, the functional properties of the system based on the execution of the model can be analyzed.
(3) Stage of system-level application: The behavioral model is extended to executable performance analysis model based on performance analysis. The performance of the system is then analyzed quantitatively according to the result of simulated execution of the model. Then, we can determine whether the functional attributes of the system meet the system requirements and apply the analysis to practice.

4.3.9 Quantitative Modeling Based on EMC Gray System Theory

When the relationship between the input and output of the system is available, we can build EMC behavioral model at system level and subsystem/equipment level (EMC information can be extracted from test data during the development stage, and the model can be further revised).

However, in many cases, the key input and output parameters of some equipment in the system are unavailable, which makes it difficult to establish an EMC behavioral model. In order to obtain the key input and output parameters of the equipment, the EMC characteristics of some equipment, such as the harmonic interference characteristics and the noise spectrum of the power supply, can be extracted through testing. Then, we can summarize the characteristics and derive regular formulas to build analytical model of equipment/module with incomplete design parameters. This is the EMC quantitative modeling method proposed by the EMC research team of Beihang University [15]. By using gray theory [17, 18], the different interference factors

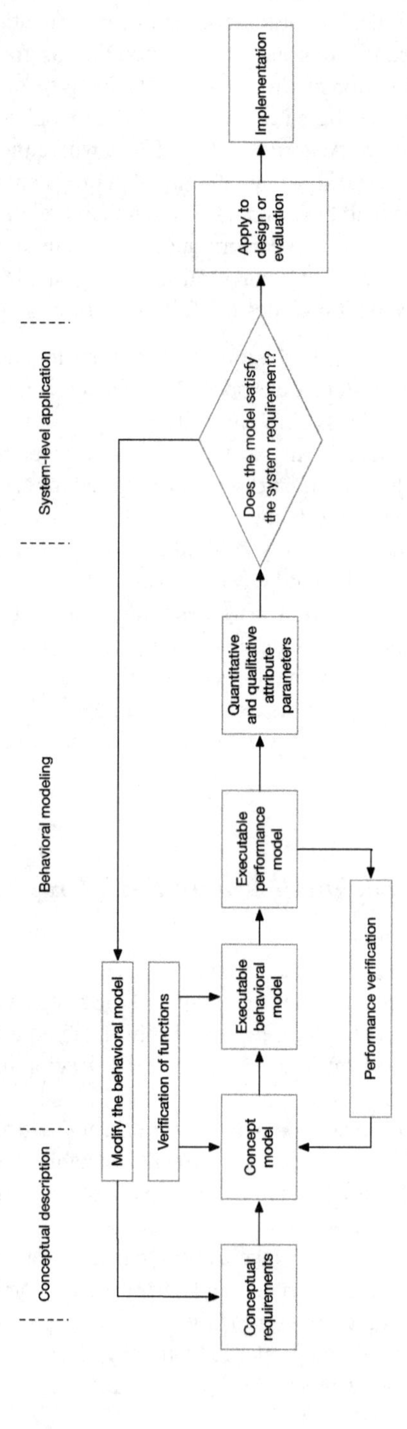

Fig. 4.18 Process of behavioral modeling, model verification, and behavioral simulation analysis

in the same equipment interference spectrum are analyzed and different harmonic interference and broadband interference of the equipment are identified based on the test data. Thus, we can establish a gray model by summarizing the emission of the equipment. Finally, we can achieve better accuracy with small testing data using the best accurate modeling method for equipment under the least square criterion.

The EMC modeling method based on gray system theory can be used to model the EMC of systems with incomplete information, thus effectively solving the problem of equipment's EMC simulation analysis and improving the accuracy of system-level EMC research.

Chapter 5
Critical Techniques of Quantitative System-Level EMC Design

The proposed quantification system-level EMC design is a top-down process. It translates the overall EMC requirements proposed in the demonstration phases into the overall EMC technical requirements, and further decomposes it into requirements for subsystems and equipment. Thus, the EMC design of system and subsystem/equipment can be implemented in the scheme phase. Then, EMC evaluation, design adjustment, performance control, and verification are conducted during the engineering development phase. Finally, the process control requirements during the production phase for EMC design and the requirements for EMC maintenance during the application phase are carried out. With this method, the advanced principle of EMC in a process from conceptual design to detailed design and finally to physical design can be achieved, through which we make continuous adjustment to achieve the optimal performance.

Quantitative system-level EMC design is a brand-new technology that incorporates multi-disciplinary knowledge including system engineering theory, electromagnetic field theory, communication theory, control theory, reliability, and theory. The advancement of this technology lies in the usage of predesign instead of post-validation, and quantitative design instead of experience-based design.

The conventional workflow of EMC design is shown in Fig. 5.1. The overall EMC requirements made by the project manager to the subsystems/equipment are mainly the test items specified by the EMC regulations, and there is little difference for a different subsystem/equipment. In this situation, if there is not enough information for subsystem/equipment about the EME where it operates, and no analysis on the requirements of the EMC test items, there will hardly be any design aiming at passing the EMC regulation test during the scheme design of subsystems/equipment. As a result, when the engineering prototype went through the EMC tests, items such as RE102, CE102, RS103, CS101, and CS114 are difficult to pass. Until now, the common method for EMC design is still to solve the problem only after it is exposed in the later phase of the development.

The root causes of a large number of EMC problems in the later phases are: (1) no top-level design of EMC in the early phase of system design; (2) no quality control

© National Defense Industry Press and Springer Nature Singapore Pte Ltd. 2019

D. Su et al., *Theory and Methods of Quantification Design on System-Level Electromagnetic Compatibility*, https://doi.org/10.1007/978-981-13-3690-4_5

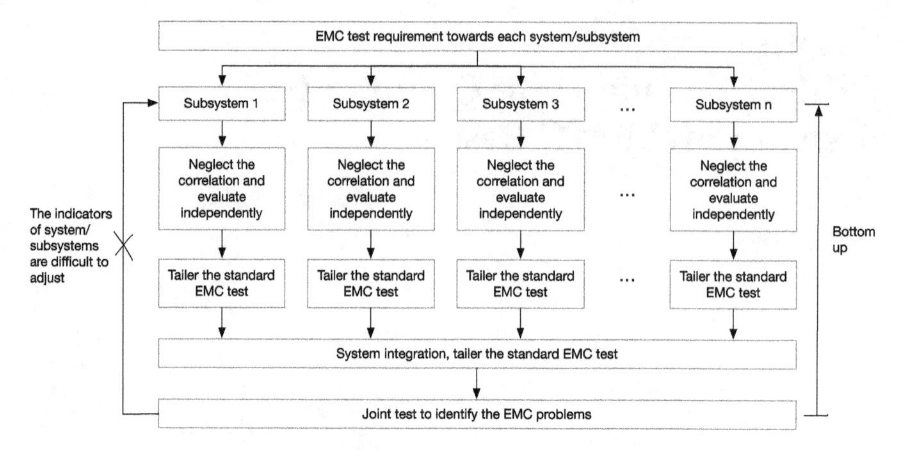

Fig. 5.1 Conventional EMC design process

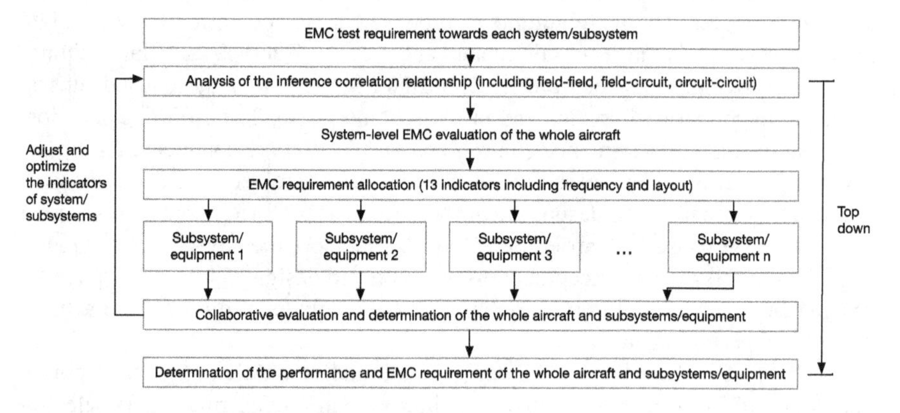

Fig. 5.2 Process of quantitative system-level EMC design

of EMC in the development process; and (3) no standard methods at industrial level for EMC demonstration and design.

Aircraft is used as an example to discuss the quantitative system-level EMC design and introduce the design methods as shown in Fig. 5.2.

5.1 Principles, Methods, and Workflow for Quantitative System-Level EMC Design

To begin with, we will introduce the basic principles of quantitative system-level EMC design. First, the system is considered as a complex system with multi-input/multi-output ports, an internal multilayer/multi-node/multi-directional coupling relationship, and frequency–time–space–code–polarization response characteristic. Second, the functional/nonfunctional field coupling relationships and circuit coupling relationship are classified into different levels, and the responses are classified into linear and nonlinear responses. Then based on the basic principle of the system, and according to the equivalent model of the input and output ports, an equivalent behavioral model of the system can be built. Thirdly, the tactical and technical requirements, functional requirements, EME requirements, and EMC requirements are taken as input of the quantitative and collaborative design methods, which also takes the system and subsystem/equipment functions into consideration. As a result, the system and subsystems/equipment can work together and the incompatible working status can be warned in advance.

According to the structure and shape (data provided by CATIA, UG, Pro-E, etc.) of the aircraft and the structure and relative layout of the installed subsystems/equipment (including electronic equipment, airborne antennas, and cables), the design principles of onboard electronics system, input and output port characteristics, and electrical characteristic pretest data, we can establish the EMC digital model of the aircraft (EMC digital prototype of the aircraft), and EMC behavioral simulation model of the airborne electronic systems and mission electronic systems. According to the operating bandwidth, receiver sensitivity, polarization characteristics, signal characteristics, nonlinear characteristics of RF front-end, receiving antenna characteristics, transmitting power, spurious emission characteristics, transmitting antenna characteristics, possible layout on the aircraft, and the attenuation characteristics of the connection cables, the EMI relationship matrix of airborne/mission electronic systems and equipment can be obtained using field–circuit coupling co-analysis method. Then, the EMC safety of the airborne electronic subsystems can be classified based on the electromagnetic emission and susceptive characteristics of airborne electronic systems. Thus, EMC prediction of the aircraft, establishment of top-level EMC indicators, decomposition of subsystem EMC indicators, quantitative evaluation, and control of the aircraft EMC can be achieved. After completing the above work, the desired EMC performance of the aircraft can be ultimately satisfied.

Now, we explain the detailed method of quantitative system-level EMC design. As an important support for the development of aircraft, the first step in system-level EMC design is to conduct EMC evaluation on the preliminary design scheme of the aircraft. The evaluation includes the rationality of the antenna layout, equipment layout, cable layout, the selection of equipment indicators, and the electromagnetic safety of the whole aircraft. Based on the result of the evaluation, electromagnetic incompatibility problems will be discovered. Then, the overall EMC technical requirements of the whole aircraft can be proposed in accordance with the overall EMC requirement for the aircraft, including the design indicator requirements and test requirements. Next, the overall technical requirements can be decomposed to subsystems and equipment.

A large number of electronic information systems are usually installed on an aircraft. Therefore, it is necessary to identify all electromagnetic emission sources and susceptive equipment of interest in order to establish an interference relationship before evaluation. Then, we can analyze the possible energy transmission channel between the electromagnetic emission sources and susceptive equipment in the form of field–field, field–circuit, and circuit–circuit. Thus, we can construct an interference correlation matrix of the whole aircraft and evaluate the EMC based on the quantitative results given by the interference correlation matrix. Layout or indicator adjustments are then performed on equipment that do not meet the compatibility requirements in the interference correlation matrix. Finally, the key parameters of EMC design that satisfy the EMC requirements of the whole aircraft can be obtained. The key parameters include: the frequency coordination relationship between the transmitters and receivers; the isolation between the electromagnetic emission source and the susceptive equipment; the layout of antennas, equipment, and cables; the EME characteristics of the key parts inside and outside the cabin; the energy matching requirements between the transmitters and the receivers (i.e., transmission power, out-of-band attenuation of transmitters, receiving sensitivity, out-of-band suppression of receivers, signal-to-noise ratio required for properly receiving), cabin shielding and resonance characteristics; tolerable degradation of equipment; equipment safety priorities; and safety margins for susceptive equipment.

Figure 5.3 illustrates the design and evaluation process for the quantitative system-level EMC design. The EMC geometric model of the aircraft is built based on the geometric model for engineering design (e.g., CATIA model, UG model, Pro-E model) Then, the EME inside and outside the cabin, the shielding effectiveness, and the resonance characteristics of the aircraft cabin are simulated based on the emission equipment and its antenna parameters, antenna layout characteristics, etc. Thus, the interference correlation matrix at the operator's position, fuel tank, etc., can be constructed and its electromagnetic security can be analyzed. Meanwhile, using the system model based on electromagnetic field method, a field–circuit coupling model based on distribution effects, a behavioral simulation model based on the circuit method [19], the isolation between transmitting antennas and receiving antennas, and the coupling between transmitters and receivers can be calculated. Furthermore, the safety of the equipment can be analyzed using the interference correlation matrix of the susceptive equipment.

According to the results of EMC evaluation, the overall technical requirements (with amendments to the indicators of the preliminary design scheme) for EMC of the whole aircraft can be proposed, and the overall technical requirements for EMC of the whole aircraft can be quantitatively allocated to subsystems and equipment. The subsystem/equipment is designed according to the technical requirements proposed by the whole aircraft. Therefore, the subsystem/equipment scheme needs to be evaluated in conjunction with the aircraft's overall design scheme and the design is subject to adjustment based on the evaluation results. EMC tests can also be conducted when necessary.

Here, we summarize the major steps of quantitative design of system-level EMC. (1) Develop the top-level technical indicators for system EMC; (2) develop EMC technical indicators for subsystem/equipment; (3) conduct subsystem/equipment's EMC design, and predict its impact on the EMC of the whole aircraft; (4) conduct collaborative design of EMC of subsystems/equipment and systems, and optimize subsystem/equipment's EMC indicators; (5) conduct subsystem/equipment's EMC tests, and improve the EMC design of subsystem/equipment based on the test results;

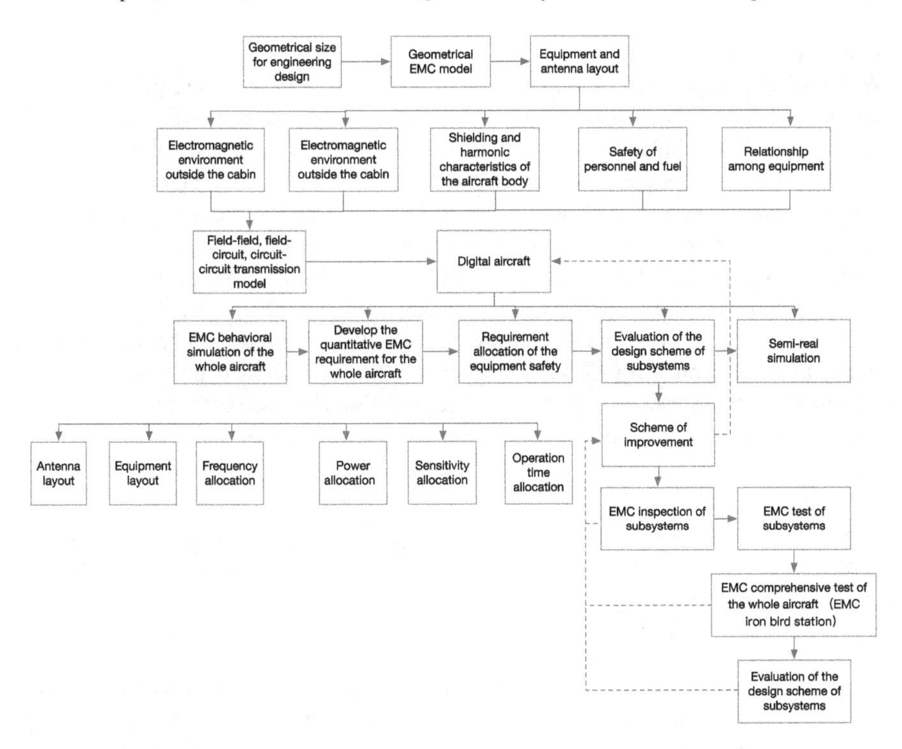

Fig. 5.3 Top-down process of quantitative EMC evaluation and design

(6) conduct joint commissioning test of EMC of subsystems/equipment, and optimize the integrated solution of the whole aircraft based on the test results; (7) perform the EMC test of the whole aircraft, and optimize its EMC. These seven steps provide a technical approach for EMC design from conceptual to detailed and then to physical design of the whole aircraft.

5.2 Solution Method for EMC Condition

Electromagnetic compatibility condition (EMC(s)) reflects the overall EMC status of aircraft after the equipment integration. EMC(s) is a metric of the contribution from EMC probability of the susceptive equipment, probability of fuel safety, and probability of personnel safety to the EMC of the whole aircraft system. It also reflects the importance of a different susceptive equipment, fuel, and personnel to the whole aircraft's system. In the process of aircraft development, two models can be used to solve the system $EMC(s)$: isolated model and serial and isolated mixed model [14].

1. Isolated model

With the isolated model, all receiving equipment and other susceptive equipment, fuel safety subsystems, and operator safety subsystems in the aircraft are considered to perform their tasks independently. In other words, there is neither serial relationship of inter-influence, nor parallel relationship for mutual backup among the equipment and subsystems, and they all affect the EMC of the entire aircraft in an isolated manner. Under a certain usage, let the receivers that operate at the same time, other susceptive equipment, fuel subsystems, and operator subsystems be R_1, R_2, R_3,..., R_n, respectively. In order to reflect the contribution of the EMC probability of various kinds of susceptive equipment after the aircraft integration, the probability of fuel safety and the probability of personnel safety to the EMC of the whole aircraft system, and to represent the importance of a different susceptive equipment, fuel, and personnel in the whole aircraft system, the EMC(s) can be defined as

$$
\begin{aligned}
EMC(s) &= W_1C_1(s) + W_2C_2(s) + W_3C_3(s) + \cdots + W_nC_n(s) \\
&= \frac{w_1}{\sum_{j=1}^{n} w_j}C_1(s) + \frac{w_2}{\sum_{j=1}^{n} w_j}C_2(s) + \frac{w_3}{\sum_{j=1}^{n} w_j}C_3(s) + \cdots + \frac{w_n}{\sum_{j=1}^{n} w_j}C_n(s) \\
&= \frac{\sum_{j=1}^{n} w_jC_j(s)}{\sum_{j=1}^{n} w_j}
\end{aligned}
\quad (5.1)
$$

where W_j is the weight of the j-th receiver (or other susceptive equipment, fuel subsystem, or operator subsystem) in the aircraft, and $0 < W_j \leq 1$; w_j is the graded weight of the j-th receiver in the aircraft (or other susceptive equipment, fuel subsystems, or operator subsystems).

Therefore, we have

$$
W_j = \frac{w_j}{\sum_{j=1}^{n} w_j} \quad (5.2)
$$

The value of the graded weights is based on the mission of the aircraft and the importance of the susceptive equipment's safety, fuel safety, and personnel safety of the whole aircraft. It is determined by the user, the program manager, and the industry experts.

$C_j(s)$ is defined as the EMC probability (safety probability) of the j-th receiving equipment (or other susceptive equipment, fuel subsystem, and personnel subsystem) in a certain operation state. Similarly, the probability (or unsafe probability) of the j-th receiving equipment (or other susceptive equipment, fuel subsystem, or personnel subsystem) is defined as $I_j(s)$; then,

$$
C_j(s) + I_j(s) = 1 \quad (0 \leq C_j(s) \leq 1; 0 \leq I_j(s) \leq 1) \quad (5.3)
$$

When analyzing the EMC probability (fuel safety probability, personnel safety probability) of a receiving equipment or a susceptive equipment, it is necessary to separately determine the susceptive ports that the equipment is likely to be interfered from. For the receiving equipment, its EMC consists of the safety of antenna ports, multiple cable ports, equivalent case shielding ports (including aperture coupling), and equivalent grounding port. For susceptive equipment, its EMC consists of multiple cable ports, equivalent shielding ports (including apertures, slots), and equivalent grounding ports. The safety of fuel system consists of the safety of multiple fuel susceptive points, and the personnel safety consists of the safety of multiple workstations. In the same equipment, these ports are mostly connected in series, which means that as long as one susceptive port is insecure, the probability of EMC of the equipment is 0. On the other hand, only if all the ports are safe, the EMC probability of the equipment will be equal to 1. Only a few backup ports with identical functions have a parallel relationship. The EMC probability of the series model is

$$C_j(s) = C_{j1}(s) \cdot C_{j2}(s) \cdot C_{j3}(s) \cdot \cdots \cdot C_{jn}(s) = \prod_{k=1}^{n} C_{jk}(s) \tag{5.4}$$

where $C_{jk}(s)$ is the probability of safety of the k-th port of the j-th receiving equipment (or other susceptive equipment, fuel subsystem, or personnel subsystem) in the series model.

In addition, the probability of interference of the parallel model is defined as

$$I_j(s) = I_{j1}(s) \cdot I_{j2}(s) \cdot I_{j3}(s) \cdot \cdots I_{jn}(s) = \prod_{k=1}^{n} I_{jk}(s) \tag{5.5}$$

And its EMC probability is

$$C_j(s) = 1 - I_j(s) = 1 - \prod_{k=}^{n} I_{jk}(s) = 1 - \prod_{k=1}^{n} [1 - C_{jk}(s)] \tag{5.6}$$

where $I_{jk}(s)$ is the probability of interference at the k-th port of the j-th receiving equipment (or other susceptive equipment, fuel subsystem, or personnel subsystem) in the series model.

In engineering practice, the EMC probabilistic model of receiving equipment (or other susceptive equipment, fuel subsystems, or personnel subsystems) is usually a pure serial structure of susceptive ports. Some equipment may have backup ports, and the EMC probability becomes a series and parallel mixed structure. In the mixed model, the quantity of serial susceptive ports is much more than the number of parallel susceptive ports, and the parallel form is simpler. The EMC of the mixed model of the entire equipment can be derived from the physical meaning of (5.4)–(5.6).

2. Series-isolated mixed model

Many aircraft have integrated high-power emission equipment, which may interfere with the fuel system and control equipment of the aero engine. Therefore, it is necessary to consider the aircraft safety problems caused by EMC. In the series-isolated mixed model for EMC prediction, the EMC condition of the whole aircraft is predicted by the probability of EMC condition of the engine control system, the fuel system, and the equipment and personnel in a series manner, while the safety of receiving equipment, other susceptive equipment, and personnel is still isolated from each other. The EMC condition of the series-isolated mixed model of the whole aircraft in application is

$$
\begin{aligned}
EMC(s) &= C_E(s) \cdot C_O(s) \cdot C_{SW}(s) \\
&= C_E(s) \cdot C_O(s) \cdot [W_1C_1(s) + W_2C_2(s) + W_3C_3(s) + \cdots + W_nC_n(s)] \\
&= C_E(s) \cdot C_O(s) \cdot \frac{\sum_{j=1}^{n} w_j C_j(s)}{n \sum_{j=1} w_j}
\end{aligned} \tag{5.7}
$$

where $C_E(s)$ is the EMC probability of the engine; $C_O(s)$ is the probability of safety of the fuel system (determined by the probability of EMC of multiple fuel susceptive points in series); $C_{SW}(s)$ is the probability of safety of the receiving equipment, other susceptive equipment, and personnel (EMC probability). The value of $C_{SW}(s)$ is determined using an isolated model (receiving equipment, susceptive equipment other than the engine, operator subsystem which are defined as $R_1, R_2, R_3, \ldots, R_n$, respectively); W_j, w_j, and $C_j(s)$ are solved in the same way as before.

The safety of the aircraft engine and the fuel system must be guaranteed first, because poor safety of the engine and fuel is the major problems in EMC design. Therefore, the safety of the aircraft is emphasized in the series-isolated mixed model. Comparing the two prediction models for EMC condition, we think the latter one is more practical.

3. Determination of the probability of interference of subsystems

The probability of interference in the subsystem is related to the following factors: the full-power emission probability $P_i^{(\max)}$ of the transmitting equipment, the probability of interference frequency between the susceptive equipment and the transmitting equipment $P_{ij}^{(f)}$, the overlapping probability of the working time between the susceptive equipment and the transmitting equipment $P_{ij}^{(t)}$, etc.

The probability of a susceptible port can be determined using the following method. Firstly, the worst-case analysis method of EMC (assuming the full-power emission of the transmitting equipment and the most susceptive receiving of the receiving equipment) can be used to analyze whether the susceptive port is interfered. If it is, then the probability of being interfered by the transmitting equipment at the i-th susceptive port is

$$
I_{ijk}(s) = \zeta_k \cdot P_i^{(\max)} \cdot P_{ij}^{(f)} \cdot P_{ij}^{(t)} = P_i^{(\max)} \cdot P_{ij}^{(f)} \cdot P_{ij}^{(t)} \tag{5.8}
$$

The corresponding EMC probability is

$$
C_{ijk}(s) = 1 - I_{ijk}(s) = 1 - P_i^{(\max)} \cdot P_{ij}^{(f)} \cdot P_{ij}^{(t)} \tag{5.9}
$$

If the susceptive port is secure, the probability of the port being affected by the i-th transmitting equipment is

$$I_{ijk}(s) = \zeta_k \cdot P_i^{(\max)} \cdot P_{ij}^{(f)} \cdot P_{ij}^{(t)} = 0 \tag{5.10}$$

The corresponding EMC probability is

$$C_{ijk}(s) = 1 \tag{5.11}$$

In Eqs. (5.8)–(5.11), $I_{ijk}(s)$ is the value of the probability of interference at the k-th port of the j-th receiving equipment (or other susceptive equipment, fuel subsystem, or personnel subsystem) under the effect of the i-th transmitting equipment. $C_{ijk}(s)$ is the safety probability of the k-th port of the j-th receiving equipment (or other susceptive equipment, fuel subsystem, or human subsystem) under the effect of the i-th transmitting equipment. ζ_k is the interference factor of the susceptive port derived from the worst-case EMC analysis method. If it is interfered, then $\zeta_k = 1$; if it is safe, then $\zeta_k = 0$; $P_i^{(\max)}$ is probability of the maximum emission during the operation of the i-th transmitting equipment; $P_{ij}^{(f)}$ is the combined probability of interference frequencies between the i-th transmitting equipment and the j-th susceptive equipment; $P_{ij}^{(t)}$ is the overlapping probability of working time between the i-th transmitting equipment and the j-th susceptive equipment.

Since the interference is not only generated in band, the interference frequency combination probability refers to the probability that the transmitting equipment and the susceptive equipment both operate in a frequency band that can generate interference (the frequencies of the two do not necessarily overlap or cover each other). The probability can be expressed as

$$P_{ij}^{(f)} = \sum_{N_i=1}^{N_E} \sum_{N_j=1}^{N_R} K_{N_i N_j} P_{N_i}^{(E)} P_{N_j}^{(R)} \tag{5.12}$$

where $P_{N_i}^{(E)}$ is the probability of usage of the N_i-th frequency of the transmitting equipment in the selected band ($N_i = 1, 2, \ldots, N_E$); $P_{N_j}^{(R)}$ is the probability of usage of the N_j-th frequency of susceptive equipment in the selected frequency band ($N_i = 1, 2, \ldots, N_R$); $K_{N_i N_j}$ indicates whether the N_i-th frequency of the transmitting equipment in the selected frequency band and the N_j-th frequency of the susceptive equipment constitute an interference relationship (N_j-th equals 1 if yes; equals 0 if no).

Overlapping probability of working time between transmitting equipment and susceptive equipment [20]: In general, assuming that susceptive equipment can be switched on at any time t_0 and work during $t_0 + t_R$, and the transmitting equipment can be switched on and off at any time. When the transmitting equipment is switched on and the time intervals are uniformly distributed, the overlapping probability of working time between the transmitting equipment and the susceptive equipment can be expressed as

$$P_{ij}^{(t)} = \begin{cases} (t_i + t_R)/(\bar{\tau}_i + \bar{t}_i); & t_R < \tau_{i\min} \\ \left[t_i + \frac{2\tau_{i\max} t_R - t_R^2 - \tau_{i\min}}{2(\tau_{i\max} - \tau_{i\min})} \right] / (\bar{\tau}_i + \bar{t}_i); & \tau_{i\min} \le t_R \le \tau_{i\max} \\ 1; & t_R > \tau_{i\max} \end{cases} \tag{5.13}$$

where $1/(\bar{\tau}_i + \bar{t}_i)$ is the average frequency of the transmitting equipment; $\bar{t}_i$ is the average radiation time of the transmitting equipment; $\bar{\tau}_i$ is the average interval of the transmitting equipment; $\tau_{i\min}$ and $\tau_{i\max}$ are the minimum and maximum of the average time interval of the transmitting equipment, respectively.

When the time interval has a normal distribution, the overlapping probability of the working time between transmitting equipment and susceptive equipment is expressed in [21] as

$$\begin{aligned} P_{ij}^{(t)} = \frac{1}{(\bar{\tau}_i + \bar{t}_i)} \Bigg\{ & \bar{t}_i + t_R - \bar{\tau}_i - (t_R - \bar{\tau}_i)\Phi\left(\frac{t_R - \bar{\tau}_i}{\sigma_{\tau i}}\right) + \bar{\tau}_i \Phi\left(\frac{\bar{\tau}_i}{\sigma_{\tau i}}\right) \\ & -0.4\sigma_{\tau i}\left[\exp\left[-\frac{1}{2}\left(\frac{t_R - \bar{\tau}_i}{\sigma_{\tau i}}\right)^2\right] - \exp\left[-\frac{1}{2}\left(\frac{\bar{\tau}_i}{\sigma_{\tau i}}\right)^2\right]\right]\Bigg\} \end{aligned} \tag{5.14}$$

The determination of the probability of interference of a subsystem is complex. In fact, it is also related to the form of the radiation signal of the transmitting equipment, the statistical characteristics of the radiation signal, and the processing method of the signal by the susceptive equipment.

When there are a limited number of transmitting equipment on the aircraft, each equipment has a unique function, their radiation power and the frequency band distribution are greatly different, and there will be basically no transmitting equipment with the same frequency band and the same power. Therefore, the probability of interference can be approximated to

$$I_{jk}(s) = \max[I_{ijk}(s)] \tag{5.15}$$

The corresponding EMC probability is

$$C_{jk}(s) = \min[C_{ijk}(s)] = 1 - \max[I_{ijk}(s)] = 1 - I_{jk}(s) \tag{5.16}$$

Substituting Eqs. (5.8)–(5.11), (5.15), and (5.16) into Eqs. (5.4)–(5.5), we can get the EMC probability of the subsystem using series model as

$$\begin{aligned} C_j(s) &= [1 - I_{j1}(s)] \cdot [1 - I_{j2}(s)] \cdots [1 - I_{jn}(s)] \\ &= \prod_{k=1}^{n} [1 - I_{jk}(s)] \\ &= \prod_{k=1}^{n} [1 - \max[I_{ijk}(s)]] \end{aligned}$$

$$= \prod_{k=1}^{n} \left[1 - \max \left[\zeta_k \cdot P_i^{(\max)} \cdot P_{ij}^{(f)} \cdot P_{ij}^{(t)} \right] \right] \tag{5.17}$$

The probability of interference and the probability of EMC of the subsystem using parallel model are

$$I_j(s) = I_{j1}(s) \cdot I_{j2}(s) \cdot I_{j3}(s) \cdot I_{jn}(s)$$
$$= \prod_{k=1}^{n} \max \left[\zeta_k \cdot P_i^{(\max)} \cdot P_{ij}^{(f)} \cdot P_{ij}^{(t)} \right]$$
$$I_j(s) = I_{j1}(s) \cdot I_{j2}(s) \cdot I_{j3}(s) \cdot \cdots \cdot I_{jn}(s)$$
$$= \prod_{k=1}^{n} \max \left[\zeta_k \cdot P_i^{(\max)} \cdot P_{ij}^{(f)} \cdot P_{ij}^{(t)} \right] \tag{5.18}$$

$$C_j(s) = 1 - \prod_{k=1}^{n} \max \left[\zeta_k \cdot P_i^{(\max)} \cdot P_{ij}^{(f)} \cdot P_{ij}^{(t)} \right] \tag{5.19}$$

The above engineering approximations can be applied to the actual EMC analysis of aircraft, but it does not have a wide range of applicability. For systems of larger size, more complex structure, and with more electronic equipment, this method cannot be directly adopted; e.g., this method cannot be used directly for electromagnetic environment performance analysis in battlefield areas.

Theoretically, by substituting the solution of (5.17)–(5.19) into (5.1) we can accurately estimate the EMC condition (EMC(s)) of the whole aircraft system. This method is called EMC accurate evaluation. However, in engineering calculations, many values in the formula are difficult to estimate accurately. Therefore, in order to quickly evaluate EMC(s) of the whole aircraft, some parameters can be further simplified and the worst-case evaluation method of EMC is adopted; i.e., when the susceptibility of the susceptive port is evaluated, the power of the transmitting equipment is considered to be the maximum radiated power and the receiving equipment is considered to operate in the most sensitive state. On this basis, safety margin of the coupled port can be analyzed using the energy criterion. When the safety margin is greater than zero, the port is considered to be safe. When the safety margin is less than zero, the port is considered to be interfered. Therefore, the probability of interference and the probability characteristics of EMC are unnecessary to be considered anymore. Then, Eqs. (5.17)–(5.19) are simplified as

$$C_j(s) = \prod_{k=1}^{n} [1 - \zeta_k] \tag{5.20}$$

$$I_j(s) = \prod_{k=1}^{n} \zeta_k \tag{5.21}$$

$$C_j(s) = 1 - \prod_{k=1}^{n} \zeta_k \tag{5.22}$$

Equations (5.20)–(5.22) indicate that in a series model, as long as one port is interfered, the entire subsystem is considered to be insecure. Similarly, in a parallel model, as long as one port is secure, the entire subsystem is considered to be safe. This kind of EMC analysis is called rough evaluation of EMC, and it is does not accept any degradation of equipment performance in EMC design. Using this method, EMC(s) can be directly evaluated based on the interference correlation matrix. The specific steps are as follows:

(1) List transmitting equipment, other radiating equipment, receiving equipment, other susceptive equipment, fuel safety subsystems, and operator safety subsystems in the aircraft system.
(2) Number the radiation source ports or susceptive ports for each equipment or subsystem, and build the EMC probability prediction model for each subsystem.
(3) Build the interference correlation matrix of the whole aircraft equipment, the fuel interference correlation matrix, and the personnel interference correlation matrix.
(4) Calculate the power of the interference through various coupling paths to different susceptive ports, fuel susceptive points, and operator workstation positions. Calculate the safety margin matrix by comparing the result with the tolerable radiation power and the susceptive ports, fuel susceptive points, and operator's position in the workstation.

According to the analysis above, the interference power matrix coupled with the susceptive port is

$$\begin{aligned}\mathbf{J_E} &= [\, J_1 \cdots J_j \cdots J_{N_2} \,] \\ &= 10\lg\left(\mathbf{T_E} 10^{\frac{(\mathbf{P_t} - \mathbf{L}_{\mathbf{t}B} - \mathbf{L_{tf}}) \otimes \mathbf{T_E} - \mathbf{I}(\mathrm{dB})}{10}}\right)(\mathrm{dBm}), \quad i \in [1, M_2], \quad j \in [1, N_2] \end{aligned} \tag{5.23}$$

where the power coupled with the receiving antenna port refers to the output power of the receiving antenna (the coupling power of the antenna port in Eq. (5.62) is the input power of the receiver).

Similarly, the radiation power matrix coupled with the fuel susceptive point and the radiation power matrix coupled with the operating station of the cabin operator can be obtained.

The radiation power matrices that the susceptive ports, fuel susceptive points, and operator workstation positions can withstand respectively are

$$\mathbf{J_E^S} = [\, J_1^S \cdots J_j^S \cdots J_{N_2}^S \,], \quad j \in [1, N_2] \tag{5.24}$$

$$\mathbf{J_O^S} = [\, J_{O_1}^S \cdots J_{O_j}^S \cdots J_{O_K}^S \,], \quad j \in [1, K] \tag{5.25}$$

$$\mathbf{J_W^S} = [\, J_{W_1}^S \cdots J_{W_j}^S \cdots J_{W_Z}^S \,], \quad j \in [1, Z] \tag{5.26}$$

The elements of the matrix in (5.24) are obtained by calculating or testing. The elements of the matrix in (5.25) and (5.26) are set according to the corresponding standards.

The safety margin matrix for the susceptive ports is

$$\begin{aligned}\mathbf{J}_{\mathbf{E}}^{\mathbf{M}} &= \mathbf{J}_{\mathbf{E}} - \mathbf{J}_{\mathbf{E}}^{\mathbf{S}} \\ &= [\, J_1 - J_1^S, \cdots J_j - J_j^S, \cdots, J_{N_2} - J_{N_2}^S \,]\end{aligned} \tag{5.27}$$

Similarly, the safety margin matrix of the fuel susceptive point and the operator table can be obtained separately as

$$\begin{aligned}\mathbf{J}_O^{\mathbf{M}} &= \mathbf{J}_O - \mathbf{J}_O^{\mathbf{S}} \\ &= [\, J_{O_1} - J_{O_1}^S, \cdots J_{O_j} - J_{O_j}^S, \cdots, J_{O_K} - J_{O_K}^S \,]\end{aligned} \tag{5.28}$$

$$\begin{aligned}\mathbf{J}_W^{\mathbf{M}} &= \mathbf{J}_W - \mathbf{J}_W^{\mathbf{S}} \\ &= [\, J_1 - J_{W_1}^S, \cdots J_{W_j} - J_{W_j}^S, \cdots, J_{W_Z} - J_{W_Z}^S \,]\end{aligned} \tag{5.29}$$

(5) Convert the safety margin matrix into a matrix of interference coefficients, extract the matrix elements, calculate $C_j(s)$ of each subsystem, and substitute (5.1) to calculate the $EMC(s)$ of the whole aircraft.

If $J_j - J_j^S > 0$, let $J_j - J_j^S = 1$, which indicates that the probability of the port being interfered is 1; i.e., the port is insecure. If $J_j - J_j^S \leq 0$, let $J_j - J_j^S = 0$, which indicates the probability of the port being interfered is 0; i.e., the port is safe. Then, we can generate the interference coefficient matrix as $\boldsymbol{\zeta}_{\mathbf{E}}^{\mathbf{I}} = \left[\zeta_1, \zeta_2, \ldots, \zeta_{N_2}\right]$. It can be seen from Eq. (5.20) that $\boldsymbol{\zeta}_{\mathbf{E}}^{\mathbf{I}}$ is the probability matrix of the susceptive port and the matrix can be written as

$$\boldsymbol{\zeta}_{\mathbf{E}}^{\mathbf{S}} = [1, 1, \cdots 1]_{1\times N_2} - [\zeta_1, \zeta_2, \cdots, \zeta_{N_2}] = [1 - \zeta_1, 1 - \zeta_2, \cdots, 1 - \zeta_{N_2}]$$

We can then extract the probability of interference and the EMC probability of all susceptive ports of a certain subsystem from the matrix. Based on the basic model of Eqs. (5.4)–(5.6), we can calculate the EMC probability $C_j(s)$ of receiving equipment and other susceptive equipment.

If $J_{O_j} - J_{O_j}^S > 0$, let $J_{O_j} - J_{O_j}^S = 1$, which indicates that the probability of inference at the fuel susceptive point is 1; i.e., the sensitive point is interfered. If $J_{O_j} - J_{O_j}^S \leq 0$, let $J_{O_j} - J_{O_j}^S = 0$, which indicates the probability of interference at the susceptive point of the fuel is 0; i.e., the sensitive point is safe. The matrix of probability of interference for fuel susceptive points is $\boldsymbol{\zeta}_{\mathbf{O}}^{\mathbf{I}} = [\zeta_{O_1}, \zeta_{O_2}, \cdots, \zeta_{O_K}]$. The EMC probability of fuel susceptive point is calculated to be $\boldsymbol{\zeta}_{\mathbf{O}}^{\mathbf{S}} = [1 - \zeta_{O_1}, 1 - \zeta_{O_2}, \cdots, 1 - \zeta_{O_K}]$. Then, we can extract the EMC probability of each element from the matrix and calculate the $C_j(s)$ of fuel subsystem according to Eq. (5.4). Similarly, the EMC probability of the operator's subsystem can be calculated.

Finally, we can substitute $C_j(s)$ of each system into Eq. (5.1) and obtain the EMC condition ($EMC(s)$) of the whole aircraft system.

5.3 EMC Modeling Methodology

5.3.1 Methodology of System-Level Modeling

The information system is a complex system with a wide frequency band, complex signal forms, and various coupling relationships. System-level EMC modeling needs to be performed from the perspective of system modeling.

1. Concept of system

Ideally, a system usually refers to a study object, an associated object, or a closed-loop system that includes a predictor. A system is independent from its "environment"; i.e., although a system is affected by the environment, it has its own characteristics and may even have an impact on the environment. The impact of the environment on the system is regarded as an input of the system. The relationship between the input and output of the system is determined by the characteristics of the system itself as shown in Fig. 5.4. The inputs and outputs of the system that change over time are called input and output variables. If the system has only one input and one output variable, it is called a single-input and single-output (SISO) system. Similarly, if the system has multiple independent input and output variables, the system is called a multi-input and multi-output (MIMO) system.

In practical applications, a system may be affected by various factors from the outside. Therefore, the three elements of the system (entities, attributes, and activities) and the environment need to be determined before we study the system any further. The system is determined only if the entities, attributes, activities, and environment are clearly defined.

2. System classification

According to different criteria, the system can be classified into linear systems and nonlinear systems; deterministic systems and stochastic systems; time-invariant systems and time-varying systems; constraint systems and unconstrained systems; continuous-state systems and discrete-state systems; continuous-time systems and discrete-time systems; time-driven systems and event-driven systems; lumped parameter systems and distributed parameter systems; systems with computer networks (intelligent systems) and systems without computer networks, etc.

Most of the systems involved in the field of EMC are characterized by nonlinearity, uncertainty, time-varying, constrained, and with distributed parameters, and they are mostly continuous-time systems. It can be predicted that the intelligent systems will become mainstream with computers taking the role of process control. Thus, complex intelligent networks will also be involved by EMC in the future.

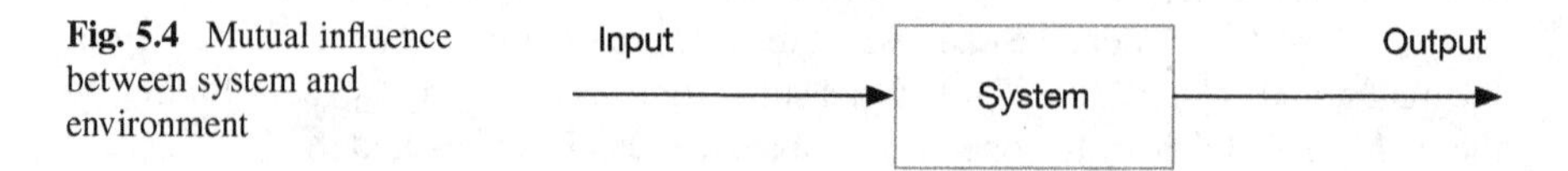

Fig. 5.4 Mutual influence between system and environment

3. Definition of system model

A model is a logical representation of a system in physics, mathematics, or other modes. It provides knowledge about the system in a certain form (such as words, symbols, diagrams, objects, mathematical formulas). The system model, on the one hand, reflects the main characteristics of the actual system, and on the other hand, is a higher-level abstraction that can be applied to similar problems.

Therefore, a good system model should have three characteristics: ① It is the abstraction or imitation of the real system; ② it is composed of the main factors that reflect the nature or characteristics of the system; ③ it is the concentration of the relationship between the main factors.

In most cases, the system model is not the system itself, but a description of the essential attributes of a certain aspect of the system. The system is complex, and its attributes are also multifaceted. Therefore, the model can be classified into static model and dynamic model according to the time dependencies; it can be classified into "black box" model and "white box" model according to whether the internal characteristics of the system are described; it can be classified into deterministic model, random model, continuous model, discrete model according to the form of variables; it can be classified into algebraic equation model, differential (difference, iterative) equation model, statistical model, and logical model according to the relationship between variables.

In engineering systems, we often encounter systems with numerous state variables, complex feedback structures, and nonlinear characteristics of inputs and outputs. These systems are called complex systems. When these characteristics are directly used as the main features of the system, they are correspondingly referred to as high-order systems, multi-loop systems, nonlinear systems, and network systems. If the number of subsystems in a complex system is large with many types and complex relationship among each other, the system is called a complex giant system. Although such systems have objectively defined characteristics, the differences in subsystems will result in variations.

4. System modeling principles

Firstly, we will introduce the concept of homomorphism and isomorphism in system modeling. Assume system A has an input signal A(t), and system S has an input signal B(t), where $t \geq 0$. Isomorphic systems are systems that have the same reaction to external excitation. For two isomorphic systems, same input will produce same output. If the input signal set and output status set of system B only correspond to a few representative inputs and outputs of system A, then system B is said to be a homomorphic system of system A. An isomorphic system must be homomorphic, but the reverse is not necessarily true.

Modeling means building the isomorphism and homomorphism of the prototype. The interaction between the entities of the system will cause its attribute to change. Entities and attributes of the system may change at different times. The change is usually described as states. At any moment, all information about entities, attributes, and activities in the system is called the state of the system at that moment. The variables representing the system state are called state variables.

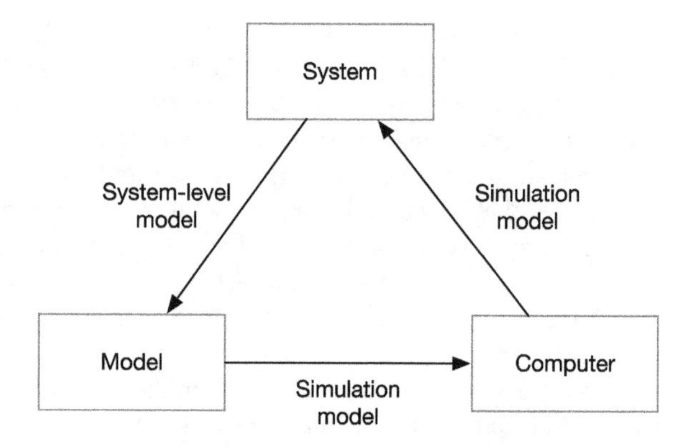

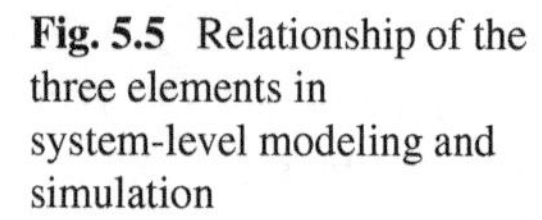

Fig. 5.5 Relationship of the three elements in system-level modeling and simulation

An ideal model for a prototype system should be a "good" homomorphic system. "Good" means that the model catches the important state that determines the system performance. Based on the analysis and research of the system, an ideal model shall be object-oriented and be built with deep understanding of the system's working environment.

System modeling and simulation include three basic activities: system modeling (primary modeling), simulation modeling (secondary modeling), and simulation testing. The three activities are connected by the three elements of the computer simulation: system, model, and computer (including hardware and software). The relationship between the three elements of system modeling and simulation is shown in Fig. 5.5.

5.3.2 *Methodology for Behavioral Modeling*

The system-level EMC problem usually contains multiple crossover EMI and multiple modes of interference coupling associating with each other. All of the subsystems, equipment, or circuit modules contained in an electronic information system have physical circuit structures. However, for a system design engineer, it is difficult to build a full simulation model for the physical structure of an electronic system. The global simulation calculation of the underlying physical circuit in the system is the bottleneck in current system simulation [22].

The top-down system-level EMC quantitative design technique is to evaluate the EMC of the system by means of simulation tests, simulation calculations, etc., and to quantitatively allocate and optimize the design [14]. This top-level EMC predesign requires a simulation technique that is in a higher level than that of the transistors and devices. The predesign method can quickly put down the indicator requirements of each subsystem or circuit module from top to bottom. Behavioral modeling technology is undoubtedly an important method for system-level EMC analysis. The simplified mathematical model is used to describe the functional characteristics of each subsystem and circuit module, simplify the underlying physical structure of the circuit, increase the simulation speed, and reduce the computational cost.

As described in Sect. 4.3.7, EMC behavioral modeling is a modeling method that targets the external behavior of the system. Therefore, behavioral modeling needs to go through conceptual description, modeling, and verification.

To build a system-level EMC behavioral model, we usually use layered simulation methods to establish a suitable model and select an appropriate simulation tool. EMC models have two categories: One is for extracting parameters; the other is for solving field and energy distribution problems. The purpose of distinguishing these two is to choose the appropriate modeling tool.

A system-level EMC behavioral model generally includes the behavioral model of subsystems/equipment/components (including susceptive unit, interference sources), coupling channel models (including radiation and conduction), and equivalent models of environmental factors. Despite the tight coupling involved in multiple domains, EMC system modeling often involves the solving of nonlinear transmission equations.

Figure 5.6 illustrates the EMC behavioral modeling method, which is an effective EMC modeling method.

5.3.2.1 Concept Description of EMC Based on System Characteristics

Before establishing a system-level EMC behavioral model, the concept description of EMC based on system characteristics should be completed first. The main purpose of the concept description is to (1) establish the overall technical framework for the system-level EMC of the object according to its intended functions; (2) formulate the principle of collaborative control and determine the overall work content; (3) summarize the major concerns; and (4) sort out the key control indicators. Then, we can determine the workflow, make a timetable, establish a staff network, and finally establish a conceptual model of system-level EMC.

(1) Establish an overall technical framework: Position the system-level EMC of the objects to be designed. Through the positioning, all parties can understand the levels, time frame, participating institutions and personnel, funding requirements, risks, etc., of the system-level EMC of the object.
(2) Formulate the principle of collaborative control: Formulate collaboration strategies in terms of time, participating institutions and personnel, and funding. In the same time, determine the collaborative interaction protocols in terms of technology, interfaces, data, and resources.
(3) Determine the overall work package: In technical, testing, controlling, managing, and archiving level, sort out and specify the lifecycle work package in terms of design, development, evaluation, production, and application.
(4) Summarize major concerns: Summarize the major technical difficulties, important control objects, and major coordination issues.

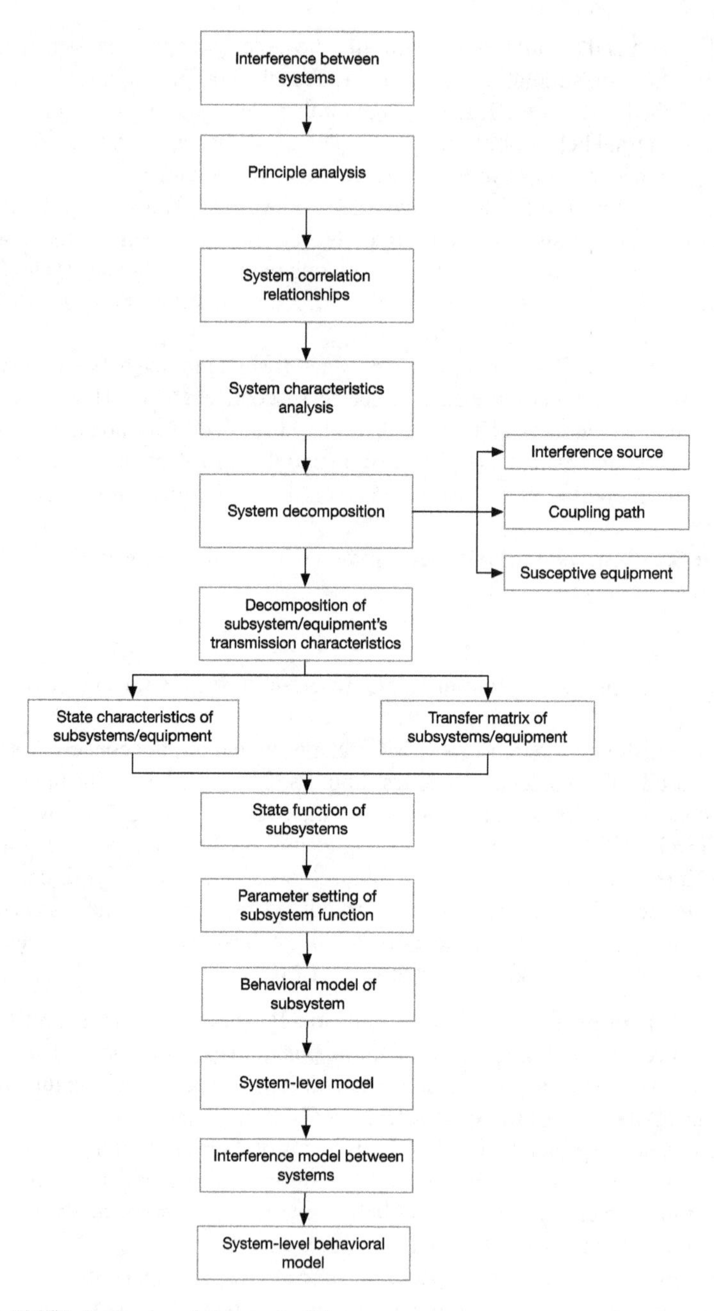

Fig. 5.6 EMC behavioral modeling methodology

(5) Sort out the key control indicators: According to the characteristics of the objects to be designed, sort out the effective technical indicators.
(6) Determine the complete workflow: Develop the EMC workflow chart and the corresponding logs.
(7) Making a timetable: Monitor the milestones through the entire developing process.
(8) Establishing a staff network: Establish the appropriate technical team and management team with clear responsibilities, work packages, and smooth information exchange.

The specific technical works involved in the concept description stage are: (1) Sort out the main radiation sources and susceptive equipment in the system, and annotate the attributes (transmitter, receiver, transceiver); (2) sort out the energy transmission relations from field, field–circuit, circuit to susceptive equipment, and build system interference relationships; (3) list the factors affecting EMC in the interference relationships; i.e., list the main EMC indicators of each equipment, for example, the overall planning of frequency resource between radiation equipment and susceptive equipment, electromagnetic environment, transmitter performance (transmission frequency, transmitting power, harmonic suppression, out-of-band suppression), receiver performance (receiving frequency, receiver susceptibility, adjacent channel suppression, out-of-band suppression), antenna layout performance (antenna patterns, polarization characteristics, antenna isolation), suppression capabilities, equipment and cable layout, interconnection characteristics (band coupling, shielding effectiveness, power usage, and grounding), and damage threshold. The output of the concept description based on the EMC of system characteristics is a conceptual model that can be used to analyze the key attributes of the system.

Assume a system with several pieces of transmitting equipment through antenna ports, other radiation equipment, receiving equipment coupled through antenna ports, and other susceptive equipment. Part of the radiation sources and susceptive equipment in a system are listed in Table 5.1, and the energy transfer relationships among the equipment are shown in Table 5.2. The main factors affecting the isolation between transmitting and receiving antennas are shown in Table 5.3.

When describing the concept of system EMC, it is important to sort out the spectral characteristics of the interference source. The EMI spectrum of the interference source can be categorized into the main interference spectrum and the out-of-band interference spectrum. It is generally believed that the radiation power of the interference source in the main interference spectrum accounts for most of its total transmission power.

The power spectral distribution of the interference source is

$$p(f) = \begin{cases} P_N, \; f < f_{\min} \\ P_B, \; f_{\min} < f < f_{\max} \\ P_T, \; f > f_{\max} \end{cases} \tag{5.30}$$

Table 5.1 Part of the radiation source and susceptive equipment

Indicator	Equipment	Operation state
1	EMI transmitter	Transmitter
2	TCAN	Transmitter/receiver
3	Shortwave radio	Transmitter/receiver
4	Radio altimeter (RADALT)	Transmitter/receiver
5	Weather radar	Transmitter/receiver
6	Transponder	Transmitter/receiver
7	Autopilot system	Transmitter/receiver
8	Radio compass	Receiver
9	VOR receiver	Receiver
10	Microwave landing system	Receiver
11	Localizer receiver	Receiver
12	Flight data recorder	Susceptive equipment
13	Power supply parameter display	Susceptive equipment
14	Fuel gauge	Susceptive equipment

where P_B is the power spectrum of the EMI emitted by the interference source in its main interference band; P_N and P_T are the power spectrum of the EMI emitted by the interference source outside its interference band; $\boldsymbol{f}_{min}$ and $\boldsymbol{f}_{max}$ are the lower band and upper band of the main interference spectrum, respectively.

The types of interference sources can be classified into broadband interference model, narrowband interference model (single-frequency interference model), harmonic interference model, and impulse interference model.

1. Broadband interference model

When the interference bandwidth is greater than the bandwidth of narrowband communication system, it belongs to the broadband interference. The spectrum of the broadband interference model is generally expressed as a function of spectral density, i.e.,

$$p(f) = \begin{cases} P_N, & f < f_{TL} \\ P_T(f), & f_{TL} < f < f_{BL} \\ P_B(f), & f_{BL} < f < f_{TH} \\ P_T(f), & f_{BH} < f < f_{TH} \\ P_N, & f < f_{TH} \end{cases} \tag{5.31}$$

where $P_T(f)$ is the power spectral density of the transition zone; f_{TL} and f_{TH} are the cutoff frequency of the lower sideband transition zone and the cutoff frequency of the upper sideband transition zone, respectively.

The output spectrum of the broadband interference model is shown in Fig. 5.7.

Table 5.2 Energy transmission among part of the equipment

Indicator	Transmitter	Type of transmitting node	Specific transmitting position	Susceptive equipment	Coupling position of susceptive equipment
1	EMI transmitter	Antenna type	Antenna	Shortwave radio	Antenna
			Antenna		Cable
			RF transmission line		Antenna
			Power line		Antenna
		Cable type	Power line		Cable
		Slot type	Case		Antenna
			Case		Cable
		Grounding system	Power ground		Power ground
			Power ground		Signal ground
2	EMI transmitter	Antenna type	Antenna	Radio altimeter	Sensor
			Antenna		Cable
			RF transmission line		Sensor
			Power line		Sensor
		Cable type	Power line		Cable
			RF transmission line		Cable
		Slot type	Case		Cable
			Case		Sensor
3	EMI transmitter	Antenna type	Antenna	Autopilot system	Cable
		Cable type	RF transmission line		Cable
			Power line		Cable
		Slot type	Case		Cable
		Grounding system	Power ground		Power ground

Table 5.3 Influencing factors for isolation between receiver and transmitting antenna of carrier system

Indicator		1	2	3
Transmitting antenna		T1	T2	T3
Receiving antenna		R1	R2	R3
Frequency	Frequency (MHz)			
	Transmit attributes			
	Receive attributes			
Isolation by testing (dB)				
Parameters of transmitting antenna	Transmit peak power (dBm)			
	Gain of transmitting antenna (dB)			
	Attenuation of transmitting band (dB)			
Parameters of receiving antenna	Gain of receiving antenna (dB)			
	Attenuation of receiving band (dB)			
	Receiving sensitivity (dBm)			
Parameters of polarization	Polarization			
	Polarization mismatch (dB)			
Spatial isolation (Ld/dB)				
Antenna isolation				

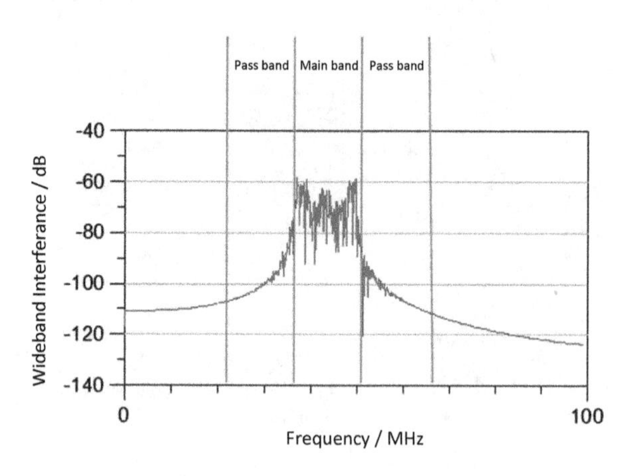

Fig. 5.7 Output spectrum of broadband interference model

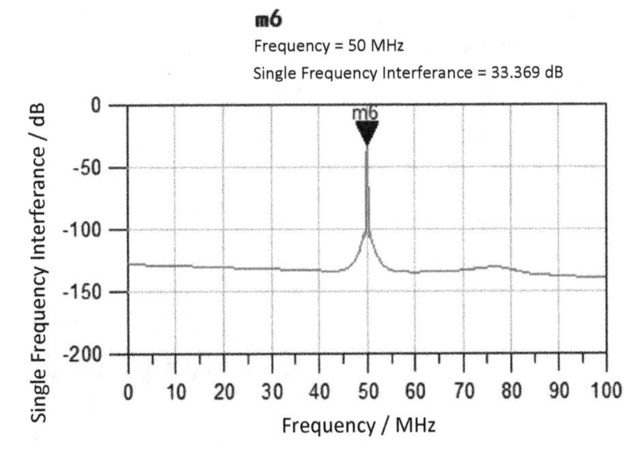

Fig. 5.8 Output spectrum of the narrowband interference model

2. Narrowband interference model

When the interference bandwidth is much smaller than the bandwidth of the communication system, it belongs to narrowband interference. In the analysis frequency range, frequency interference is similar to the single-frequency interference. The spectrum of the interference model is generally the frequency band energy function at a certain frequency resolution.

$$p(f) = \begin{cases} P_N, & f < f_B \\ P_B, & f = f_B \\ P_N, & f > f_B \end{cases} \tag{5.32}$$

where f_B is the current interference frequency and P_B is the output power of the frequency point. The output spectrum of the narrowband interference model is shown in Fig. 5.8.

3. Harmonic interference model

Interference components that occur at certain frequency intervals are harmonic interference. In the behavioral modeling, we can represent the single-frequency interference component of the frequency cycle distribution as

$$p(f) = \begin{cases} P_N, & f \neq f_H \cdot k \\ P_B(k), & f = f_H \cdot k \\ P_N, & f \neq f_H \cdot k_B \end{cases} \tag{5.33}$$

where f_H is the fundamental frequency of the harmonic interference and $P_B(k)$ is the output power of the k-th harmonic frequency.

Figure 5.9 shows the output spectrum of the harmonic interference model.

4. Pulse interference model

Common pulse sources include nuclear pulses, lightning pulses, and static pulses. Electronic equipment also generates pulse currents when they are switched on/off and when their working conditions change. Due to its wide spectrum, flexibility, and

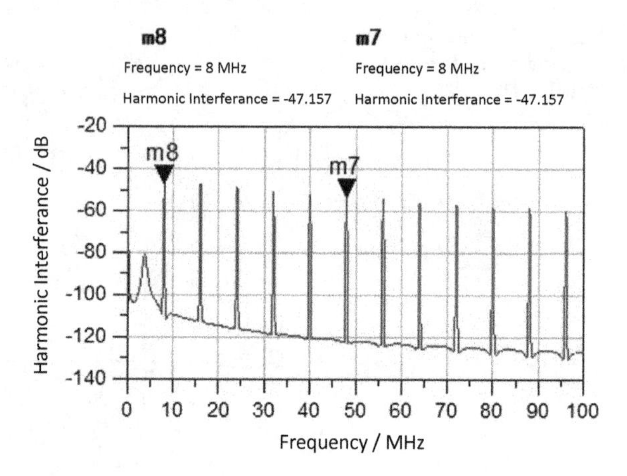

Fig. 5.9 Output spectrum of the harmonic interference model

pertinence, pulse interference is usually high frequency and high energy that instantly exceeds the safety threshold of the electronic devices and equipment. Therefore, the pulse interference has a very strong ability to destroy electronic systems such as communications, radar, and computer networks. The electromagnetic pulse is random and abrupt, and it becomes a major research direction for electromagnetic protection.

The characteristics of the pulse interference are related to its form in time domain. Thus, to build a behavioral model of the pulse interference, we need to establish a time-domain model first and then re-analyze its time–frequency characteristics. In this book, the typical pulse interference models are the rectangular pulse model, exponential decay pulse model, and damped sinusoidal pulse model. The main parameters of the pulse model are start time, delay time, end time, and decay time (for attenuation-type pulses only).

For example, the rectangular pulse model has a time-domain representation function as

$$V(t) = \begin{cases} 0, & t < 0 \\ V_0, & 0 < t < T \\ 0, & t > T \end{cases} \tag{5.34}$$

where V_0 is the pulse amplitude and T is the pulse end time, respectively.

The typical output of time domain and frequency domain for the rectangular pulse model, exponential decay pulse model, and damped sinusoidal pulse model is showed in Fig. 5.10, Fig. 5.11, and Fig. 5.12, respectively.

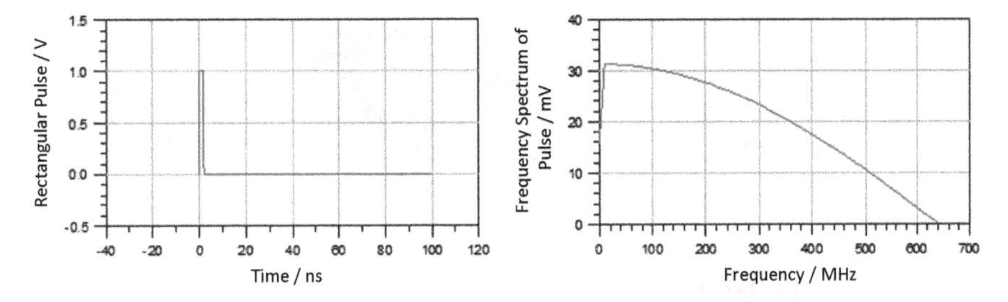

Fig. 5.10 Time-domain and frequency-domain output of rectangular pulse model

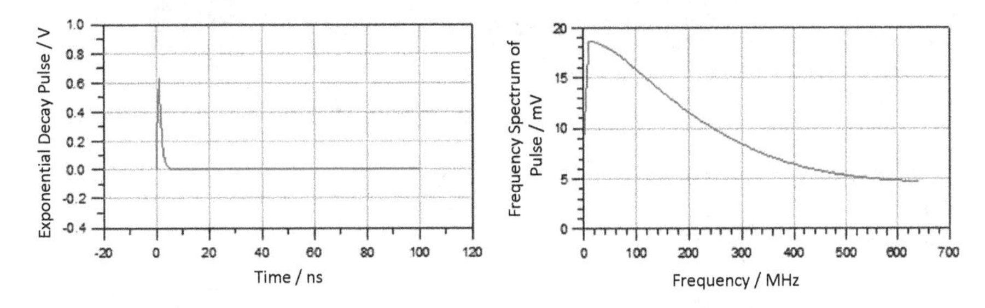

Fig. 5.11 Time-domain and frequency-domain output of exponential decay pulse model

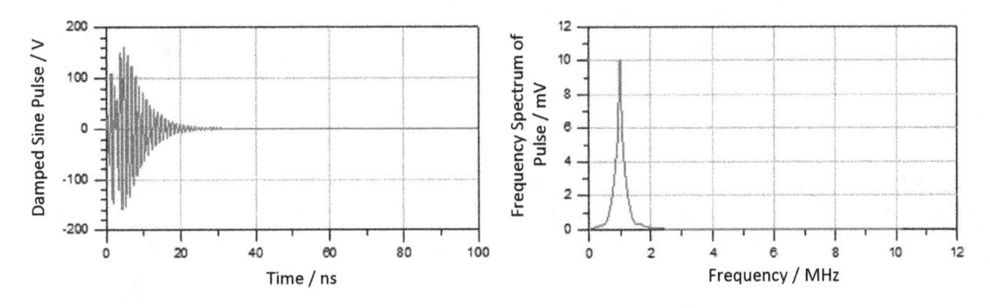

Fig. 5.12 Time-domain and frequency-domain output of damped sinusoidal pulse model

5.3.2.2 EMC Behavioral Modeling

A behavioral model is the quantification of a conceptual model. Formal modeling languages and various modeling tools can be used to build an executable abstract model. Behavioral modeling can be further divided into behavioral modeling based on equipment characteristics and behavioral modeling based on interference correlation matrix. Since the nonlinearity is an important cause of the complexity of EMC problems, we will specifically analyze the behavioral modeling methods for nonlinear systems at the end of this section.

1. Behavioral modeling based on equipment characteristics

EMC behavioral modeling based on equipment characteristics is to directly integrate the equipment behavioral models of each system or subsystem together to form a

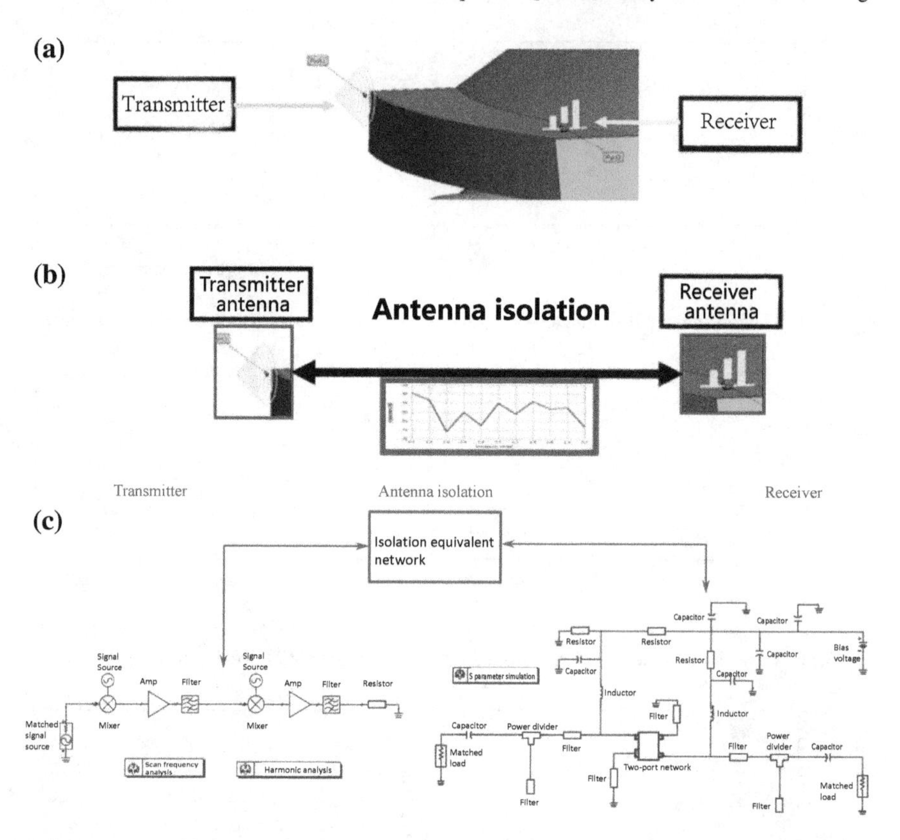

Fig. 5.13 EMC behavioral model of an airborne transceiver system based on equipment characteristics. **a** Geometry model of airborne transmitters and receivers; **b** isolation between transmitting and receiving antennas; and **c** EMC behavioral model of an airborne transceiver system

system-level or subsystem-level behavioral model. In the behavioral model based on equipment characteristics, we only consider the one-way energy flow between equipment, without considering the complex multiple coupling relationships between equipment. Although the model built this way might suffer from some errors, the modeling process is simple and clear. Therefore, we still consider the model to be effective for the first rough analysis.

We will use Fig. 5.13 as an example to illustrate the EMC behavioral modeling based on the equipment characteristics: Fig. 5.13a is the geometry model of an airborne transceiver system. The coupling relationship between the transmitting and the receiving antennas can be determined during concept description. Only the transmitting system needs to be considered in the EMC behavioral modeling based on equipment characteristics.

The behavioral model of the receiving system and the antenna coupling model (frequency characteristics of the antenna isolation) between the transmitting and the receiving antenna is shown in Fig. 5.13b. When the relationship between the antenna isolation and the frequency is substituted by the equivalent magnitude–frequency network, we obtain the behavioral model as shown in Fig. 5.13c. Thus, we can see that the behavioral model shown in Fig. 5.13c is a direct integration of the transmitter-side behavioral model, the receiver-side behavioral model, and the isolation model. The complex coupling relationship between the transmitting and the receiving antennas is not taken into account into the behavioral modeling.

To establish the EMC behavioral model for equipment as shown in Fig. 5.13c (such as the transmitting system and the receiving system behavioral model), we can adopt the idea of node analysis: Based on the principle of circuit configuration of EMI source and electromagnetic susceptive object, the equivalent circuit model can be simulated to obtain the external behavioral response of the equivalent circuit model. Then, we can build the EMC behavioral model of the equipment based on the functional indicator correction.

We will use the behavioral model of shortwave radio as an example to explain the method of behavioral modeling of the equipment's EMC in the transmitter or the receiver as illustrated in Fig. 5.13c.

Shortwave radios generally have high transmission power and are often the main source of interference to other equipment.

Based on the analysis of the EMI characteristics among various pieces of equipment in the system, when modeling the behavior of a shortwave radio station, the focus is to simulate its interference characteristics of externally generated radiation.

(1) Requirement analysis for shortwave radio station

As a transmitting equipment, the fundamental frequency of a shortwave radio station will affect other equipment operating in the same frequency in the system. The harmonics and broadband noise generated by the radio frequency power amplifier will interfere with other highly susceptive receiving equipment.

Since the load of the shortwave radio power amplifier's last transistor cannot reach the optimum at each frequency in the wide band, in order to obtain the same large output power, the transistor may be working in saturation region or cutoff region when the load is not in a frequency band of optimal state. This will cause large nonlinear distortion, and a series of harmonics and broadband noise.

If harmonics exist in a shortwave radio station, the efficiency of energy transmission and energy utilization will be reduced, the equipment will be overheated, noise will be generated, the insulation of the radio stations will be deteriorated, and the life of the equipment will be shortened. Harmonics can even result in the radio station failure or burnout. What's more, harmonics may cause partial parallel resonance or series resonance of the system; thus, the harmonic content increases and the capacitor and other equipment might be burned. Therefore, when modeling the harmonic characteristics of a shortwave radio station, it is necessary to consider both the simulation of the normal operation of the radio station itself and the simulation of the external interference. CE106 project requirements in GJB 151 can be referred to modeling.

The wideband noise generated by the shortwave radio stations is an important cause of the interference from the radio stations toward the external sensitive receiving equipment. Since the broadband noise exists in a frequency band between the main frequency and the harmonics and it is close to the fundamental and harmonic waves, it is difficult to use filters or other technical means to separate or eliminate the noise. The broadband noise produced by shortwave radios is mainly due to nonlinearity, which mainly are the harmonics and intermodulation signal produced by the transmitting equipment.

(2) Modeling process

The superheterodyne structure is used in the shortwave radio transmission equipment, and an equivalent model can be established on the circuit simulation software platform according to the circuit principle of the shortwave radio station: Firstly, build an emission model to analyze its external radiation characteristics; secondly, build models for the typical sub-modules including AM module, mixer module, and nonlinear transmission module.

(1) Establish a transmission model. The basic parameters needed to establish the transmission model are frequency range, frequency interval, work type, output power, harmonic attenuation, IF selectivity, harmonic rejection ratio, etc. The transmission model scheme is as follows. Firstly, the radio frequency carrier signal is modulated by the modulation signal and the impedance of modulator is set. Secondly, the signal is up-converted by the first local oscillator and the first IF signal is obtained after the filter. Then, after up-converted by the second local oscillator, the output frequency point is moved to the transmission carrier frequency. Finally, after amplified by a small-signal amplifier and a power amplifier, the output fundamental wave signal in frequency domain can be obtained which satisfies the power demand. The shortwave radio transmission model is shown in Fig. 5.14.
(2) Establish models for typical sub-modules. The main modules in the transmission model are AM module, mixer module, and nonlinear transmitter module. The model of the nonlinear transmitter module is the most important, since it determines the core functional indicators of the shortwave radio RF port.

   ① AM module. Since the original signal is in low frequency, it is generally not suitable for direct transmission. Therefore, it is necessary to perform signal modulation. The most common analog modulation methods are amplitude

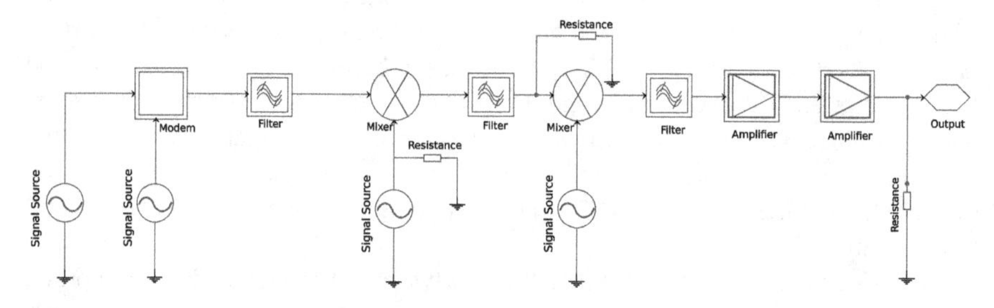

Fig. 5.14 Transmission model of a shortwave radio station

modulation and angle modulation using a sine wave as carrier. The short-wave radio behavioral model exploits amplitude modulation methods such as amplitude modulation (AM), double-sideband modulation (DSB), vestigial sideband modulation (VSB), and single-sideband modulation (SSB).
Amplitude modulation: The amplitude of the sinusoidal carrier changes linearly with the modulation signal, i.e.,

$$S_m(t) = F[s_m(t)] = A[M(\omega - \omega_c) + M(\omega + \omega_c)]/2 \tag{5.35}$$

Double-sideband modulation: If the input baseband signal has no DC component and $h(t)$ is an ideal band-pass filter, then the resulting output signal is a double-sideband modulated signal without a carrier component, which can also be called a double-sideband modulation signal. If the input baseband signal has a DC component and it is also assumed to be an ideal band-pass filter, then the resulting output signal is a double-sideband signal with a carrier component.
Single-sideband modulation: The double-sideband modulated signal contains two sidebands, the upper and lower sidebands. Since these two sidebands contain the same information, from the perspective of information transmission, it is enough to transmit only one sideband. Single-sideband modulation is a modulation method that produces only one sideband.
Residual sideband modulation: It is a linear modulation between the double sideband and single sideband. It not only overcomes the shortcomings of bandwidth occupied by the double-sideband modulated signal, but also solves the problem of implementing single-sideband signal. This method partially suppresses one sideband instead of completely suppressing; thus, a small portion of the sideband remains.
When establishing the AM sub-module model, we need to consider the RF carrier signal frequency and the modulation signal frequency. The AM sub-module behavioral model is shown in Fig. 5.15.

② Mixer module. The basic functions of the mixer are up-conversion and down-conversion. Up-conversion mixes the IF signal with the RF oscillation signal into the new RF signal and transmits it through the antenna; the down-converter is used to mix the RF signal received by the antenna with the local carrier signal and filter it into IF signal. Then, the IF signal is sent to the IF processing module. In the conversion process, the modulation type (regardless of amplitude modulation, frequency modulation, or phase modulation) and the modulation parameters (such as modulation frequency, number of modulations, etc.) do not change, and only the signal carrier frequency is changed. According to the principle of the mixer, the mixer behavioral model can be built as shown in Fig. 5.16.
③ Nonlinear emission module. Equipment that causes nonlinearity in short-wave radios usually has mixers, filters, amplifiers, and so on. We will focus on the behavioral model of the power amplifier module.

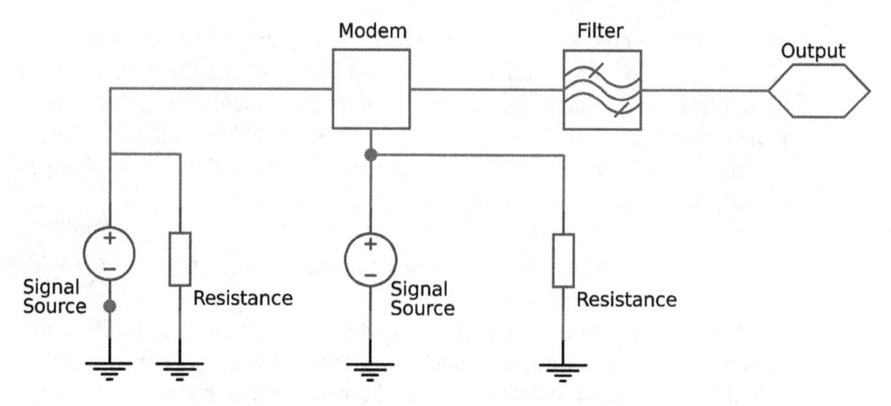

Fig. 5.15 Behavioral model of AM sub-module

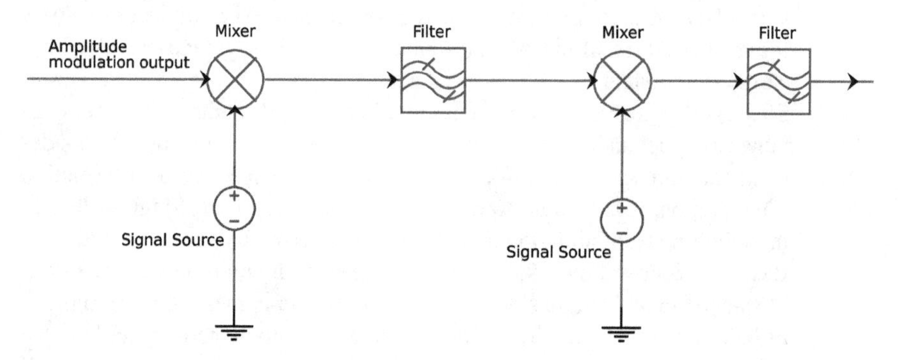

Fig. 5.16 Behavioral model of the mixer

RF power amplifier is an important part of nonlinear transmission equipment. In the transmitter's prestage circuit, the power of the RF signal generated by the modulation circuit is very small, and it needs to go through a series of amplification, such as the buffer level, the intermediate amplifier level, and the final power amplifier level. After sufficient RF power is obtained, it can be sent to the antenna. Therefore, RF power amplifiers must be used to obtain sufficient RF output power.

In the transmission system, the output power of the RF power amplifier mainly refers to the output power of the final-stage power amplifier. In order to achieve high-power output, the prefinal stage must already have sufficient power levels for excitation. Its main technical indicators are output power and efficiency. At the same time, the harmonic components in the output should be as small as possible to avoid generating interference toward other channels.

According to the different current conduction angles, the RF power amplifier can be categorized into three classes, which are Class A, Class B, and Class C. In addition, there are Class D and Class E amplifiers in which the electronic devices operate in the on–off state.

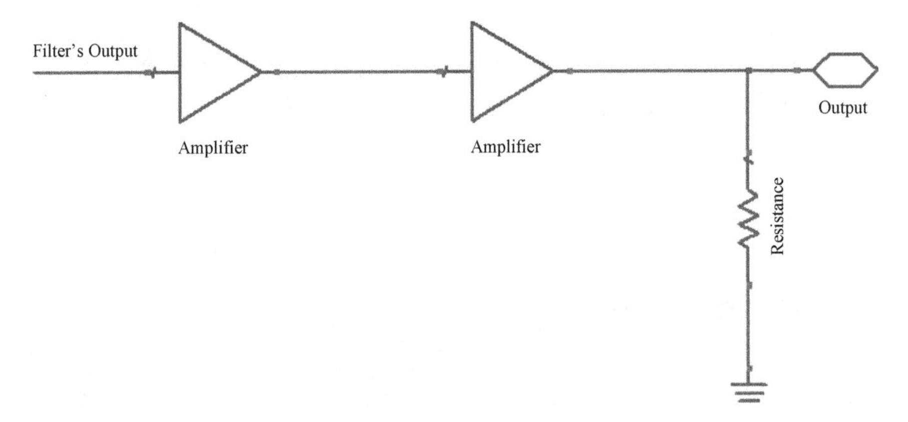

Fig. 5.17 Behavioral model of the nonlinear transmission part

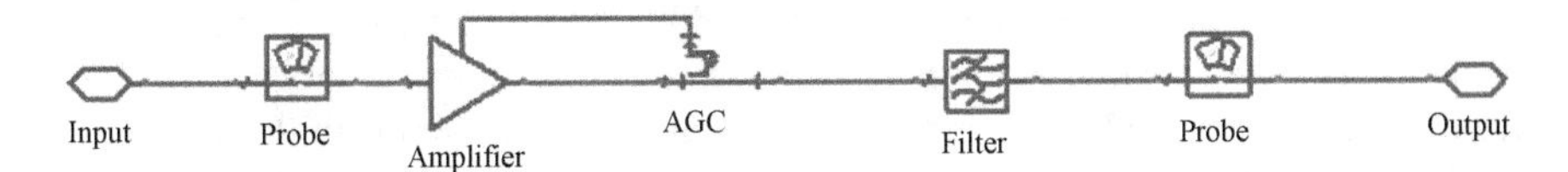

Fig. 5.18 Behavioral model of final-stage power amplifier

In terms of the characteristics of the amplifier, the key parameters are gain and gain flatness, operating frequency and bandwidth, output power, DC input power, input/output reflection coefficient, and noise coefficient. In addition, other parameters such as intermodulation distortion (IMD), harmonics, feedback, and thermal effects are often considered. All of these parameters can seriously affect the performance of the amplifier.

The modeling scheme is designed as follows: The signal input to the final power amplifier is firstly passed through a small-signal amplifier, and the transmitted small signal is amplified to the required magnitude; after that, the signal is adjusted through the final power amplifier, which is the main component of the nonlinear emission model to meet the requirement of shortwave radio transmission power. The behavioral model of the nonlinear emission part is shown in Fig. 5.17. The behavioral model of the final-stage power amplifier is shown in Fig. 5.18.

When the first-stage amplifier in the model shown in Fig. 5.17 fails to meet the requirement of the actual transmitting power, the final-stage power amplifier can be designed using automatic gain control and automatic power control technology. The automatic gain control can automatically control the amplitude of the gain by changing the input and output compression ratio. It maintains the amplitude of the output signal or keeps the signal changing within a small range when the amplitude of the input signal varies greatly. Therefore, the system can operate properly with very low input signal, and the receiver will not be saturated or jamming due to the input signal being too large. The behavioral model of automatic gain control amplifier is shown in Fig. 5.19.

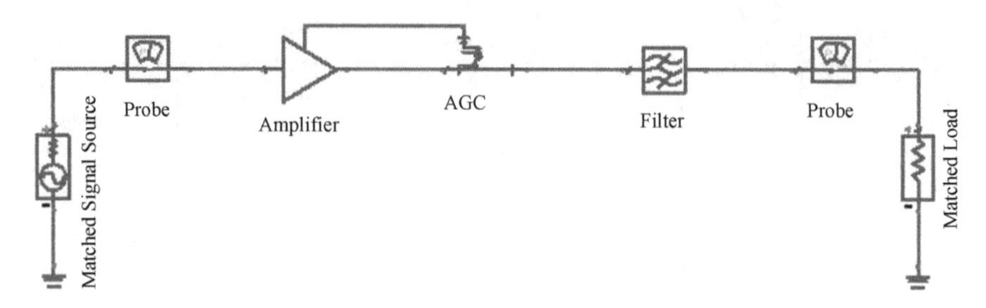

Fig. 5.19 Behavioral model of automatic gain control amplifier

Based on the behavioral models presented above, we can simulate and predict the spurious emissions (including harmonic emission and broadband noise) generated by a shortwave radio.

Before modeling subsystem or systems with multiple equipment, we need to consider the interference coupling problems among equipment; in other words, we need to establish a model of interference coupling transmission path. In behavioral modeling based on equipment characteristics, the model of interference coupling transmission path only includes one-way energy flow between equipment, regardless of the multiple and complex coupling relationships between equipment.

Now, we will discuss an example of the model of interference coupling transmission path which is shown in Fig. 5.20. The system is composed of a susceptive object and an interference source (the susceptive object and the interference source are subsystems which are composed of several lower-level subsystems. We can get more detailed coupling relationships by identifying the input and output variables of these lower-level subsystems). Each of the input and output variables is marked with an arrow indicating its direction of interference. The EMI generated by the interference source acts on the susceptive object and is an input variable for the susceptive object. In addition to the internal sources of interference, external environmental interference also affects the entire system and acts on the susceptive object, which can be considered as another input variable for the susceptive object. At the same time, the response of a susceptive object may have a counter-effect on the interference source. This effect is not necessarily interference, but it may determine whether the interference source or even the whole aircraft can continue to work normally or not.

When building the model of interference coupling transmission path, special attention must be paid to the direction of the transmission path between subsystems and the change in energy in the transmission direction. The EMC behavioral model describes the dynamic process of interference coupling by a path transfer function. The transfer function can be determined by frequency response test or numerical calculation.

Figure 5.21 shows the interference coupling model based on the transfer function, where $T(\Theta, \Phi, t, f)$ represents the interference source model; $H(\Theta, \Phi, t, f)$ represents the interference coupling path model; $S(\Theta, \Phi, t, f)$ represents the suscep-

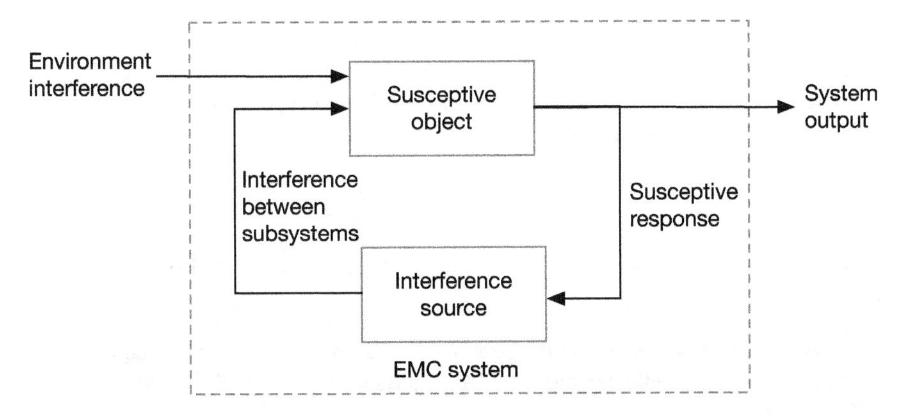

Fig. 5.20 Model of interference coupling transmission path of an electronic system

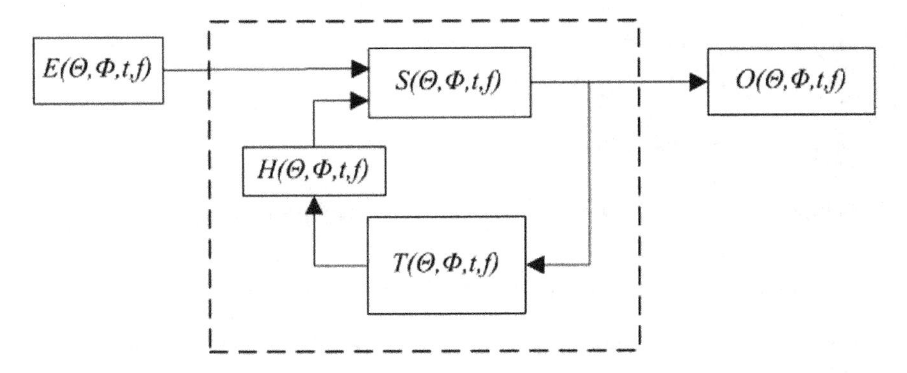

Fig. 5.21 Interference coupling model based on the transfer function

tive object model; $E(\Theta, \Phi, t, f)$ represents the environmental EMI source model; $O(\Theta, \Phi, t, f)$ represents the interference output model.

Figure 5.21 The interference coupling model based on the transfer function. where t and f represent the change of system characteristics with time and frequency; Θ and Φ represent other factors that affect the system characteristics.

EMC can be predicted based on the interference source model, interference coupling path model, susceptive object model, environmental EMI source model, etc. For the sake of simplicity, Θ and Φ are not considered for now. Then, the interference model $T(t,f)$ represents any interfering equipment with radiation or conduction emission characteristics in the system. The susceptive model $S(t, f)$ represents any equipment with radiation- or conduction-susceptive characteristics within the system. The coupling function $H(t,f)$ represents the radiation coupling coefficient and the conduction coupling coefficient, which are mainly the coupling of the radiation field to the cable, the coupling between the cables, and the common power coupling. The signal coupled with the susceptive equipment is

$$R(t, f) = T(t, f) \times H(t, f) \tag{5.36}$$

If $I(t,f)$ represents the interfering state discriminant function (safety margin function) of the susceptive equipment, then $I(t,f)$ can be expressed as

$$I(t, f) = R(t, f) - S(t, f) \tag{5.37}$$

The performance of the susceptive equipment can be evaluated by the value of $I(t, f)$:

a. If $I(t, f) > 0$, it means that the susceptive equipment will be interfered, and the value of $I(t, f)$ indicates the strength of the interference at the same time.
b. If $I(t, f) = 0$, it means that the susceptive equipment is in a critical state of interference. The equipment may suffer from interference, and the safety margin is 0.
c. If $-6 < I(t, f) < 0$, it means that the susceptive equipment has not been interfered but is vulnerable to interference.
d. If $I(t, f) < -6$, it means that the susceptive equipment will not be interfered and can work safely and stably.

2. Behavioral modeling based on interference correlation matrix

(1) **System-Level Energy Coupling Model**

The system's energy coupling model is used to study the interference signal energy coupled with the susceptive equipment.

Assume the number of radiation source ports in the system is M_2, in which there is M_1 transmitting equipment whose front end is connected to the antenna. Since each transmitting equipment may have multiple radiation source ports (including the emission ports formed by the radiation of the antenna), usually $M_2 \geq M_1$. The radiation sources that emit radiation emission through the transmitting antenna ports are denoted as $E_1, E_2, \ldots, E_{M_1}$,and the other radiation source ports are denoted as $E_{M_1+1}, \ldots, E_{M_2}$. Similarly, the number of the equivalent susceptive ports to be analyzed in the system (including fuses for pyrotechnics) is N_2, and the number of receiving equipment whose front ends are connected to the antenna is N_1. Since the energy transmitted by the emission source may be coupled with each receiving equipment through multiple ports (including the energy radiated by the radiation source through the receiving antenna port), it can be derived that $N_2 \geq N_1$. The susceptive ports formed by the coupling of the receiving antenna ports are labeled as $R_1, R_2, \ldots, R_{N_1}$,and other susceptive ports are labeled as $R_{N_1} + 1, \ldots, R_{N_2}$.

In the environment of a system, the total power [23, 24] from the radiation source port coupled with the susceptive port in the frequency range $f_a \sim f_b$ is

$$P_{Rji} = \int_{f_a}^{f_b} \frac{\eta_i(f)\beta_j(f)}{I_{ij}(f)} df \tag{5.38}$$

where $\eta_i(f)$ is the output power spectral density of the radiation source port i; $\beta_j(f)$ is the response function of the susceptive port to different modulation signals; $I_{ij}(f)$ is the isolation between the radiation source port i and the susceptive port j.

The total power received by the susceptive port j from the radiation source is

$$P_{Rj}\Big|_{f_a}^{f_b} = \sum_{i=1}^{M_2} \int_{f_a}^{f_b} \frac{\eta_i(f)\beta_j(f)}{I_{ij}(f)} df \tag{5.39}$$

The electromagnetic radiation received by fuel, ordnance, and personnel is an important topic in system-level EMC analysis and prediction. Therefore, it is necessary to study the energy coupling model at the fuel inlet, ordnance, and work chamber of the aircraft.

The total power from the radiation source port to the fuel susceptive point in the frequency range $f_a \sim f_b$ is

$$P_{Oji} = \int_{f_a}^{f_b} \frac{\eta_i(f)\xi_j(f, T)}{T_{ij}(f)} df \tag{5.40}$$

where the values of f_a and f_b may refer to the corresponding national military standard [25, 26]; $T_{ij}(f)$ is the power transfer function of the radiation port i to the fuel inlet j; $\xi_j(f, T)$ is the electromagnetic power absorption rate of the fuel and gas mixture, which is determined by the frequency, time, temperature, density of the fuel and gas mixture, the flow speed of air, and other factors.

The total power received by the electromagnetic radiation source at the fuel susceptive point j is

$$P_{Oj}\Big|_{f_a}^{f_b} = \sum_{i=1}^{M_2} \int_{f_a}^{f_b} \frac{\eta_i(f)\xi_j(f, T)}{T_{ij}(f)} df \tag{5.41}$$

The total power from the radiation source port to the operator's workstation in the frequency range $f_a \sim f_b$ is

$$P_{Wji}\Big|_{f_a}^{f_b} = \int_{f_a}^{f_b} \frac{\eta_i(f)}{T_{ij}(f, t)} df \tag{5.42}$$

where the values f_a, f_b may refer to the corresponding military standards [27]; $T_{ij}(f)$ is the power transfer function from the transmitting equipment i to the operator workstation j. It is a function of frequency and time.

The total electromagnetic radiation power at the operator's workstation j is

$$P_{Wj}\Big|_{f_a}^{f_b} = \sum_{i=1}^{M_2} \int_{f_a}^{f_b} \frac{\eta_i(f)}{T_{ij}(f, t)} df \tag{5.43}$$

(2) System-Level Interference Correlation Matrix

The interference correlation matrix describes the energy coupling relationship among the equipment in the system. In this section, we will establish the isolation matrix, antenna isolation matrix, transmitting power matrix of the transmitters, attenuation matrix of the transmitters at the frequency point to be analyzed, transmitting feeder loss matrix of the transmitters, sensitivity matrix of the receivers, and reception suppression matrix of the receivers at the analysis frequency point, receiving feeder loss matrix of the receivers, signal-to-noise ratio matrix required when the receivers operate normally, EMC safety margin matrix required by the receiving equipment, transmitting conversion matrix, receiving conversion matrix, safety threshold matrix of antenna isolations, interference power matrix of the receiver antenna ports.

Assuming the one-to-one correspondence between the radiation source port and the susceptive port, the relationship matrix between the radiation source port and the susceptive port of the entire system is

$$\begin{bmatrix} (E_1, R_1) & \cdots & (E_1, R_{N_1}) & \cdots & (E_1, R_{N_2}) \\ \vdots & & \vdots & & \vdots \\ (E_{M_1}, R_1) & \cdots & (E_{M_1}, R_{N_1}) & \cdots & (E_{M_1}, R_{N_2}) \\ \vdots & & \vdots & & \vdots \\ (E_{M_2}, R_1) & \cdots & (E_{M_2}, R_{N_1}) & \cdots & (E_{M_2}, R_{N_2}) \end{bmatrix}, \quad i \in [1, M_2], \quad j \in [1, N_2] \tag{5.44}$$

where (E_i, R_j) represents the energy transfer relationship between the radiation source port E_i and the susceptive port R_j.

The energy transfer function between the radiation source port and the susceptive port describes the interference relationship and the strength of interference between the two ports.

The isolation between the radiation source port and the susceptive port is a one-to-one correspondence, and it exhibits as a power transfer function. Therefore, considering the safety of susceptive equipment, the interference matrix of the susceptive equipment is constructed with isolation $I_{ij}(f)$ as elements. The antenna isolation matrix is

$$\boldsymbol{I} = (I_{ij}(f)) = \begin{bmatrix} I_{11}(f) & \cdots & I_{1j}(f) & \cdots & I_{1N_2}(f) \\ \vdots & & \vdots & & \vdots \\ I_{i1}(f) & \cdots & I_{ij}(f) & \cdots & I_{iN_2}(f) \\ \vdots & & \vdots & & \vdots \\ I_{M_21}(f) & \cdots & I_{M_2j}(f) & \cdots & I_{M_2N_2}(f) \end{bmatrix}$$

$$= \begin{bmatrix} I(E_1, R_1) & \cdots & I(E_1, R_{N_1}) & \cdots & I(E_1, R_{N_2}) \\ \vdots & & \vdots & & \vdots \\ I(E_{M_1}, R_1) & \cdots & I(E_{M_1}, R_{N_1}) & \cdots & I(E_{M_1}, R_{N_2}) \\ \vdots & & \vdots & & \vdots \\ I(E_{M_2}, R_1) & \cdots & I(E_{M_2}, R_{N_1}) & \cdots & I(E_{M_2}, R_{N_2}) \end{bmatrix}, \quad i \in [1, M_2], \quad j \in [1, N_2] \tag{5.45}$$

We can rewrite it in the form of dB

$$\boldsymbol{I}(\text{dB}) = 10\boldsymbol{lg}\ \boldsymbol{I} \tag{5.46}$$

The sub-matrix $\boldsymbol{A}$ in the matrix $\boldsymbol{I}$ is the antenna isolation matrix, which represents the degree of isolation between the transmitter and the receiver when energy is transmitted through the antenna ports. We can take ideal values for the isolation of the equipment used for both transmitting and receiving, i.e.,

$$\mathbf{A} = \begin{bmatrix} I(E_1, R_1) & \cdots & I(E_1, R_j) & \cdots & I(E_1, R_{N_1}) \\ \vdots & & \vdots & & \vdots \\ I(E_i, R_1) & \cdots & I(E_i, R_j) & \cdots & I(E_i, R_{N_1}) \\ \vdots & & \vdots & & \vdots \\ I(E_{M_1}, R_1) & \cdots & I(E_{M_1}, R_j) & \cdots & I(E_{M_1}, R_{N_1}) \end{bmatrix}, i \in [1, M_1], j \in [1, N_1] \tag{5.47}$$

where the isolation $I(E_i, R_j)$ between the transmitting antenna i and the receiving antenna j is determined by factors such as the radiation characteristics of the transmitting and the receiving antennas, the mounting position in the system, the electrical size between the transmitting and the receiving antennas, the shielding condition, and the polarization matching between the transmitting and the receiving antennas. The antenna isolation matrix is an important topic in antenna layout optimization.

We can rewrite (5.47) in the form of dB

$$\boldsymbol{A}(\text{dB}) = 10\mathbf{lg}\,\boldsymbol{A} \tag{5.48}$$

The transmitting power matrix of the transmitters is

$$\mathbf{P_t} = \begin{bmatrix} P_{t1} \\ \vdots \\ P_{ti} \\ \vdots \\ P_{tM_1} \end{bmatrix} (dBm), \quad i \in [1, M_1] \tag{5.49}$$

where P_{ti} is the transmitting power of the i-th transmitting equipment.

The emission attenuation matrix of the transmitters at the analysis frequency is

$$\mathbf{L_{tB}} = \begin{bmatrix} L_{tB1}(f) \\ \vdots \\ L_{tBi}(f) \\ \vdots \\ L_{tBM_1}(f) \end{bmatrix} (dBc), \quad i \in [1, M_1] \tag{5.50}$$

where $L_{tBi}(f)$ is the amount of emission attenuation of the i-th transmitting equipment at the analysis frequency point and it is a function of the frequency. We can assume $L_{tBi}(f) = 0\text{dB}$ within the emission bandwidth. Out of operation bandwidth, $L_{tBi}(f)$ represents the out-of-band attenuation and harmonic attenuation, and the data can be estimated using theoretical methods or obtained from tests.

The transmitting feeder loss matrix of the transmitters is

$$\mathbf{L_{tf}} = \begin{bmatrix} L_{tf1}(f) \\ \vdots \\ L_{tfi}(f) \\ \vdots \\ L_{tfM_1}(f) \end{bmatrix} (\text{dB}), \quad i \in [1, M_1] \tag{5.51}$$

where $L_{tfi}(f)$ is the RF transmission loss of the i-th transmitter. It indicates the feeder loss between the transmitter output port and the input port of the transmitting antenna. Similarly, it is a function of frequency.

Transmission line loss consists of transmission line length loss, waveguide discontinuity loss, rotating joint loss, connection failure loss, and transceiving switching loss (transceiving switching loss exists if the same antenna is used for both transmitting and receiving). At the same time, transmission line loss can only be relatively determined after system installation. The value should be 0.5–3.5 dB under most conditions.

The sensitivity matrix of the receivers is

$$\mathbf{P_{smin}} = [\, P_{s1} \cdots P_{sj} \cdots P_{sN_1} \,](\text{dB}m), \quad j \in [1, N_1] \tag{5.52}$$

where P_{sj} is the design sensitivity of the j-th receiving equipment.

The reception suppression matrix of the receivers at the analysis frequency is

$$\mathbf{L_{rB}} = [\, L_{rB1}(f) \cdots L_{rBj}(f) \cdots L_{rBN_1}(f) \,](\text{dB}c), \quad j \in [1, N_1] \tag{5.53}$$

where $L_{rBj}(f)$ is the reception suppression of the j-th receiving equipment at the analyzing frequency and it is a function of the frequency. We can assume $L_{rBj}(f) = 0\text{dB}$

within the receiver's operating bandwidth. Outside the operating bandwidth, $L_{rBj}(f)$ represents the out-of-band suppression of the receiver, harmonic suppression, etc. As with the data for $L_{tBi}(f)$, the data for $L_{rBj}(f)$ can also be estimated using the methods provided in Chap. 3, but the test data is the best to use here.

The receiving feeder loss matrix of the receivers is

$$\mathbf{L_{rf}} = [\, L_{rf1}(f) \cdots L_{rfj}(f) \cdots L_{rfN_1}(f) \,](\mathrm{dB}), \quad j \in [1, N_1] \tag{5.54}$$

where $L_{rfj}(f)$ is RF transmission loss of the j-th receiving equipment. It is the transmission loss of the transmission line between the transmitting antenna output port and the receiver input port. It is also a function of frequency.

The signal-to-noise ratio (SNR) matrix required when the receivers operate normally is

$$\frac{\mathbf{S}}{\mathbf{N}} = \left[(\tfrac{S}{N})_1 \cdots (\tfrac{S}{N})_j \cdots (\tfrac{S}{N})_{N_1} \right](\mathrm{dB}), \quad j \in [1, N_1] \tag{5.55}$$

where $(\frac{S}{N})_j$ is the SNR of the j-th receiving equipment.

The EMC safety margin required by the receiving equipment is

$$\mathbf{S_m} = [\, S_{m1} \cdots S_{mj} \cdots S_{mN_1} \,](\mathrm{dB}), \quad j \in [1, N_1] \tag{5.56}$$

where S_{mj} is the EMC safety margin required for the j-th receiving equipment to operate normally, and the value may be 6 dB or 16.5 dB.

The transmitting conversion matrix is a $1 \times M_1$ matrix. It can be expressed as

$$\boldsymbol{T_E} = [1,\ 1,\ \cdots,\ 1] \tag{5.57}$$

The receiving conversion matrix is an $N_1 \times 1$ matrix. It can be expressed as

$$\mathbf{T_R} = \begin{bmatrix} 1 \\ 1 \\ \vdots \\ 1 \end{bmatrix} \tag{5.58}$$

Considering the one-to-one correspondence between the transmitting equipment and the receiving equipment, if the entire system needs to fully meet the EMC requirements, it must be ensured that the receiver will not be interfered or desensitized. The safety threshold matrix of antenna isolations is then defined as

$$\begin{aligned} \boldsymbol{A}_{\mathrm{limit}}(\mathrm{dB}) &= (A_{\lim it}) \\ &= (\boldsymbol{P}_{\mathrm{t}} - \boldsymbol{L_{tB}} - \boldsymbol{L_{tf}}) \otimes \boldsymbol{T_E} - (\boldsymbol{P_{smin}} + \boldsymbol{L_{rB}} + \boldsymbol{L_{rf}} - \boldsymbol{S_m}) \otimes \boldsymbol{T_R} \end{aligned} \tag{5.59}$$

where $\otimes$ is the Kronecker product; $\boldsymbol{P_t}$ is the transmitting power matrix of the transmitters [Eq. (5.49)]; $\boldsymbol{L_{tB}}$ is the emission attenuation matrix of the transmitters at the analysis frequency [Eq. (5.50)]; $\boldsymbol{L_{tf}}$ is the transmitting feeder loss matrix of the transmitters [Eq. (5.51)]; $\boldsymbol{P_{smin}}$ is the sensitivity matrix of the receivers [Eq. (5.52)]; $\boldsymbol{L_{rB}}$ is the reception suppression matrix of the receivers at the analysis frequency (Eq. 5.53); $\boldsymbol{L_{rf}}$ is the receiving feeder loss matrix of the receivers (Eq. 5.54); $\boldsymbol{S_m}$ is the EMC safety margin matrix required by the receiving equipment (Eq. 5.56); $\boldsymbol{T_E}$ is the transmitting conversion matrix (Eq. 5.57); $\boldsymbol{T_R}$ is the receiving conversion matrix (Eq. 5.58).

It is worth mentioning that Eq. (5.95) is the optimal threshold for EMC design. Satisfying this equation means that the signals emitted by other transmitting equipment in the system will not interfere with the receiver through the antenna ports. Of course, the interference phenomenon does not necessarily occur if (5.95) is not satisfied. For a receiving equipment, as long as (5.60) and (5.61) are satisfied at the same time, the receiver can work normally. But it might be desensitized, and the corresponding tactical and technical indicators might also decline.

$$\frac{P_{sj}}{J_j + N_j} \geq (\frac{S}{N})_{mj} \tag{5.60}$$

$$P_{sj} \geq P_{s\min} \tag{5.61}$$

where P_{sj} is the signal power received by the receiver j; J_j is the external interference power received by the receiver j; N_j is the internal noise power of the receiver j. For a particular receiver, the value of N_j is relatively stable.

In fact, for each receiver, any spectral component will be affected by any transmitting equipment in the system, but the degree of influence will be different. Considering that the receiver is affected by all transmitting equipment (through antenna-radiated power), the interference power matrix coupled with the receiver antenna port is

$$\begin{aligned} \boldsymbol{J_A} &= [\, J_1 \cdots J_j \cdots J_{N_1} \,] \\ &= 10\boldsymbol{lg}\left(\boldsymbol{T_E} 10^{\frac{(\mathbf{P_t} - \mathbf{L_{t}}_B - \mathbf{L_{tf}}) \otimes \mathbf{T_E} - (\mathbf{L_{r}}_B + \mathbf{L_{rf}}) \otimes \mathbf{T_R} - \mathbf{A}(\mathrm{dB})}{10}}\right) (i \in [1, M_1], j \in [1, N_1]) \quad (\mathrm{dB\ m}) \end{aligned} \tag{5.62}$$

where $10^{\mathrm{A}} = 10^{[I(E_i, R_j)]} = [10^{I(E_i, R_j)}]$ (a matrix power of a constant).

Considering that the power radiated from all sources is coupled with the susceptive equipment, besides the antenna port, the coupled port can also be a slot, apertures, cable, or other interference channels. Based on different coupling approaches, the interference power coupled with other susceptive ports can be calculated using the same method to obtain $J_{N_1+1}, \cdots, J_{N_2}$. The safety threshold of corresponding port isolation is $\boldsymbol{I_{limit}}$.

The energy transfer relationship between the radiation source port and the susceptive port described in Eq. (5.44) can also be described in the following way.

It is assumed that the number of equipment in the system is M, and each equipment is regarded as a radiation source $E(f)$ with a constant radiation state or a periodic change. For example, the i-th equipment is expressed as $E_i(f)(i = 1, 2, \ldots, M)$, which indicates that the number of ports that can interfere with the outside is M. When the system is relatively large, several pieces self-contained equipment can be combined into subsystems to simplify the decomposition of the system. However, when the equipment is in an unstable state, the equipment should be further decomposed until it reaches a stable state.

Similarly, assuming that there are N receiving ports in the system and each port is denoted by $R(f)$, we define the j-th receiving port as $R_j(j = 1, 2, \ldots, N)$and the coupling function between the j-th receiving port and the i-th interference port as $H_{i,j}(f)$. Then, the component of the total interference that may be received by the j-th receiving port is

$$R_j(f) = \begin{bmatrix} E_1(f) \cdots E_i(f) \cdots E_M(f) \end{bmatrix}_{1\times \mathrm{M}} \times \begin{bmatrix} H_{1,j}(f) \\ \vdots \\ H_{i,j}(f) \\ \vdots \\ H_{M,j}(f) \end{bmatrix}_{\mathrm{M}\times 1}$$

$$= \sum_{i=1}^{M} E_i(f) H_{ij}(f) \tag{5.63}$$

We can derive that the total interference received by all receiving ports within the entire electronic information system is

$$\begin{bmatrix} R_1(f) \\ \cdots \\ R_j(f) \\ \cdots \\ R_N(f) \end{bmatrix} = \begin{bmatrix} E_1(f) \cdots E_i(f) \cdots E_M(f) \end{bmatrix}_{1\times M} \times \begin{bmatrix} H_{1,1}(f) & \cdots & H_{1,j}(f) & \cdots & H_{1,N}(f) \\ \vdots & & \vdots & & \vdots \\ H_{i,1}(f) & & H_{i,j}(f) & & H_{i,N}(f) \\ \vdots & & \vdots & & \vdots \\ H_{M,1}(f) & \cdots & H_{M,j}(f) & \cdots & H_{M,N}(f) \end{bmatrix}_{M\times N}$$

$$= \begin{bmatrix} \sum_{i=1}^{M} E_i(f) H_{i,1}(f) \\ \vdots \\ \sum_{i=1}^{M} E_i(f) H_{i,j}(f) \\ \vdots \\ \sum_{i=1}^{M} E_i(f) H_{i,N}(f) \end{bmatrix} \tag{5.64}$$

where the $M \times N$ matrix is called the interference correlation matrix I. It represents the coupling among different ports in the system and can be expressed as

$$I = \begin{bmatrix} H_{1,1}(f) & \cdots & H_{1,j}(f) & \cdots & H_{1,N}(f) \\ \vdots & & \vdots & & \vdots \\ H_{i,1}(f) & & H_{i,j}(f) & & H_{i,N}(f) \\ \vdots & & \vdots & & \vdots \\ H_{M,1}(f) & \cdots & H_{M,j}(f) & \cdots & H_{M,N}(f) \end{bmatrix} \tag{5.65}$$

The coupling function between the i-th interference port and the j-th receiving port is $H_{i,j}(f)$.

When there are multiple coupling relationships (such as multi-path effects in communication) among subsystems and equipment in the system, there is no longer a one-to-one correspondence between the interference ports and the receiving ports; in other words, there is correlation among different interfering ports or receiving ports. Under this circumstance, we need to take the coupling factors brought by the correlation into consideration. In the following paraphrase, we will analyze the coupling relationship between multiple interference sources and a single receiving port in the system.

Assuming an electronic information system, the number of transmitting equipment (ports) is M and the number of susceptive equipment (ports) is N, and if there is correlation between the interference sources, the EMI coupled with the j-th susceptive object's input port can be written as

$$R_j(f) = \underline{\underline{\sum_{i=1}^{M} H_{ji}(f) \times \mathrm{E}_i(f)}} \quad j = 1 \cdots N \tag{5.66}$$

If the susceptive limit of the j-th receiving port is $S_j(f)$, its interference state function can be written as

$$\begin{aligned} I_j(f) &= R_j(f) - S_j(f) \\ &= \underline{\underline{\sum_{i=1}^{M} H_{ji}(f) \times E_i(f) - S_j(f)}} \end{aligned} \tag{5.67}$$

where $\underline{\underline{\sum_{i=1}^{M}}}$ means that the transmission value is not directly arithmetically added, and the interference calculation at the port cannot be simply calculated using the sum of the system functions. We also need to take into account the correlation function between the interference variables and the correlation function between the coupling transfer functions.

In EMC analysis, it is usual to find that multiple interference sources are uncorrelated, multiple transmission paths are uncorrelated, or the interference source functions and transmission path functions are uncorrelated. Since the output functions of the independent systems are uncorrelated, $R_j(f)$ can be expressed by the energy of the interference signal that the port couples with as

$$R_j(f) = \left\| \sum_{i=1}^{M} H_{ji}(f) \times \mathrm{E}_i(f) \right\| = \sum_{i=1}^{M} \| H_{ji}(f) \times \mathrm{E}_i(f) \| \quad j = 1 \cdots N \tag{5.68}$$

If the energy value of the largest component of multiple interference components is much larger than the other components, the total received interference energy can be approximately replaced by the maximum component. Then, (5.68) can be simplified as

$$R_j(f) = \boldsymbol{max}\left(\| H_{ji}(f) \times E_i(f) \|\right), \quad j = 1 \cdots N \tag{5.69}$$

where $I_j(f)$ represents the working condition of the j-th receiving port and the interference state function $I_j(t, f)$ can be written as

$$\begin{aligned} I_j(f) &= R_j(f) - S_j(f) \\ &= \sum_{i=1}^{M} \| H_{ji}(f) \times E_i(f) \| - S_j(f) \\ &\approx \boldsymbol{max}\left(\| H_{ji}(f) \times E_i(f) \|\right) - S_j(f) \quad j = 0, 1, \cdots N \end{aligned} \tag{5.70}$$

① If $I_j(f) < 0$, it can be regarded that the j-th equipment will not be interfered in the environment.

② If $I_j(f) = 0$, it can be regarded that the j-th equipment is in the interference critical state in the environment; that is, the equipment may be interfered and the safety margin is 0.

③ If $I_j(f) > 0$, then the j-th equipment can be considered to be subject to interference in the environment, and the value of $|I_j(f)|$ indicates the strength of the interferences.

The practical applications show that through the EMC behavioral simulation, we can analyze the influence of various interferences toward the electronic system quantitatively and provide an effective design for the reinforcement of the EMC of the electronic system.

(3) **Analysis Model of Each Parameter of the Interference Matrix**

According to the interference relationship between the radiation source and the receiving port of the system, a quantitative analysis model of each element of the interference correlation matrix is constructed (if possible, the data of these elements can also be obtained through testing). The analysis model mainly includes radiation source characteristic model, receiver response model, and energy coupling model.

(1) Radiation source characteristics model. The important characteristics of radiation sources in EMC design mainly include radiation power, signal modula-

tion method, operating frequency band, instantaneous bandwidth, out-of-band characteristics of transmitting equipment, radiation characteristics of airborne antennas, input impedance, and out-of-band characteristics of airborne antennas. Specifically, the out-of-band characteristics of the transmitting equipment, the radiation characteristics of the airborne antennas, and the out-of-band characteristics analysis of the airborne antennas are problems that are easily overlooked and difficult to quantify in the EMC analysis.

① The spectral distribution model of the transmitting equipment. Common transmitters usually have large transmitting power and weak antenna directivity, such that the spectral distribution characteristics of the transmitter have an important influence on the electromagnetic environment of the aircraft. In the analysis and design of the aircraft's EMC, we need to understand not only the spectral power distribution within the band of the transmitter, but also the spectral power distribution of the out-of-band spurious and harmonics. Therefore, in order to quantitatively allocate the indicators of the whole aircraft, it is necessary to determine the spectral distribution of the transmitter.

The functions and structures of different transmitters are different, and in a different equipment, the circuit for the same function may have different structures. Therefore, the rules of spectral distribution of different transmitters are also different from each other. Ideally, the power spectral distribution of the transmitting equipment can be expressed as [21]

$$P_t(f) = \begin{cases} \bar{P}_B + a\delta_B, & f_L \leq f \leq f_H \\ \bar{P}_N, & f < f_L,\ f > f_H \end{cases} \tag{5.71}$$

where f_L is the low end of the transmitter's operating frequency band; f_H is the high end of the transmitter's operating frequency band; a is the confidence coefficient; δ_B is the standard deviation; $\bar{P}_N$ is the statistical mean of the transmitter's noise power; $\bar{P}_B$ is the statistical mean of the fundamental radiation power; $\bar{P}_B$ and δ_B can be written as

$$\bar{P}_B = \frac{1}{m}\sum_{i=1}^{m} P_i \tag{5.72}$$

$$\delta_B = \sqrt{\frac{1}{m-1}\sum_{i=1}^{m}(\bar{P}_B - P_i)^2} \tag{5.73}$$

where P_i is the measured value of a single transmitter's output power of the fundamental wave; m is the sampling number of the transmitter.

Equation (5.71) is significant for the accurate determination of the fundamental radiation power of the existing transmitter. In engineering practice, due to the nonlinear effects of signal sources, power amplifiers, filters, and other electronic com-

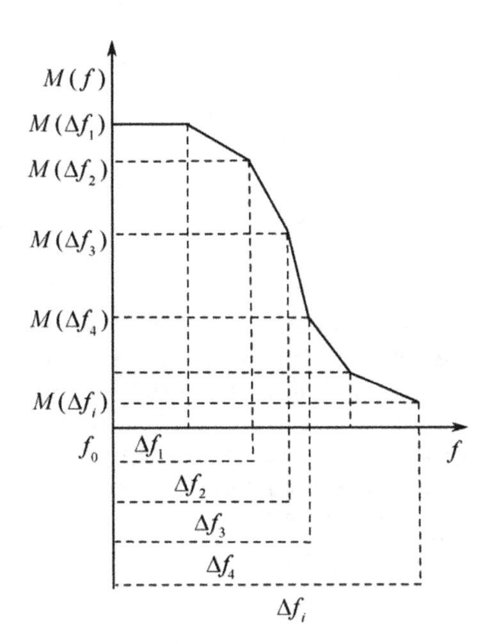

Fig. 5.22 Spectrum distribution model near the fundamental wave of the transmitter

ponents in the transmitter, the time-domain characteristics of the radiation signal of the transmitter, the modulation method of the signal, the type of power amplification, and the filtering performance are different. As a result, the spectral power distribution of the transmitter is quite different from (5.71). The output power of the transmitter is not completely limited to a certain frequency or a narrow band, but is mainly distributed in the frequency band near the fundamental wave. In the process of EMC analysis and design, we need to know not only the spectrum power value within the frequency band of the transmitter, but also the out-of-band spurious and harmonic power values. The spectrum of the transmitter is described in [21].

The spectral distribution envelope near the fundamental frequency is described by a polyline segment. The power of each frequency point is determined by the maximum emission power of the fundamental wave, the frequency value of the point, and the position and slope of the polyline segment where the frequency point is located. The spectrum distribution model near the fundamental frequency of the transmitter is shown in Fig. 5.22.

Assuming that the central operating frequency of the transmitter at a certain moment is f_0, the relative value of a certain spectral component and the maximum transmitting power of the fundamental wave is $M(f)$ (dBc), and then $M(f)$ can be expressed as

$$M(f) = M_i + N_i \lg(\frac{|f - f_0|}{\Delta f_i}), \quad (f \in [\Delta f_{i-1}, \Delta f_i]; \quad i \in [1, n]) \tag{5.74}$$

where Δf_i is the frequency bandwidth between the low-end part of the i-th spectral component and the central operating frequency f_0; M_i is the constant term of the

Table 5.4 Harmonic suppression constants of transmitters

Fundamental frequency (MHz)	A (dB/dec)	B (dB)	Standard deviation δ (dB)
<30	−70	−20	10
30~300	−80	−30	15
>300	−60	−40	20

baseband modulation characteristic in the i-th frequency band; N_i is the slope of the spectral envelope within the frequency band $[\Delta f_{i-1}, \Delta f_i]$,and it is a constant term; n represents the number of segments that the spectral envelope covers at the one side of the center frequency, and $n \geq 1$. When $n = 0$, the characteristics of the frequency spectrum can be simplified as (5.71).

According to Eq. (5.71), the spectrum power distribution near the fundamental wave of the transmitter is

$$P(f) = \bar{P}_{\text{max}} + M(f) \tag{5.75}$$

where $\bar{P}_{\text{max}}$ is the statistical mean of the maximum radiation power of the transmitter.

The spectral power of the harmonics can be statically calculated as

$$\overline{P_h}(N_h f_0) = \bar{P}_B + A \lg N_h + B \tag{5.76}$$

where $\bar{P}_B$ is the statistical mean of the fundamental emission power; f_0 is the fundamental frequency; N_h is the harmonic number; A, B are the harmonic suppression constants of the transmitter. For common transmitter, the value of A, B can be determined using Table 5.4.

Both Eqs. (5.74) and (5.76) are based on a large amount of priori data. They have a certain degree of universality, but the valuation bias is relatively large (as shown in Table 5.4). The prerequisite for using the two functions is that a large amount of test data has been obtained for the product. In addition, the functions only estimate the spectrum and harmonic power in the vicinity of the fundamental wave, but they are incapable of evaluating the out-of-band spurious power. In fact, if a mature product or similar system has already existed, we can directly test the full-band spectral power distribution data of the transmitting equipment and then substitute it into the EMC design.

In the engineering practice of EMC analysis and design, if the transmitter of the aircraft system is a mature product or an improved version of a mature product, we will use a test method similar to CE106 instead of (5.74) and (5.76) to measure the power distribution characteristics of the spectrum, which is denoted as $P_t(f)$. Figure 5.23 shows the partial test data for the spectral power distribution of a transmitter. It is obvious that the spectral distribution shown in the figure is quite different from the spectrum distribution represented by Eqs. (5.74) and (5.76). Therefore,

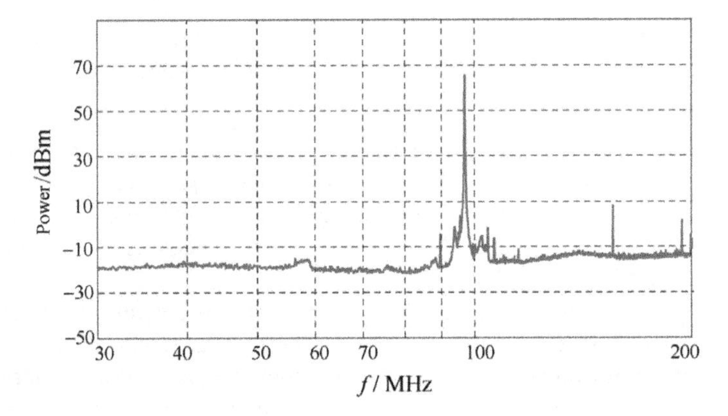

Fig. 5.23 Partial test data of the spectral power distribution of a transmitter

when the spectral power distribution characteristics of the product can be obtained by testing, test data should be preferred; when the test data cannot be obtained, the results should be predicted first, and as soon as the test data can be obtained, the prediction results should be modified with the test data.

In the engineering practice of EMC analysis and design, if the transmitter selected for the aircraft system is a product to be developed, the maximum power value of the emission spectrum can be estimated using Eq. (5.77).

$$P_t(f) = P_{signal}(f) \cdot G_{amp}(f) \cdot P_{amp}(f) \cdot L_{filter} \cdot P_{filter}(f) \tag{5.77}$$

where $P_{signal}(f)$ is the spectral power distribution of a weak signal input for a high-power microwave power amplifier of a transmitter. The signal's characteristics in time domain can be simulated and calculated according to the circuit design scheme, and $P_{signal}(f)$ can be obtained by a Fourier transform [19, 28]; $G_{amp}(f)$ is the power gain of a high-power microwave power amplifier. It is a design input of the power amplifier which can also be provided by the product supplier. $P_{amp}(f)$ is the spectral distribution characteristics of the power amplifier. For individual power amplifier components, the data can be measured before the development of the transmitter. Considering the nonlinear factor, the power amplifier should be in the maximum output state during the test. L_{filter} is the attenuation of the emission power of the microwave filter; $P_{filter}(f)$ is the spectral power distribution characteristic of the microwave filter.

For separate filter components, L_{filter} and $P_{filter}(f)$ can be provided by the product supplier or be tested before the development of the transmitter. Considering the nonlinear factors, the input power of the microwave filter needs to simulate the maximum output of the power amplifier. And for filters integrated into a specific circuit, these two data can be simulated and calculated during the design phase [28].

② In-band and out-of-band radiation characteristics of airborne antennas. The radiation characteristics (directionality, direction power gain) of the airborne antenna are important parts of the energy transmission from the transmitter to the susceptive port. The accuracy of this data has a great influence on the analysis and design of the system EMC.

When the antenna is installed on the surface of the fuselage, the boundary conditions formed by the installation environment of the antenna greatly affect the radiation characteristics of the antenna. The aircraft will induce electromagnetic current and generate secondary radiation, such that the radiation characteristics of the antenna will be remodeled.

In general, the radiation characteristics of the airborne antenna can be obtained by measurement or calculation. For small aircraft, if the antenna is installed and the test environment is available (in the darkroom), then the radiation characteristics of the airborne antenna can be obtained through measurement. For large and medium-sized aircraft, the test environment is difficult to build, and the design scheme for antennas is usually not fully determined during the demonstration phase. Therefore, the only way to obtain the radiation characteristics of the airborne antenna is through simulation calculations.

There is a large number of the research literatures on the radiation characteristics of structured antennas in free space, such as array antennas, horn antennas, parabolic antennas, and loop antennas, and relatively accurate calculation result has been achieved. After the installation, the antenna's radiation characteristics are much more difficult to calculate. The major calculation methods are approximation and numerical methods. Our readers can refer to [10, 29–37] for more research under this topic.

Regarding the radiation characteristics of the antenna, we would like to point out in particular that the out-of-band characteristics of the antenna are nonnegligible in the study of EMC. In the traditional engineering development, only the in-band parameters such as power gain, pattern, characteristic impedance, and radiation efficiency are considered for the selection of the airborne antenna. In recent years, many problems have been found in the implementation of EMC engineering, which forced the engineers to pay more attention to the effects from the aircraft platforms on the antenna's radiation characteristics. However, rare attentions have been paid to the out-of-band characteristics of antennas. In fact, the antenna at the front end of the equipment is equivalent to a filtering system, and its out-of-band response characteristics have an important influence on the out-of-band radiation characteristics of the transmitter and the out-of-band response of the receiver. There are plenty of transmitting and receiving antennas locating on the airborne platform, and the out-of-band of some antennas falls inside the in-band of other antennas. The out-of-band radiation characteristics of the antenna may cause mutual interference between the transmitting and receiving antennas. In practical applications, the authors have performed theoretical analysis and practical testing on the out-of-band characteristics of a large number of antennas. Generally, the antenna under study shows that: The directionality of the antenna may be split at the high frequency band, and the gain may become large. Although the radiation efficiency of the antenna is degraded, then the intensity of radiation is still relatively great, the energy transmitted by the transmitting antenna out-of-band can still cause interference to susceptive equipment, and the out-of-band of the susceptive equipment may also respond to the interference signal. In practical applications, the authors have incorporated the out-of-band characteristics of the antenna into the overall EMC design, which has significantly improved the accuracy of the EMC design [14].

(2) Receiver response model. The aircraft system is equipped with a large number of receiver for aeronautical navigation, detection, reconnaissance, warning, and communication. The normal operation of the receivers is of great significance for flight safety and performance, and it is also an important goal in aircraft EMC design. The receiver and its front-end antenna together constitute the response to electromagnetic energy in a specific spatial, polarization, time, and frequency domain. Among them, we mainly need to select the time-domain and the frequency-domain characteristics of the electromagnetic energy for the receiver. In the design of aircraft system EMC, the frequency-domain characteristics of the receiver are mainly considered.

For an ideal receiver, the frequency selectivity should exhibit a rectangular characteristic; i.e., the receiver is only susceptive to the electromagnetic energy in a specific frequency range, and all the energies outside the frequency range of the spectrum are rejected from entering the receiving channel. Due to the nonlinear effects of the functional circuits such as preselectors, mixers, filters, and IF amplifiers in receiver, the frequency selectivity and rectangular selection characteristics of the receiver are inconsistent. What's more, the frequency selectivity and rectangular characteristics of some receivers are far from each other. In the EMC design process, the frequency selectivity of the receiver must be quantified.

Similar to the representation of the spectral distribution near the fundamental wave of the transmitter, the frequency selectivity of the receiver near the operating bandwidth or the major receiving channel can also be represented by a piecewise straight-line approximation [21].

$$S(\Delta f) = S(\Delta f_i) + S_i \lg(\Delta f / \Delta f_i) \tag{5.78}$$

where $S(\Delta f)$ is the selectivity of the receiving equipment when the frequency is deviated from the tuned frequency as Δf; $S(\Delta f_i)$ is the value of the selectivity of the receiver on the frequency boundary Δf_i; S_i is the slope of the selectivity of the receiver on the line segment of frequency Δf_i.

Signals located within the frequency points and frequency range outside the main receiving channel can also enter the receiver due to the system's nonlinear characteristics, affecting the in-band of the receiving system. These frequencies or frequency bands are called additional receiving channels. They mainly include IF receiving channel, image receiving channel, and harmonic receiving channel. Using the same method, the selective model of the receiver in the main additional receiving channels can be obtained

$$S(f_a) = I \lg(f_a / f_0) + J \tag{5.79}$$

Table 5.5 Harmonic suppression constants of transmitters

Frequency (MHz)	I (dB/dec)	J (dB)	$\delta(f_a)$ (dB)
<30	25	85	15
30~300	35	85	15
>300	40	60	15

where f_0 is the central operating frequency of the receiver; f_a is the operating frequency of the additional receiving channel; I and J are linear approximation coefficients of the frequency selective curve.

Selective models often use a statistical model, which exploits the mean $\bar{S}(f_a)$ and the mean squared error $\sigma(f_n)$, where I, J, and $\sigma(f_n)$ are obtained from data averaged over different types of receivers. Table 5.5 lists the mean of I,J, and $\sigma(f_n)$ for different types of receivers.

Both Eqs. (5.78) and (5.79) are based on a large amount of priori data. They have a certain degree of universality, but the bias of the predicted value is relatively large (as shown in Table 5.5). The prerequisite for using the two functions is that a large amount of test data has been obtained for the product. Different types of receivers have different mechanisms when generating major additional channels, so the error in the estimation of additional channel selectivity may be greater using (5.79). If a mature product or similar system has already existed, we can also directly test the full-band selectivity data of the receiver and then substitute it into the EMC design. The accuracy of the analysis and evaluation will be improved in this way.

In a sense, the frequency range outside the working frequency band of the receiver (main receiving channel) can all be called additional receiving channels. Besides the IF receiving channel, the image receiving channel, and the harmonic receiving channel, if the input energy is large enough, the nonlinearity of the system may cause the receiver to respond to signals at any frequency outside the band. The full-band response characteristics of the receiver can be analyzed using the behavioral simulation method of the circuit [19, 28]. For mature equipment, it is best to use the test methods to accurately obtain the response characteristics: The sensitivity of the receiver can be measured by in-band method, and the intermodulation suppression characteristics of the testing receiver can be measured by out-of-band method.

(3) Energy coupling model. The coupling model mainly involves antenna coupling, field–line coupling, cable coupling, and aperture coupling.

① Antenna coupling. A large number of transceiver antennas are installed within the limited space of the aircraft, and it is likely that these antennas are difficult to work together. The energy coupling between airborne transceivers of aircraft is one of the main ways for airborne receivers to be interfered. When performing EMC design and quantitative allocation of indicators, it is necessary to quantitatively analyze and control the energy of transmitting equipment coupled with the airborne receiving antenna through the antenna port.

In the engineering design of EMC, when analyzing the coupling between the transmitting and receiving antennas, the electrical size, the shielding condition, the polarization state of the transmitting and receiving antennas, the operating frequency band, the out-of-band characteristics, and the mounting mode should be fully considered.

According to the different electrical sizes of the transmitting/receiving antennas, the coupling can be categorized into near-field coupling, far-field coupling, and near-far mixing field. Among them, near-field coupling and near-far mixing field energy transfer can be solved by numerical methods such as moment method (MoM) and finite element method (FEM). When analyzing the coupling between antennas, we also need to consider whether there are obstacles between the antennas: For near-field conditions, we can use low-frequency numerical methods no matter there are obstacles or not; for the far-field situation, if there is no obstacle, we can use the numerical method and the geometrical optics method to do hybrid analysis. The numerical method is used to calculate the radiation characteristics of the antenna, and the geometrical optics method is used to calculate the attenuation of the electromagnetic field in space. For cases where the transmitting and receiving antennas are in the far field of each other, and there is shielding between airborne antennas from the fuselage or the wing, we can use numerical method and geometrical theory of diffraction (GTD) or uniform geometrical theory of diffraction (UTD) to do hybrid analysis. The numerical method is used to calculate the radiation characteristics of the antenna, and GTD or UTD is used to calculate the attenuation, diffraction attenuation, and occlusion loss of the electromagnetic field in space.

The airborne antenna isolation is an indicator describing the antenna coupling. It fully reflects the directionality, gain, polarization state, in-band and out-of-band characteristics of the antenna, and the contribution of space between antennas to the energy coupling between transmitting and receiving antennas. Antenna isolation refers to the ratio of the transmitting power P_{ta} of the transmitting antenna to the power P_{ra} received by the receiving antenna (P_{ra} is the power P_{ta} received by the receiving antenna after various attenuations), i.e.,

$$L = \frac{P_{ta}}{P_{ra}} \text{ or } L(\text{dB}) = 10\lg \frac{P_{ta}}{P_{ra}} \tag{5.80}$$

Antenna isolation is only one isolation item between the transmitter and the receiver. In engineering applications, although it is not possible to estimate the interference relationship between the transmitter and the receiver simply by the value of the antenna isolation, the antenna isolation analysis is the key technology in isolation of the transmitter and receiver.

In the following paragraphs, we will discuss the antenna isolation at far field without obstacle coupling, the antenna isolation at near-field coupling, and the antenna isolation at far field with obstacles coupling separately. Then, we will introduce the concept of isolation between transmitters and receivers.

(a) Antenna isolation at far field without obstacle coupling. Since the two antennas are far from each other, their energy coupling is mainly through the radiation field. Assuming the transmitting power of the transmitting antenna is P_{ta}, and the gain is $G_t(\theta_t, \varphi_t)$; the receiving power of the receiving antenna is P_{ra}, and the gain is $G_r(\theta_r, \varphi_r)$; the distance between the receiving antenna and the transmitting antenna is D; normally, the antenna isolation can be solved by Eq. (5.80) when the transmitting/receiving antennas are in direct view. When the size of the transmitting/receiving antennas is relatively small, they can be approximately regarded as a point source with a certain directionality; thus, the electromagnetic wave transmitted by the transmitting antenna can be approximated as a spherical wave and further approximated to a plane wave at the position of the receiving antenna. The antenna isolation can be expressed as

$$\begin{aligned} L_{antenna}(\text{dB}) &= L_d - G_t(\theta_t, \varphi_t) - G_r(\theta_r, \varphi_r) \\ &= 20\lg[\frac{4\pi D}{\lambda}] - G_t(\theta_t, \varphi_t) - G_r(\theta_r, \varphi_r) \end{aligned} \tag{5.81}$$

where $L_d = 20\lg[\frac{4\pi D}{\lambda}]$ is the spatial isolation of the transmitting and the receiving antennas in direct view. It is determined by factors including the distance D between the transmitting and the receiving antennas and the analysis wavelength λ. $G_t(\theta_t, \varphi_t)$ is the antenna gain of the transmitting antenna in the receiving direction, and it can be looked up from the radiation plot of the airborne transmitting antenna gain based on the relative positions of the antenna. The specific angle is determined by the position of the line-of-sight segment between the transmitting and the receiving antennas on the aircraft coordinate system. $G_r(\theta_r, \varphi_r)$ is the antenna gain of the airborne receiving antenna in the transmitting direction. It can be looked up from the radiation plot of the airborne receiving antenna gain based on the relative positions of the antenna. The specific angle is determined by the position of the line-of-sight segment between the transmitting and the receiving antennas on the aircraft coordinate system.

When the polarization of the transmitting/receiving antennas does not completely match, we need to consider the loss L_P caused by the mismatch polarization; i.e., the entire isolation of antennas is

$$\begin{aligned} L_{antenna}(\text{dB}) &= L_d - G_t(\theta_t, \varphi_t) - G_r(\theta_r, \varphi_r) + L_P \\ &= 20\mathbf{lg}\left[\frac{4\pi D}{\lambda}\right] - G_t(\theta_t, \varphi_t) - G_r(\theta_r, \varphi_r) + L_P \end{aligned} \tag{5.82}$$

Now, we will discuss the solution of the loss caused by the polarization mismatch. Assume the amplitude ratio of the orthogonal components E_θ and E_φ of the transmitting polarized wave to be ρ_1, and their phase difference is β_1. Similarly, assume the amplitude ratio of the orthogonal components E_θ and E_φ of the receiving polarized wave to be ρ_2 and their phase difference to be β_2. When the amplitude and phase of the two polarized waves are different, the loss of the polarization mismatch can be calculated as [38]

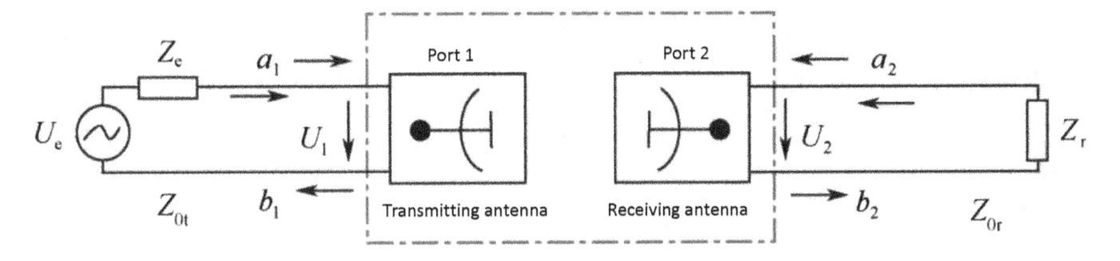

Fig. 5.24 Equivalent two-port network of near-field antenna coupling

$$L_P = 20\lg \frac{[1+\rho_1\rho_2\cos(\beta_1+\beta_2)]^2 + [\rho_1\rho_2\sin(\beta_1+\beta_2)]^2}{(1+\rho_1)^2+(1+\rho_2)^2} \tag{5.83}$$

When the magnitude of the two polarized waves is not identical, and $\beta_1 = \beta_2 = 0$, the loss of polarization mismatch is

$$L_P = 20\lg \frac{(1+\rho_1\rho_2)^2}{(1+\rho_1)^2+(1+\rho_2)^2} = \frac{(1+tg\alpha_1 tg\alpha_2)^2}{(1+tg\alpha_1)^2+(1+tg\alpha_2)^2} \tag{5.84}$$

(b) Antenna isolation at near-field coupling. If the receiving antenna is not in the far-field region of the transmitting antenna or vice versa, the mutual interference between the two antennas does not couple with the radiation field but couples with the near-field bound field or the near-field induction field. This situation often occurs in isolation between the linear antenna arrays. Since the concept of antenna power gain is established in the far field, Eq. (5.80) does not apply to near-field antenna isolation analysis. Even if the concept of power gain is extended to the near field, it still describes the power gain of radiation field. However, in this case, the main factor of interference between antennas is not the radiation field.

In the near-field scenario, the system consisting of the transmitting/receiving antennas can generally be viewed as a two-port network as shown in Fig. 5.24. Assuming the maximum output power of the signal source to be P_{tmax} and the power absorbed by the load impedance Z_r to be P_r, then the antenna isolation is [3, 39, 40]

$$L_{antenna} = \frac{P_t\max}{P_r} \tag{5.85}$$

As shown in Fig. 5.24, the transmitter is equivalent to the signal source U_e, and its internal impedance is Z_e; the receiver is equivalent to the load impedance Z_r; the radiating antenna and its port, and the receiving antenna and its port together constitute a two-port network, in which the Port 1 is connected to the transmitter through the transmission line with characteristic impedance Z_{0t} and the Port 2 is connected to the receiver through the transmission line with characteristic impedance Z_{0r}. Then, we introduce the normalized complex voltage inward waves a_1 and a_2, and outward waves b_1 and b_2. They are defined as

$$a_1 = \frac{U_1 + I_1 Z_{0t}}{2\sqrt{Z_{0t}}} \tag{5.86}$$

$$a_2 = \frac{U_2 + I_2 Z_{0r}}{2\sqrt{Z_{0r}}} \tag{5.87}$$

$$b_1 = \frac{U_1 - I_1 Z_{0t}}{2\sqrt{Z_{0t}}} \tag{5.88}$$

$$b_2 = \frac{U_2 - I_2 Z_{0r}}{2\sqrt{Z_{0r}}} \tag{5.89}$$

Then, the normalized matrix S can be obtained as

$$\begin{bmatrix} b_1 \\ b_2 \end{bmatrix} = \begin{bmatrix} s_{11} & s_{12} \\ s_{21} & s_{22} \end{bmatrix} \begin{bmatrix} a_1 \\ a_2 \end{bmatrix} \tag{5.90}$$

The reflection coefficient of signal source impedance Z_e relative to Z_{0t} is defined as

$$\Gamma_1 = \frac{Z_e - Z_{0t}}{Z_e + Z_{0t}} \tag{5.91}$$

The reflection coefficient of load impedance Z_r relative to Z_{0r} is

$$\Gamma_2 = \frac{Z_r - Z_{0r}}{Z_r + Z_{0r}} \tag{5.92}$$

Then, we can obtain the power absorbed by the load as

$$P_r = \frac{1}{2}|b_2|^2\left(1 - |\Gamma_2|^2\right) \tag{5.93}$$

Assume the output power of the signal source is $\frac{1}{2}|b_s|^2$, where b_s is

$$b_s = \frac{U_e/\sqrt{Z_{ot}}}{Z_e/Z_{0t} + 1} = \frac{U_e}{Z_e + Z_{0t}}\sqrt{Z_{0t}} \tag{5.94}$$

The maximum output power of the signal source is

$$P_t\max = \frac{1}{2}\frac{|b_s|^2}{1 - |\Gamma_1|^2} \tag{5.95}$$

Since $a_1 = b_s + \Gamma_1 b_1$, thus

$$b_s = a_1 - \Gamma_1 b_1 = a_1 - \Gamma_1(s_{11}a_1 + s_{12}a_2) = a_1(1 - \Gamma_1 s_{11}) - s_{12}\Gamma_1 a_2 \tag{5.96}$$

Since $b_2 = s_{21}a_1 + s_{22}a_2$, $a_2 = b_2\Gamma_2$, thus

$$L_{antenna} = \frac{|b_s|^2}{|b_2|^2\left(1 - |\Gamma_1|^2\right)\left(1 - |\Gamma_2|^2\right)} = \frac{|(1 - \Gamma_1 s_{11})(1 - \Gamma_2 s_{22}) - s_{12}s_{21}\Gamma_1\Gamma_2|^2}{|s_{12}|^2\left(1 - |\Gamma_1|^2\right)\left(1 - |\Gamma_2|^2\right)} \tag{5.97}$$

In the near field, since the operating wavelength of the antenna is relatively long and does not satisfy the calculation conditions of the GTD, the antenna isolation in the near-field case can be calculated using (5.97). Using the FEM, the feed port of the transmitting/receiving antennas is used as the two ports of the two-port network. Then, the parameter S between the two can be calculated, and the VSWR of the antenna is calculated for each port. Substituting the parameters obtained by the numerical method into Eq. (5.97) will result in antenna isolation in the case of a near field.

When the FEM is used to calculate the antenna isolation, the optimization algorithm is used to calculate the matching impedance of the transmitting and receiving antenna ports. Thus, the impedance of the transmitting and the receiving antenna ports can be matched and we have $\Gamma_1 = 0$, $\Gamma_2 = 0$. Since the impedance matching has a frequency characteristic, when the operating frequency bands of the transmitting and the receiving antennas do not overlap with each other, Γ_1 and Γ_2 cannot equal to zero at the same time. In the emission frequency band, we have

$$L_{antenna} = \frac{|1 - \Gamma_2 s_{22}|^2}{|s_{12}|^2\left(1 - |\Gamma_2|^2\right)} \tag{5.98}$$

In the receiving operating frequency band,

$$L_{antenna} = \frac{|1 - \Gamma_1 s_{11}|^2}{|s_{12}|^2\left(1 - |\Gamma_1|^2\right)} \tag{5.99}$$

If the receiving and the transmitting bands are overlapped, then in the operating frequency band, there is

$$L_{antenna} = \frac{1}{|s_{12}|^2} \tag{5.100}$$

(c) Antenna isolation at far field with obstacle coupling. If the transmitting and the receiving antennas satisfy far-field area condition, and the fuselage or the wing is obstructed between the antennas, we then need to consider the attenuation or obstruction effect of the fuselage.

When there is diffraction, we can obtain the far-field antenna isolation according to (5.101).

$$L_{antenna}(\text{dB}) = L_P + L_d - G_t(\theta_t, \varphi_t) - G_r(\theta_r, \varphi_r)$$

$$= L_P + 20\lg[\frac{4\pi D}{\lambda}] + L_{diffraction} + SF_W - G_t(\theta_t, \varphi_t) - G_r(\theta_r, \varphi_r) \tag{5.101}$$

where L_P is the polarization attenuation between the transmitting and the receiving antennas; L_d is the spatial isolation between the transmitting and the receiving antennas; L_d is determined by the distance D between the transmitting and the receiving antennas, the diffraction attenuation $L_{diffraction}$, the shielding attenuation SF_W, the analysis wavelength λ, etc. Therefore, L_d can be expressed as $L_d = 20\lg[\frac{4\pi D}{\lambda}] + L_{diffraction} + SF_W$. Diffraction and block attenuation can be obtained by analyzing the surface diffraction or edge diffraction attenuation based on the diffraction model of the fuselage surface or the airfoil.

From the diffraction model, we can see that in order to obtain the diffraction attenuation of the field on the surface of the fuselage or the airfoil, it is necessary to accurately solve the trajectory of the diffraction ray. In the following paragraphs, we introduce a new algorithm for ray tracing "minimum included angle algorithm," which was proposed by the EMC research team of Beihang University. This method improves the speed of tracing, the accuracy, and flexibility of calculation, and guarantees the uniqueness of ray tracing. This method can be used to perform calculation of ray tracing and isolation between any pair of antennas on the surface of the fuselage.

(d) Isolation between airborne transmitters and receivers. The ratio of the transmitting power P_t to the receiving power P_r is defined as the isolation between the transmitter and the receiver (where P_r refers to the power P_t that reaches the receiver after various attenuations), i.e.,

$$L = \frac{P_t}{P_r} \text{ or } L(\text{dB}) = 10\lg\frac{P_t}{P_r}(\text{dB}) \tag{5.102}$$

Antenna isolation only takes into account the isolation between the transmitting and the receiving antennas and does not fully reflect the degree of isolation between the transmitter and the receiver. The factors affecting the energy transmission between the transmitter and the receiver are shown in Fig. 5.25. When analyzing the degree of isolation between a certain transmitter and a certain receiver, the frequency band attenuation of the transmitter, the loss of the transmission line, and the frequency band suppression characteristics of the receiver must be considered in addition to the isolation between the transmitting and the receiving antennas. Equation (5.103) is usually used for this analysis

$$\begin{aligned} L_{TR}(\text{dB}) &= L_{tB}(f) + L_{rB}(f) + L_{tf} + L_{rf} + L_P + L_d - G_t - G_r \\ &= L_{tB}(f) + L_{rB}(f) + L_{tf} + L_{rf} + L_{antenna} \end{aligned} \tag{5.103}$$

Equation (5.103) is the basis for the matrix elements of (5.95) and (5.62). From (5.103), it can be seen that antenna isolation is an important part of the isolation between the transmitter and the receiver. At the same time, the layout of the transmitting antenna has a relatively great impact on other susceptive ports, so the overall antenna layout design and optimization are quite important to the compatibility

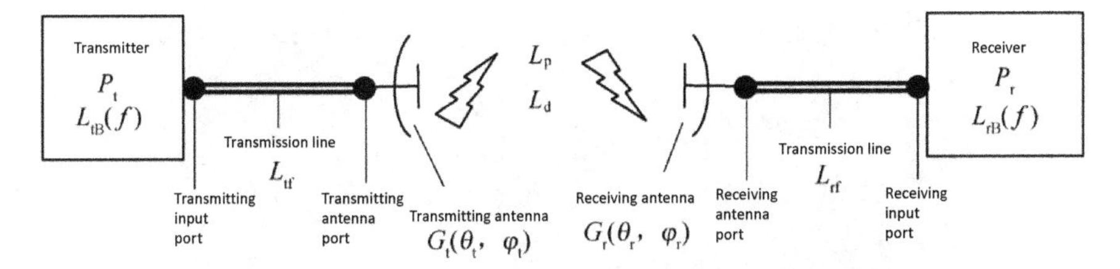

Fig. 5.25 Factors influencing the isolation between the transmitter and the receiver

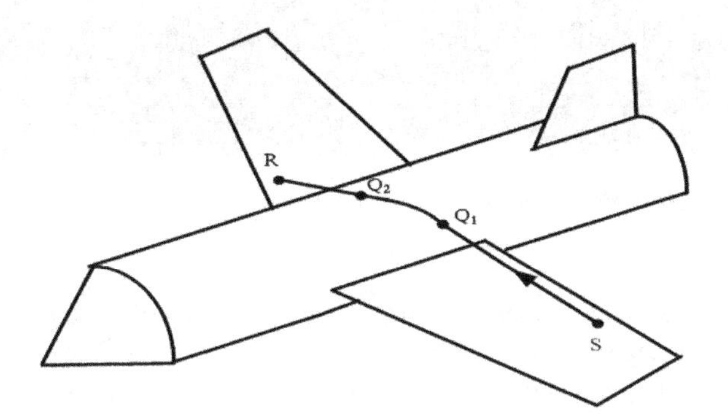

Fig. 5.26 Illustration of ray tracing on the surface

between the transmitter and the receiver. However, the antenna isolation is not completely equivalent to the isolation between the transmitter and the receiver, so other EMC indicators must be designed synergistically during the EMC design and optimization.

Constructing a three-dimensional geometric mesh model of the aircraft surface is an important basis for solving antenna isolation of airborne transmitters and receivers. Antenna coupling path is an important parameter in isolation calculation. When the mesh is dense enough, the computational cost to calculate the antenna coupling path will be very high, and it will be difficult even for a supercomputer to finish in a short time. Here, we will introduce the antenna coupling path calculation method based on the minimum included angle algorithm [41].

Calculating the coupling path between the transmitting and the receiving antennas on an arbitrary surface is essentially solving the problem with a geodesic [42]. In a well-defined area of the curve, there is only one curve between the two points to ensure that the path length is the minimum. This particular curve is called a geodesic. The creeping wave travels along the surface of the curved surface and can be seen as a geometric optical ray. Its propagation satisfies Fermat's principle that the path of propagation is the shortest. As shown in Fig. 5.26, the path of the surface ray between the two points of Q1 and Q2 is the geodesic connecting the two points. Q1 is the point of tangency of the ray SQ1 on the surface of the aircraft, and the point Q2 is the point of tangency of the ray SQ2 on the surface of the aircraft.

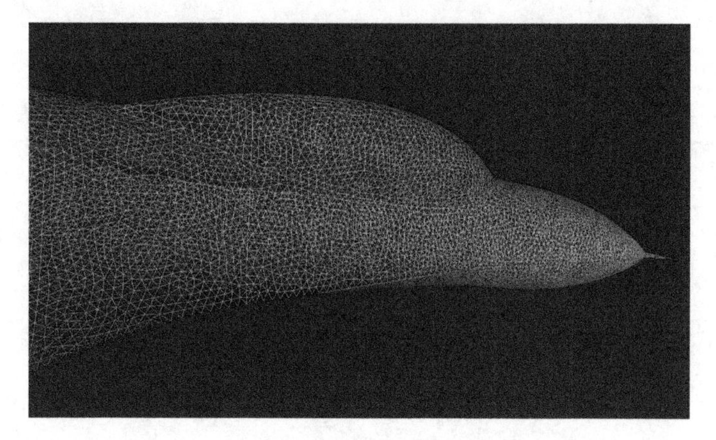

Fig. 5.27 Mesh subdivision of the model surface

In order to track the surface rays, it is assumed that [43]: ① The creeping wave inside a single triangle is a straight line; ② when two adjacent triangles develop into a plane, the creeping wave trajectories on the two triangles are a straight line when connected.

Any surface can be split into multiple triangular planes, as shown in Fig. 5.27, and the coordinates of the vertices of each triangle can be obtained. Taking vertex (B) and all the triangles that contain vertex (B), as shown in Fig. 5.28, v is the vector from the source point to the destination point. Starting from point B, the triangular edge BC_i which forms the smallest angle with v is the creeping track we are looking for. In Fig. 5.28, assuming that the angle between BC_2 and v is the smallest, the wave propagates along BC_2. Then, with C_2 as the new vertex, the creeping wave propagation trajectory is found using the principle of minimum included angle, and so on. With the minimum included angle method, we only need to analyze and compare the triangular edges within a certain range in the direction of v in the tracking process, which greatly improves the tracking efficiency. The included angle formula is:

$$\theta = \min\{\theta_1, \theta_2, \theta_3\} = \min\{\arccos(\vec{\upsilon} \cdot \vec{BC_i} /(\left|\vec{\upsilon}\right| \cdot \left|\vec{BC_i}\right|))\} \tag{5.104}$$

Curve 1 in Fig. 5.29 can be obtained as a sawtooth waveform using the minimum included angle method. On this basis, a NURBS curve [44] is constructed. The rational fractional expression of the NURBS curve is:

$$p(u) = \frac{\sum_{k=0}^{n} \omega_k p_k B_{k,d}(u)}{\sum_{k=0}^{n} \omega_k B_{k,d}(u)} \tag{5.105}$$

where p_k is the $n + 1$-th control point; the parameter ω_k is the weighting factor of the control point; the larger the ω_k, the closer the curve is to the control point p_k; $B_{k,d}(u)$ is the basis function of the B-spline function; it is defined as

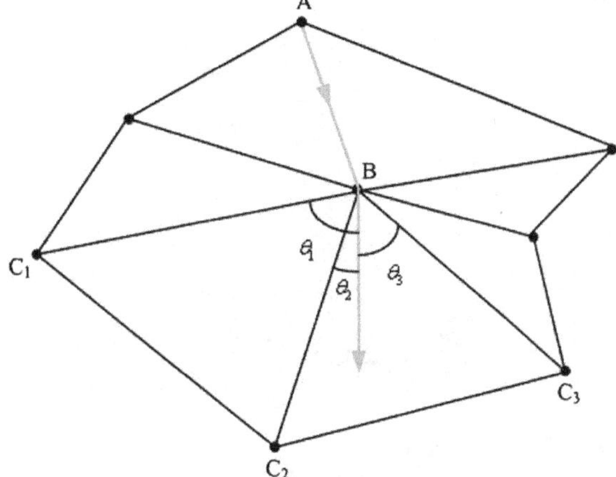

Fig. 5.28 Illustration of the algorithm based on the triangle mesh

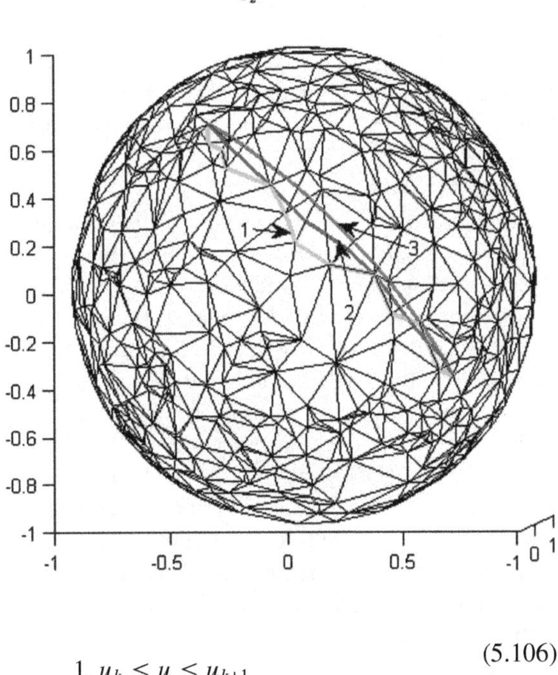

Fig. 5.29 Ray tracing process

$$B_{k,d}(u)= \begin{cases} 1, u_k \le u \le u_{k+1} \\ 0, other \end{cases} \tag{5.106}$$

Therefore, as long as the control point, the weighting factor and the weight vector are extracted from the sawtooth wave Curve 1, and a NURBS curve can be defined. The NURBS curve can be further interpolated and fitted. The interpolation fitting method is to compare the curvature of the NURBS curve at each pre-interpolation point with the given radius of curvature threshold R_{max}: If $R_i < R_{max}$, it indicates that the curvature of the interval is large and it is suitable to use the arc segment fitting; if $R_i > R_{max}$, it indicates that the bending degree of this interval is small. Based on

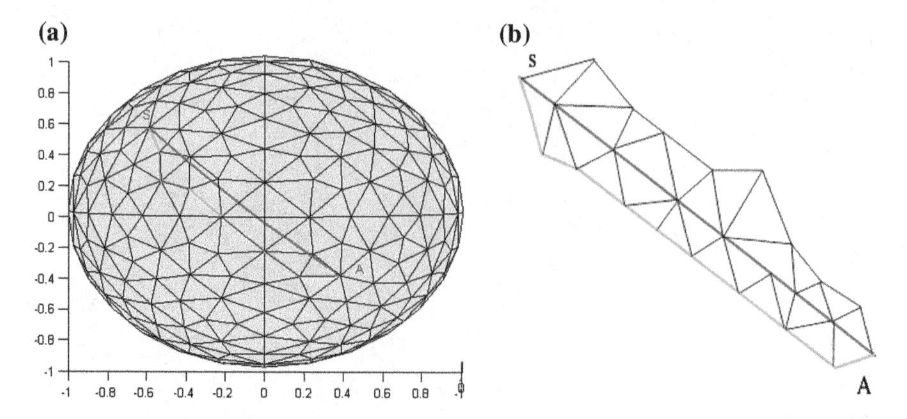

Fig. 5.30 Convergence of minimum included angle ray tracing. **a** Ray tracing on sphere and **b** projection of sawtooth wave on a plane

this idea, only the triangles near Curve 1 are more densely divided to obtain the Curve 2 in Fig. 5.29. Then, we keep interpolating using the same method until a two-dimensional smooth Curve 3 in Fig. 5.29 is formed. Therefore, the computing time can be greatly saved using the minimum included angle algorithm.

In the tracking process, the wave follows the direction which forms the smallest angle with the direction vector $\boldsymbol{v}$, as shown in Fig. 5.30a. Therefore, a sawtooth wave that is creeping along the direction vector $\boldsymbol{v}$ is obtained on the curved surface. When the sawtooth wave is projected on the plane of the direction vector and perpendicular to the plane of the tracking ray and the direction vector $\boldsymbol{v}$ (Fig. 5.30b), the included angle will be the smallest, such that the sawtooth wave is guaranteed to connect to the endpoint A, which ensures it convergence.

In order to verify the accuracy and versatility of the algorithm, the algorithm was applied to developable surfaces and nondevelopable surfaces, respectively. The accuracy of the algorithm was verified by comparing the programmed trace results with the numerical results. Considering the direct wave and creeping wave, according to the new algorithm, the ray tracing path between the left and right of the cylinder can be obtained as shown in Fig. 5.31.

The geodesic line from the emission point S to the receiving point R satisfies Fermat's principle. The tracing rays are divided into three segments: RP1, P1P2, and SP2. RP1 and SP2 can be viewed as direct waves on a concave surface. P1P2 is the creeping waves on the cylinder, as shown in Fig. 5.32. The geodesic on the cylinder can be calculated by the following formula

$$|P_1P_2| = \sqrt{h^2 + |P_2Q|^2} \tag{5.107}$$

Therefore, the numerical calculation length is

$$L = \overline{SP_2} + \widehat{P_1P_2} + \overline{P_1R} = 5.284$$

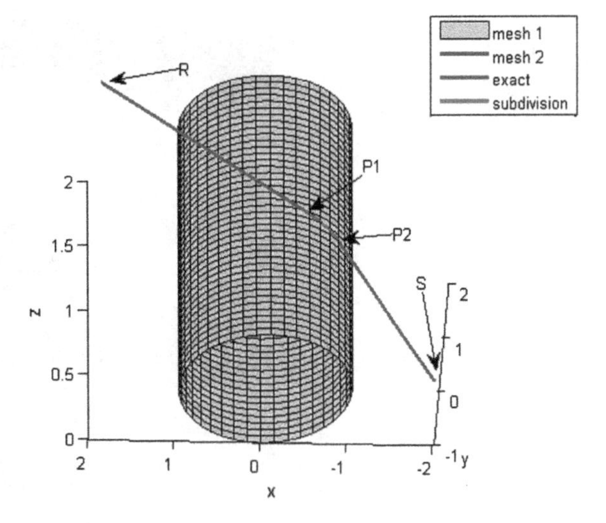

Fig. 5.31 Ray tracing between two sides of the cylinder

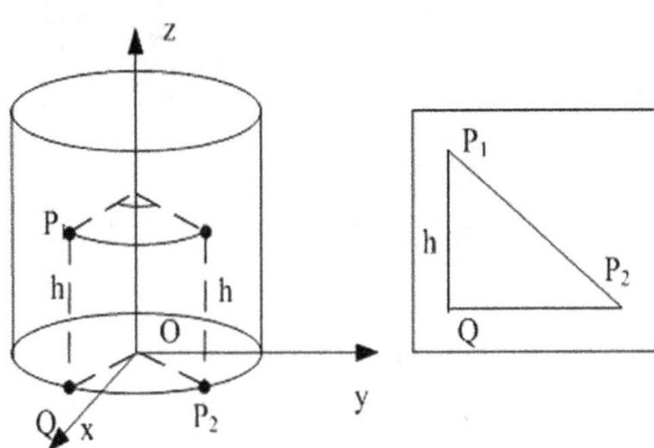

Fig. 5.32 Calculation of the geodesic line on the cylinder

Table 5.6 Comparison of calculation results and tracing results on arbitrary surfaces

Coordinates of the emission point S (m)	Coordinates of the receiving point R (m)	Coordinates of the point of tangency (P1) (m)	Coordinates of the point of tangency (P2) (m)	Tracing result (m)	Numerical distance (m)
2	−2	0.511	−0.973	5.245	5.284
2	−1	0.860	−0.231		
1.5	0.5	1.163	0.750		

The numerical calculation results can then be compared with the tracing results, as listed in Table 5.6.

Using the new algorithm, the relative error between the tracing result and the ground truth is: a = (5.284 − 5.245)/5.245 = 0.74%. It can be seen that the minimum included angle method is very accurate in result tracing on the developable surface, so the algorithm is applicable to the developable surface.

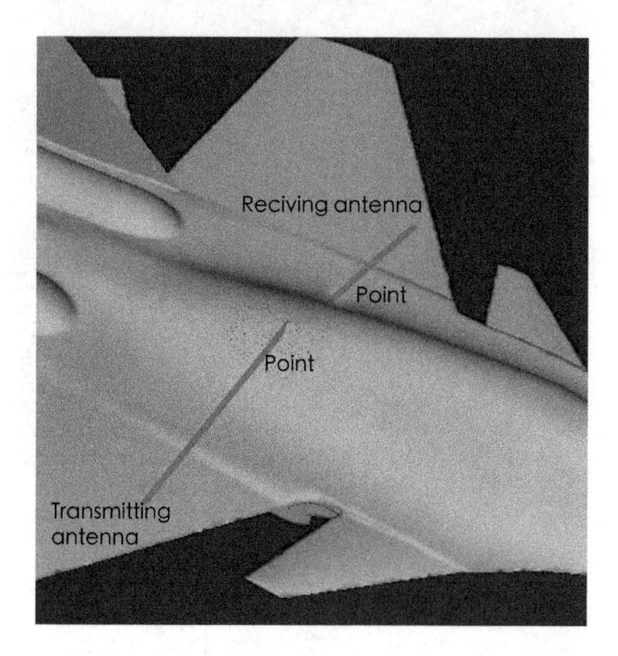

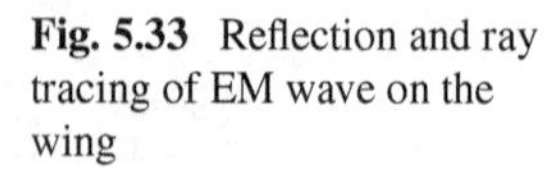

Fig. 5.33 Reflection and ray tracing of EM wave on the wing

The reflection and ray tracing of EM wave on the convex–concave surface: Reflected waves are also one of the factors that affect antenna isolation. A simple model is constructed for the reflected waves as shown in Fig. 5.33. The transmitting antenna on the left plane emits a wave, which is reflected by the left plane and directly hits the cylinder. After creeping on the cylinder for a certain distance, the reflection occurs on the right plane and then arrives at the receiving antenna.

As can be seen from Fig. 5.33, after importing the model into the program, the geodesic reflection from the transmitting antenna to the receiving antenna can be obtained. Therefore, the minimum included angle method is suitable for convex—concave surfaces.

Ray tracing on nondevelopable surfaces: Taking an ellipsoid as an example, we will illustrate that the minimum included angle method can be applied to the solution of a geodesic line on a nondevelopable surface. Taking two points at the end of the x-axis, according to the algorithm, the ray tracing path on the ellipsoid can be obtained, as shown in Fig. 5.34.

The approximation of the circumference of an ellipse is

$$L \approx \pi\left[1.5(a+b) - \sqrt{ab}\right] \tag{5.108}$$

Table 5.7 compares the numerical result with the tracing result.

Using the new algorithm, the relative error between the tracking result and the ground truth is: a = (13.4015 − 13.1534)/1.4015 = 0.36%. It can be seen that the

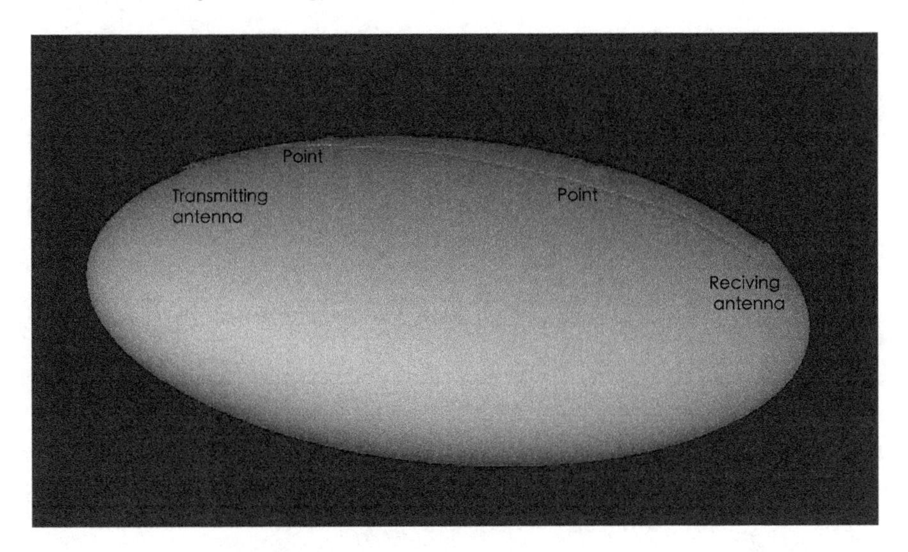

Fig. 5.34 Ray tracing on the ellipsoid

Table 5.7 Comparison between numerical result and tracing result on the ellipsoid

Coordinate of the emission point S (m)	Coordinate of the receiving point R (m)	Tracing result (m)	Numerical distance (m)
−6	6	13.3534	13.4015
0	0		
0	0		

tracing result is very accurate. Therefore, the minimum included angle method is applicable to nondevelopable surfaces.

Ray tracing on a combination of solids: We use a combination of cone and cylinder to verify that the algorithm is also feasible for combination of solids. The lateral surface of the cone is a circular sector (*SBC*), as shown in Fig. 5.35. Thus, we can calculate the length of the geodesic line between any two points on the cone using Eq. (5.109)

$$BD^2 = BS^2 + SD^2 - 2BS * SD * \cos\theta \tag{5.109}$$

The ray path on the combination of solids as shown in Fig. 5.36 can be obtained using the algorithm. The ray path includes *SD* on the cone and *DR* on the cylinder.

So, the length calculated numerically is

$$L = \widehat{SD} + \overline{DR} = 5.942$$

Table 5.8 compares the numerical result with the tracing result.

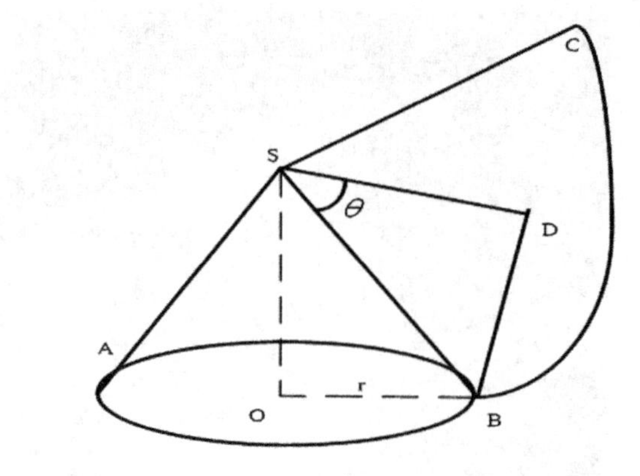

Fig. 5.35 Lateral surface of the cone

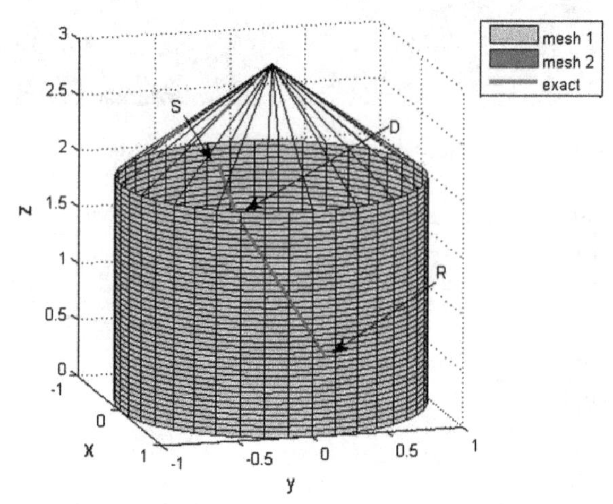

Fig. 5.36 Ray tracing on a combination of solids

Table 5.8 Comparison between numerical result and tracing result on a combination of solids

Emission point S (m)	Receiving point R (m)	Inflection point D (m)	Emission point S (m)	Tracing result (m)	Numerical distance (m)
0.489	0.996	0.85	0.489	5.942	5.897
−0.49	0.091	−0.481	−0.49		
2.308	0.689	1.997	2.308		

Using the new algorithm, the relative error between the tracking result and the ground truth is: a = (5.942 − 5.897)/5.897 = 0.76%. It can be seen that the tracing result is very accurate, such that the minimum included angle method is applicable to assemblies. Figure 5.37 shows the result for an aircraft obtained from the ray tracing.

Result analysis: During tracing, the surface of the model is split into triangles. Therefore, the geodesic line obtained using the minimum included angle method is

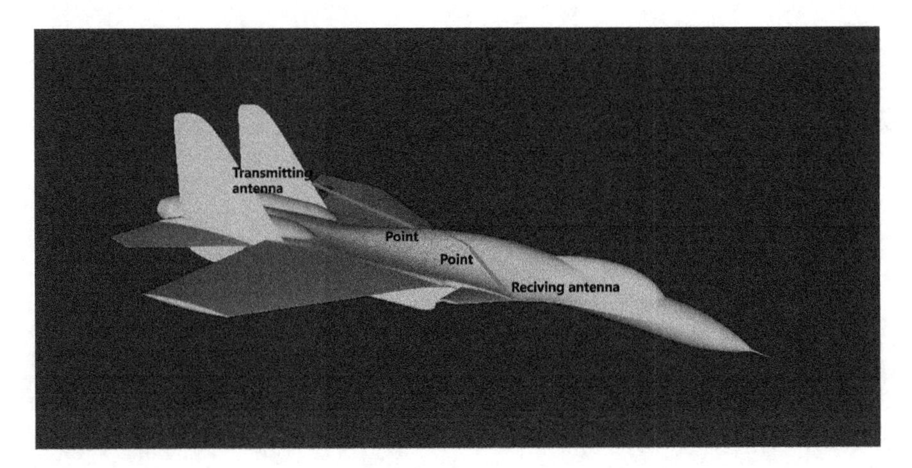

Fig. 5.37 Ray tracing result of an aircraft

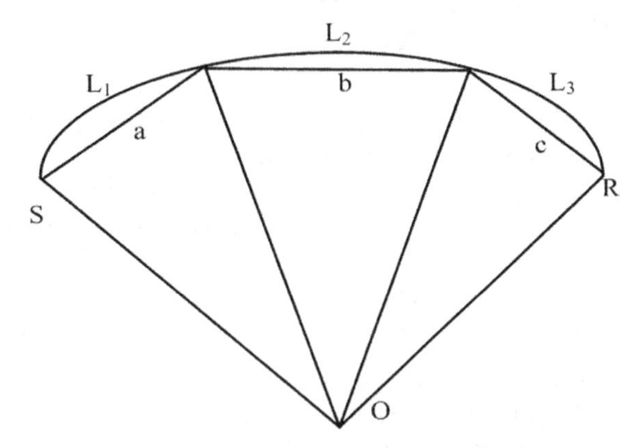

Fig. 5.38 Tracing result analysis

not a smooth curve. In fact, the geodesic line is composed of three segments, which are a, b, and c in Fig. 5.38, i.e.,

$$L = a + b + c \tag{5.110}$$

However, the actual length of geodesic line is the arc length:

$$\widehat{SR} = \widehat{L}_1 + \widehat{L}_2 + \widehat{L}_3 \tag{5.111}$$

It can be seen that the length of the geodesic line obtained by the new algorithm is very different from the ground truth. The error of the ray tracing is mainly determined by the density of the curved surface. The denser the split, the more accurate the result, and the more computing time it takes.

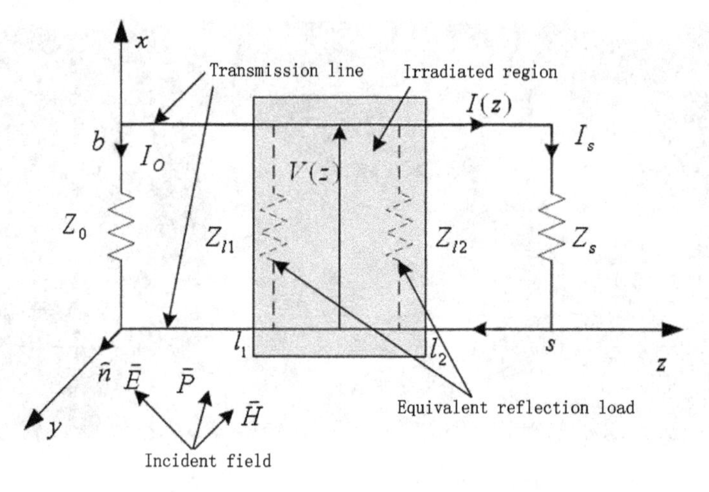

Fig. 5.39 Transmission line radiated by electromagnetic fields and the equivalent parameter settings

② Field–line coupling. The electromagnetic field radiated by the aircraft's airborne transmitters may affect the susceptive equipment through antenna coupling and cable coupling. Besides the distribution of the space, the coupling of the cable to the space field is also affected by the laying mode, the connection mode (such as single-wire grounding, double-line reciprocation), and the protection mode (unshielded single wire, single-shielded single wire, double-shielded single wire, twisted pair, shielded twisted pair, etc.) of the cables.

The connection between equipment often uses parallel two-conductor lines. Let the length of the parallel two-conductor lines be l, the spacing be b, and the height from the reference ground be h. The transmission line is exposed to the electromagnetic field as shown in Fig. 5.39. The area of the transmission line being exposed is A_l, and then

$$\int_{A_l} (\nabla \times \vec{E}) \cdot \hat{n} dS = -\frac{\partial}{\partial t} \int \vec{B} \cdot \hat{n} dS \tag{5.112}$$

There is an induced voltage for a given transmission line impedance, which is

$$V(z) = -\int_{0}^{b} E_x(x, z) dx \tag{5.113}$$

Then, we can derive the differential equations for the transmission line as [23, 45]

$$\frac{\partial V(z)}{\partial z} + ZI(z) = j\omega \int_0^b B_y^i(x, z)dx \tag{5.114}$$

$$\frac{\partial I(z)}{\partial z} + YV(z) = j\omega Y \int_0^b E_x^i(x, z)dx \tag{5.115}$$

where E^i and B^i are incident field; Y and Z are the admittance and impedance of the transmission line, respectively.

If the equivalent impedance Z_0 of the transmission line shown in Fig. 5.39 is not affected by the electromagnetic field, and the load of the transmission line is Z_s, then the currents through Z_0 and Z_s that can be calculated using (5.114) and (5.115) are

$$I_0 = \frac{Z_C}{D[Z_C \cosh(\gamma l_1) - Z_{l1}\sinh(\gamma l_1)]} \cdot \left[\int_{l_1}^{l_2} [E_z^i(b, z) - E_z^i(0, z)][Z_c \cosh\gamma(l_2 - z) + Z_{l2}\sinh\gamma(l_2 - z)]dz + [Z_C \cos\gamma(l_2 - l_1) + Z_{l2}\sinh\gamma(l_2 - l_1)] \int_0^b E_x^i(x, l_1)dx - Z_C \int_0^b E_x^i(x, l_2)dx \right] \tag{5.116}$$

$$I_s = \frac{Z_C}{D[Z_C \cosh(s - l_2) + Z_{l2}\sinh\gamma(s - l_2)]} \cdot \left[\int_{l_1}^{l_2} [E_z^i(b, z) - E_z^i(0, z)][Z_c \cosh\gamma(z - l_1) + Z_{l1}\sinh\gamma(z - l_1)]dz - [Z_C \cosh(\gamma l_2) + Z_{l1}\sinh(\gamma l_2)] \int_0^b E_x^i(x, l_2)dx + Z_C \int_0^b E_x^i(x, l_1)dx \right] \tag{5.117}$$

where Z_C is the characteristic impedance of the transmission line; γ is the complex propagation constant; b is the line spacing; s is the total length of the transmission line; l_1 and l_2 are the positions of the electromagnetic radiation transmission line. And there is

$$D = (Z_C Z_{l1} + Z_{l2} Z_C)\cosh\gamma(l_2 - l_1) + (Z_C^2 + Z_{l2} Z_{l1})\sinh\gamma(l_2 - l_1) \tag{5.118}$$

where Z_{l1} and Z_{l2} are equivalent reflection loads at l_1 and l_2, which can be written as

$$Z_{l1} = Z_C \frac{Z_0 - Z_C \tanh\gamma l_1}{Z_C - Z_0 \tanh\gamma l_1} \tag{5.119}$$

$$Z_{l2} = Z_C \frac{Z_s + Z_C \tanh \gamma(s - l_2)}{Z_C + Z_s \tanh \gamma(s - l_2)} \tag{5.120}$$

Equations (5.116) and (5.117) are the induced currents obtained by irradiating an electromagnetic field to a certain area. For cases where multiple areas on the transmission line are irradiated, the superposition method can be used to solve the problem.

Because all the cables laying on the aircraft are affected by the distribution of the space field, the workload of analysis and calculation is huge, such that the influence is difficult to be fully predicted during predesign. To solve this problem, we can exploit the knowledge of engineering development, test, and usage of other similar equipment, to screen, forecast, and calculate the key parts of the cable.

③ Aperture coupling. There are many high-power transmitters and susceptive equipment in the cabins of aircraft systems. For susceptive equipment, it is required that the enclosures have good shielding capabilities. In fact, many pieces of susceptive equipment have to include apertures of different sizes and numbers; thus, the electromagnetic field in the cabin can be coupled with the apertures and acts on the cables and susceptive components inside the casing of the equipment. At the same time, the high-power radiation field outside the fuselage can also be coupled with the inside of the fuselage through apertures, thereby affecting the environment electromagnetic fields, cables, and equipment inside the cabin.

The analysis of the aperture-coupled electromagnetic field is complex, and it is difficult to obtain an accurate analytical solution. Generally, diffraction theory or electromagnetic dual theory is used to obtain approximate results [46].

The theoretical basis for diffraction in physical optics is the Huygens principle, which assumes that every point on the wave front can be seen as a secondary light source that radiates wavelets. The wavelets are then stacked in front of the transmission direction; thus, the field quantity in the transmission direction can be obtained. Based on Huygens principle and the Kirchhoff formula derived from Green's theorem, the field quantity $\psi(r)$ (omitting the time factor $e^{-j\omega t}$) at any point P in a certain area V after the aperture coupling can be expressed as ψ and $\frac{\partial \psi}{\partial n}$ of the boundary surface S from V as

$$\psi(r) = -\frac{1}{4\pi} \oint_S \frac{e^{-jkR}}{R} n \left[\nabla' \psi + \left(jk - \frac{1}{R} \right) \frac{R}{K} \psi \right] dS' \tag{5.121}$$

where R is the distance from the field point (P) to the center of the aperture.

The center of the aperture is set to be the origin of the Cartesian coordinate system. The field intensity at the center of the aperture is ψ_0. The incident wave is projected onto the conductive plate at an angle θ_0. The angle between the radius vector from the field point to the center of the aperture and the normal direction of the conductive plate is θ. For apertures of any shapes, we have

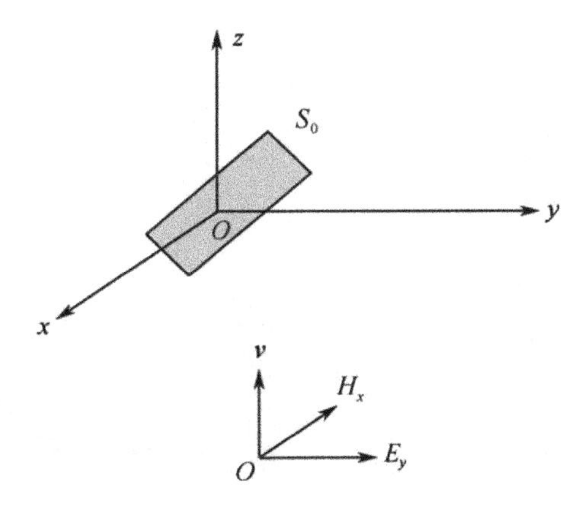

Fig. 5.40 Rectangle aperture coupling

$$\psi(r) = j\frac{ke^{-jkR}}{4\pi}(\theta_0 + \theta)\int\limits_{S_0} \psi_0 e^{-jkr'} dS' \tag{5.122}$$

For a rectangle aperture S_O as shown in Fig. 5.40, let the length of the aperture be $2a$, and the width be $2b$. The incident ray is along the direction of z-axis and $\theta_0 = 0^0$. The incident field is $E_y = E_0 e^{-jkr'}$; thus, there is a coupling field and we have

$$\begin{aligned} E_P &= -\frac{jke^{-jkR}}{4\pi R}(1 + \boldsymbol{cos}\theta)\int\limits_{S_0} e^{-jkr'} dS' \\ &= -\frac{jke^{-jkR}}{4\pi R}(1 + \boldsymbol{cos}\theta)\int\limits_{S_0} E_0 e^{-jk(x'\boldsymbol{sin}\,\theta\,\boldsymbol{cos}\varphi + y'\boldsymbol{sin}\,\theta\,\boldsymbol{sin}\,\varphi)} dx'dy' \end{aligned} \tag{5.223}$$

where R is the distance from the field point to the origin; θ is the angle between R and the z-axis; E_0 is the field strength of the incident wave at the rectangle aperture plane S_0; φ is the angle between the projection of R on xoy and the x-axis; x', y' are the coordinates of any point on the rectangle aperture plane S_0.

For circular aperture S_0 as shown in Fig. 5.41, let the aperture radius be a. When the incident field is a constant, i.e., $E_y = E_0$, the coupling field is

$$E_P = -\frac{jke^{-jkR}}{4\pi R} E_0 S(1 + \mathbf{cos}\theta)\frac{2J_1(ka\ \mathbf{sin}\theta)}{ka\ \mathbf{sin}\theta} \tag{5.124}$$

where J_1 is a first-order Bessel function and $S = \pi a^2$.

In addition, when the conductive plate is thin, and the size of the aperture is small compared to the wavelength, the distribution of the aperture coupling field can be approximated using the electromagnetic dual principle.

④ Coupling between cables. Due to space limitations, large number of equipment, and difficulty in equipment layout adjustment, the internal cables of the aircraft

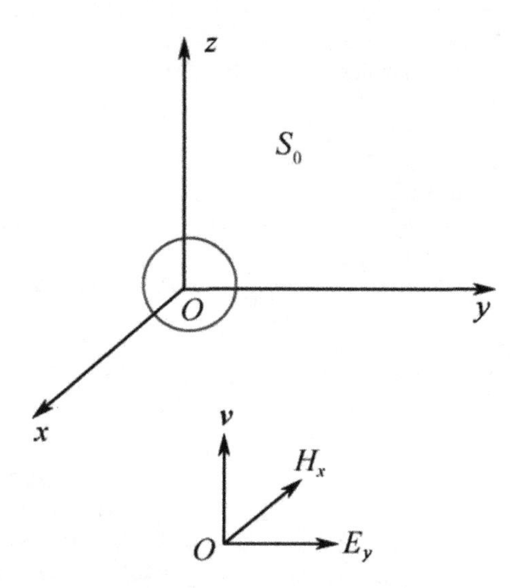

Fig. 5.41 Circular aperture coupling

are densely populated. Sometimes, high-power RF cable accessories are installed with a large number of weak signal transmission cables. In this situation, mutual coupling between cables may cause interference response or decrease of safety margins of the susceptive equipment.

Cable coupling can be capacitive or inductive. There are many kinds of coupled cables, such as unshielded line coupling, shielded line coupling, unshielded twisted pair coupling, and shielded twisted pair coupling. Among them, the radiation coupling relationship between two parallel wires located above the ground plane is very typical (the height from the ground, the radius of the insulation layer, and the radius of the wire layers of the two wires are referred to as h_1, R_1, r_1 and h_2, R_2, r_2 respectively; the distance between the two wires is denoted by s; the length of the two parallel wires is denoted by l_s; the dielectric constant of the insulating layer is denoted by ε).

The equivalent distributed parameters of the capacitive coupling between the radiating cable and the receiving cable are shown in Fig. 5.42. The coupling capacitance between the two parallel wires is

$$C_{12} = -\frac{2\pi\varepsilon_{eff} P_{12}}{\cosh^{-1}\left(\frac{h_1}{r_1}\right) - \cosh^{-1}\left(\frac{h_2}{r_2}\right) P_{12}^2} \tag{5.125}$$

$$\begin{aligned} P_{12} = 1/2 \Bigg[&\cosh^{-1}\left(\frac{s^2 + (r_1^2 - r_2^2)}{2sr_1}\right) + \cosh^{-1}\left(\frac{s^2 - (r_1^2 - r_2^2)}{2sr_2}\right) \\ &- \cosh^{-1}\left(\frac{4h_1h_2 + s^2 + (r_1^2 - r_2^2)}{2r_1\sqrt{4h_1h_2 + s^2}}\right) - \cosh^{-1}\left(\frac{4h_1h_2 + s^2 - (r_1^2 - r_2^2)}{2r_2\sqrt{4h_1h_2 + s^2}}\right) \Bigg] \end{aligned} \tag{5.126}$$

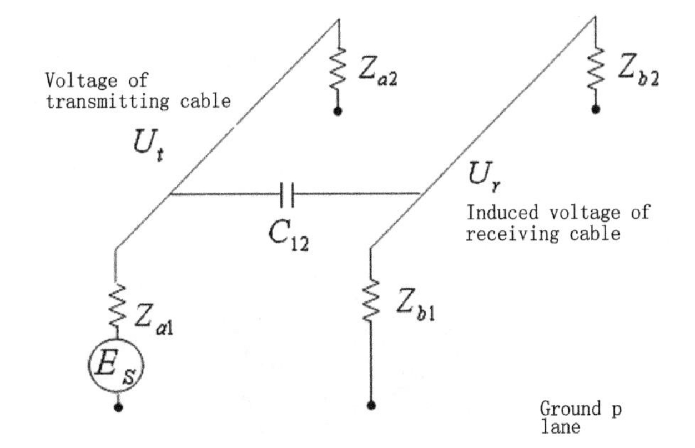

Fig. 5.42 Equivalent distributed parameters of capacitive coupling of parallel laying between radiating and receiving cables

$$\varepsilon_{eff} = \varepsilon_0 + 2\frac{\left(\frac{R_1+R_2}{r_1+r_2}\right)^2 - 1}{\left(\frac{2s+R_1+R_2}{r_1+r_2}\right)^2}(\varepsilon - \varepsilon_0) \tag{5.127}$$

The equivalent capacity of capacitive coupling of parallel laying between radiating and receiving cables can be used to solve the coupling voltage, i.e.,

$$U_2 = \left(\frac{1}{1 + \frac{1}{j\omega Z C_{12} l_s}}\right)U_1 \tag{5.128}$$

where $Z = \frac{Z_{b1}Z_{b2}}{Z_{b1}+Z_{b2}}$.

If the radiating cable is a shielded cable, let t_e be the thickness of the shielding layer, δ_e be the skin depth of the emitting shielding line, r_{si} be the radius of the inner conductor of the shielding layer, and F_e be the constant related to the coverage of the shielding layer. Then, we have

$$U_2 = \left(\frac{1}{1 + \frac{1}{j\omega Z C_{12} l_s}}\right)\left(e^{-\frac{t_e}{\delta_e}} + F_e\right)\left(\frac{1}{1 + \frac{1}{j\omega R_s C_{ws} l_s}}\right)U_1 \tag{5.129}$$

where $C_{ws} = \frac{2\pi\varepsilon}{\ln(r_{si}/r_1)}$ is the capacitance value between the core and shielding layer of the radiating wire per unit length; R_s is the shielding resistance which depends on frequency. The expression of R_s is

$$R_s = R_0\left(\frac{t_e}{\delta_e}\right)\left(\frac{\sinh\frac{2t_e}{\delta_e} + \sin\frac{2t_e}{\delta_e}}{\cosh\frac{2t_e}{\delta_e} - \cos\frac{2t_e}{\delta_e}}\right)$$

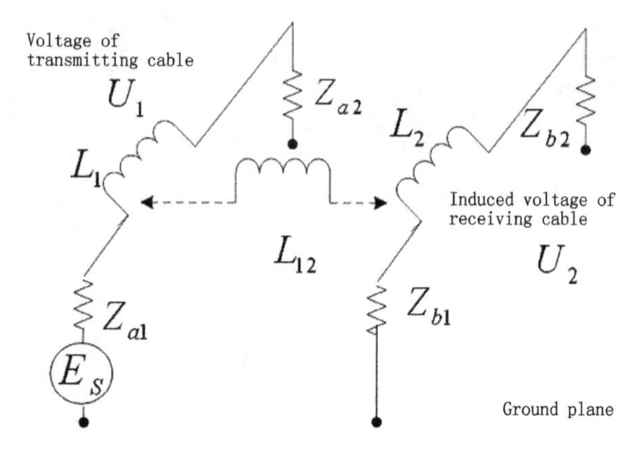

Fig. 5.43 Equivalent distributed parameters of inductive coupling of parallel laying between radiating and receiving cables

where $R_0 = \frac{l_{es}}{2\pi\sigma r_{si}t}$ is the DC resistance of the shielding layer, l_{es} is the length of the shielding wire, σ is the electric conductivity of the shielding layer, and t is the thickness of the shielding layer.

For double-shielded radiation cables, there is

$$U_2 = \left(\frac{1}{1 + \frac{1}{j\omega Z C_{12} l_s}}\right)\left(e^{-\frac{t_e}{\delta_e}} + F_e\right)\left(\frac{1}{1 + \frac{1}{j\omega R_s C_{ws} l_s}}\right)\left(e^{-\frac{t_{e2}}{\delta_{e2}}} + F_{e2}\right)\left(\frac{1}{1 + \frac{1}{j\omega R_{s2} C_{s1s2} l_s}}\right)U_1 \tag{5.130}$$

where C_{s1s2} is the capacitance value between the inner and outer shielding layers per unit length; R_{s2}, t_{e2}, δ_{e2}, and F_{e2} are the impedance, thickness, skin depth, and shielding constant of the outer shielding layer, respectively.

The equivalent distributed parameters of the inductive coupling between the radiating cable and the receiving cable are shown in Fig. 5.43. The mutual inductance between parallel cables is

$$L_{12} = \frac{\mu_0}{4\pi} \ln\left[\frac{(h_1 + h_2)^2 + s^2}{(h_1 - h_2)^2 + s^2}\right] \tag{5.131}$$

The induced voltage of the receiving cable due to the inductive coupling at the terminal load Z_{b2} is

$$U_L = j\omega L_{12} I_1 l_s \frac{Z_{b2}}{Z_{b1} + Z_{b2} + j\omega L_2 l_s} \tag{5.132}$$

where $L_2 = \frac{\mu_0}{2\pi} \ln\left[\frac{2h_2}{r_2} - 1\right]$; I_1 is the radiating cable current; L_2 is the self-inductance per unit length of the receiving cable.

The expression for current I_1 is

$$I_1 = \frac{U_1}{Z_{a2} + j\omega L_1 l_s} \tag{5.133}$$

where U_1 is the voltage of the radiating cable; L_1 is the self-inductance per unit length of the radiating cable, and its expression is

$$L_1 = \frac{\mu_0}{2\pi} \ln\left[\frac{2h_1}{r_1} - 1\right] \tag{5.134}$$

If the radiating cable is shielded and grounded at multiple points, the radius of the shield is r_s, the self-inductance per unit length of the shield is L_{sw}, the resistance of the radiating cable shield is R_s, and the induced voltage of the shield is U_s, then the induced current of the shield can be calculated as

$$I_{sh} = \frac{U_s}{R_s + j\omega L_{sw} l_s} = \frac{j\omega L_{sw} I_1 l_s}{R_s + j\omega L_{sw} l_s} = \frac{j\omega I_1 l_s \left[\frac{\mu_0}{2\pi} \ln(\frac{2h_1}{r_s} - 1)\right]}{R_s + j\omega L_{sw} l_s \left[\frac{\mu_0}{2\pi} \ln(\frac{2h_1}{r_s} - 1)\right]} \tag{5.135}$$

Correspondingly, the equivalent radiation inference is

$$I_{eff} = I_1 - I_{sh} = \frac{R_s I_1}{R_s + j\omega L_{sw} l_s} = \frac{R_s I_1}{R_s + j\omega l_s \left[\frac{\mu_0}{2\pi} \ln(\frac{2h_1}{r_s} - 1)\right]} \tag{5.136}$$

The induced voltage of the receiving cable due to the inductive coupling to the terminal load Z_{b2} is

$$\begin{aligned} U_L &= j\omega L_{12} I_{eff} l_s \frac{Z_{b2}}{Z_{b1} + Z_{b2} + j\omega L_{sw} l_s} \\ &= j\omega L_{12} l_s \frac{Z_{b2}}{Z_{b1} + Z_{b2} + j\omega l_s \left[\frac{\mu_0}{2\pi} \ln(\frac{2h_1}{r_s} - 1)\right]} \frac{R_s I_1}{R_s + j\omega l_s \left[\frac{\mu_0}{2\pi} \ln(\frac{2h_1}{r_s} - 1)\right]} \end{aligned} \tag{5.137}$$

If the receiving cable is shielded and grounded at multiple points, L_{s2w2} is the mutual inductance between the shielding layer and the receiving core per unit length, R_{rs} is the resistance of the receiving cable's shielding layer, L_{es2} is the mutual inductance between the radiation cable and the receiving cable's shielding layer per unit length, L_{ew2} is the mutual inductance between the radiating cable and the receiving cable per unit length, L_{rs} is the self-inductance of the receiving cable shielding layer per unit length, and U_2 is the induced voltage of the receiving cable core, then the induced current and the induced voltage of the shielding layer are

$$I_{rs} = \frac{U_{rs}}{R_{rs} + j\omega L_{rs} l_s} = \frac{j\omega L_{es2} l_s I_1}{R_{rs} + j\omega L_{rs} l_s} \tag{5.138}$$

$$U_2 = j\omega L_{ew2} l_s I_1 - j\omega L_{s2w2} l_s I_{rs} = j\omega L_{ew2} l_s I_1 - j\omega L_{s2w2} l_s \left(\frac{j\omega L_{es2} l_s I_1}{R_{rs} + j\omega L_{rs} l_s} \right) \tag{5.139}$$

5.3.2.3 Nonlinear System Behavioral Modeling Method

In the system-level EMC study, it is necessary to build a behavioral model of the nonlinear system in order to analyze the nonlinear factors that affect the characteristics of the system. On this basis, we need to linearize the nonlinear subsystems and equipment, and improve the accuracy of the equipment's EMC behavioral simulation modeling.

This book takes the power amplifier, a most common nonlinear device in communication systems, as an example to explain the modeling process. The nonlinear description method is used to establish a behavioral model of the nonlinear system. With this method, we can further study the spectrum regeneration or expansion outside the useful signal bandwidth among the communication systems, and the mutual interference between adjacent channels [16].

1. The nonlinearity of electronic systems

Nonlinearity in electronic systems is mainly reflected in the following aspects.

(1) **The generation of harmonics**

One of the most obvious characteristics of a nonlinear system is the generation of harmonics of the excitation frequency. If a nonlinear system is represented by a Taylor series expansion, when the input is a single-frequency signal $u(t) = A\cos\omega_1 t$, the output contains not only the fundamental frequency component but also the harmonic component at frequency $N\omega_1$ (where N is a positive integer), as shown in Fig. 5.44. In most cases, the harmonics can be filtered out in narrowband receiving systems, but in the broadband system, the harmonics may cause serious interference to the system. All amplifiers will produce harmonics of different strengths. The stronger the harmonics, the worse the nonlinearity of the amplifier is.

(2) **Intermodulation**

Intermodulation is the recombination of several frequencies. When the input is a dual-tone signal, the output frequency can be expressed as $m\omega_1 + n\omega_2$ $(m, n = \cdots, -3, -2, -1, 0, 1, 2, 3, \cdots)$. The intermodulation can cause serious interference

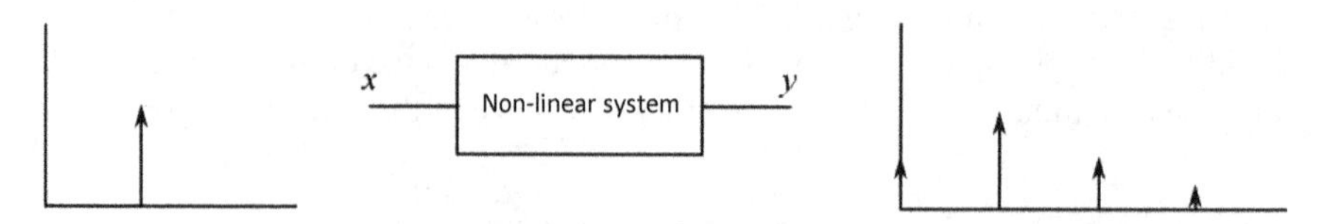

Fig. 5.44 Input and output spectrum diagram of a single signal through the nonlinear system

to the receiver. The intermodulation caused by the third power term of the nonlinear device is called the third-order intermodulation, and the intermodulation caused by the fifth power term is called the fifth-order intermodulation. The third-order intermodulation signal is the strongest signal in the odd-order terms. When the frequencies of the two input signals are very close, the third-order intermodulation signal is difficult to filter out.

(3) **Cross-modulation**

Cross-modulation is also caused by the nonlinearity. When the modulation of one signal is converted into another signal, it will result in cross-modulation distortion, which is likely to interfere with the adjacent channels.

(4) **Saturation and dynamic range reduction**

When the power amplifier operates in the nonlinear region, the response current does not increase proportionally with the increase of the excitation voltage. Under certain conditions, the amplifier saturates, and the amplification factor and the dynamic range of the amplifier decrease, making the amplifier unable to meet the design requirements. The power consumption will also be increased in the same time.

(5) **Conversion of amplitude modulation/phase modulation (AM/PM)**

AM–PM conversion refers to the phase deviation caused by the change of signal amplitude in a nonlinear circuit. In RF power amplifiers, phase distortion can also cause the third-order intermodulation distortion.

2. **Nonlinear system modeling**

In system-level EMC studies, amplifier (including power amplifier) is a common and important electronic device of all electronic systems. The amplifier's inherent nonlinearity usually leads to spectrum regeneration or expansion outside the useful signal bandwidth, which causes interference to adjacent channels or signals. As a result, the bit error rate of the corresponding receiving communication systems will increase. The main products of the nonlinear distortion of the power amplifier include high harmonics at main frequency, intermodulation distortion, cross-modulation distortion, AM–PM conversion, and wideband spurious. Among them, the first three occupy multiple channels and may cause some frequencies to be unusable, and the wideband spurs can cause interference in a very wide frequency band, which greatly reduces the receiving capacity of the communication system, or even block the RF front end of the entire system. From the above analysis, we can see that establishing the characteristic model accurately is the key in analyzing the system's self-compatibility, due to the power amplifier's nonlinearity.

The nonlinearity of the amplifier (including the mixer with conversion gain) can usually be represented by the characteristic parameters shown in Fig. 5.45, namely the output power $P_{1\text{dB}}$ of the 1 dB gain compression point, the input power P_{IIP3} or the output power P_{OIP3} of the third-order intercept point, the output power of the saturation point P_{sat}, and the gain compression G_{sat} of the saturation point.

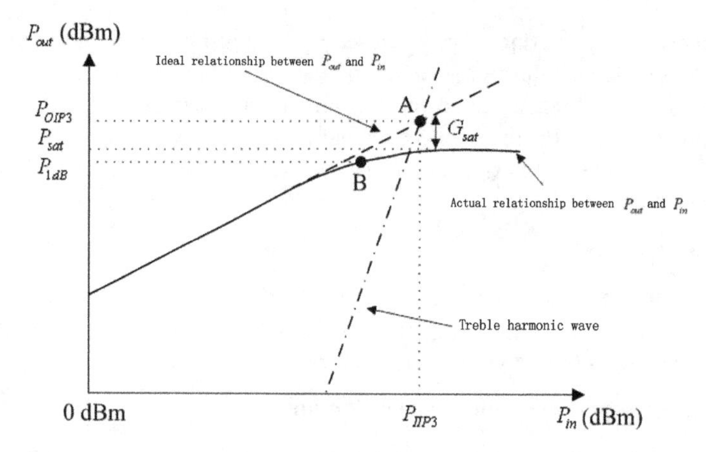

Fig. 5.45 Input/output transfer characteristics of a nonlinear amplifier

The 1 dB gain compression point is defined as the point where the difference between the amplifier linear gain and the actual nonlinear gain is 1 dB. It is used to measure the power capacity of the amplifier and indicates amplifier deviating constant. In the mixer, $P_{1\text{dB}}$ is the output power corresponding to the frequency when the functional relationship between the intermediate frequency (IF) signal and the RF input signal starts to deviate from the linear characteristic.

The third-order intercept (TOI) point is the intersection of the linear power extension line of the third-order intermodulation product and the linear power extension line of the first-order product. They correspond to the input intercept point and the output intercept point, respectively. The TOI point describes the degree of the third-order intermodulation distortion of the nonlinear amplifier. The further the TOI point is from the 1 dB compression point, the smaller the amplitude of the third-order intermodulation component.

The saturation point is the output power point at which the amplifier's maximum output power reaches saturation. At this power value, the linearity of the amplifier is poor, but the efficiency is high.

The difference between the amplifier gain and the ideal linear gain is defined as the saturation point gain compression.

(1) **Nonlinear behavioral modeling method**

In general, the amplifier model has three types, namely the physical model for circuit simulation (device model), the behavioral model for system-level simulation, and the behavioral model based on the partial physical model (the gray box model). The amplifier model built based on equivalent circuit and device is more suitable for circuit-level simulations and can provide accurate simulation results. However, the system will be too complex, or the implementing cost will be too high due to the usage of device-level or transistor-level models. The behavioral model only depends on the input and output test data of the amplifier, such that the internal components of the amplifier are not required for modeling. Using nonlinear mathematical expressions can also simulate the nonlinear characteristics and memory effect of the actual amplifier, which is very suitable for system-level EMC simulation.

The nonlinear behavioral model of an amplifier is usually a nonlinear expression. A behavioral model describing the nonlinear characteristics and memory effect can be obtained by solving the coefficients of the mathematical expression using the input and output data. This behavioral model includes memoryless nonlinear behavioral models and memory nonlinear behavioral models. For narrowband systems, it is generally not necessary to consider the memory effect of the amplifier. However, in the case of a wideband signal input, the amplifier has a nonnegligible memory effect due to the existence of the bias network and the matching network in the amplifying circuit. The following section discusses the memoryless behavioral model and the memory nonlinear behavioral model of the amplifier. We also provide the applicable situations and performance characteristics of various different models.

(1) Memoryless behavioral model. For an amplifier, the memory effect is negligible when the input is a narrowband signal. The memoryless behavioral model can be represented by two algebraic functions of the instantaneous envelope amplitude, which describe the real and imaginary parts or the magnitude and phase of the output envelope component. Let the amplifier input signal be

$$x(t) = A(t)e^{j\phi(t)} \tag{5.140}$$

The common memoryless Taylor polynomial is

$$y(t) = \sum_{n=1}^{\infty} a_n x^n(t) \tag{5.141}$$

where a_n is a constant.

When considering the complex coefficients and ignoring even orders, the model is in the form

$$y(t) = \sum_{n=1}^{N} a_{2n-1} |x(t)|^{2(n-1)} x(t) \tag{5.142}$$

where N is the order of the polynomial.

The advantage of the Taylor series behavioral model is that the relative magnitude of each order distortion is reflected by the coefficient. The model is suitable for systems with smaller nonlinear distortion order and memoryless.

Another memoryless behavioral model is the Saleh function polynomial model [47]. Two equations are used to describe the AM–AM and AM–PM characteristics of the system, respectively

$$A_y[A(t)] = \frac{\alpha_A A(t)}{1 + \beta_A [A(t)]^2} \tag{5.143}$$

$$\phi_y[A(t)] = \frac{\alpha_\phi A(t)^2}{1 + \beta_\phi [A(t)]^2} \tag{5.144}$$

where α_A, β_A, α_ϕ, and β_ϕ are the measured fitting parameters for the AM–AM characteristics $A_y[A(t)]$ and AM–PM characteristics $\phi_y[A(t)]$ of the amplifier. The Saleh model is suitable for memoryless quasi-linear systems with weak nonlinearity.

(2) Memory nonlinear behavioral model. The memory nonlinear model can be categorized into linear (static) memory effect behavioral model and nonlinear (dynamic) memory effect behavioral model based on the different memory effects. So far, the linear memory behavioral model of an amplifier mainly refers to the Wiener model and Hammerstein model.

The Wiener model consists of a linear filter followed by a nonlinear function model. The output is expressed as [48]

$$y(k) = \sum_{n=1}^{N} b_n \left\{ \sum_{m=0}^{M-1} a_m x(k-m) \right\}^n \tag{5.145}$$

where $y(k)$ and $x(k)$ are the output and input of the model, respectively; M and N are the memory length and highest order of the polynomial, respectively.

The Hammerstein model is a cascade of nonlinear models without memory effects and linear time-varying modules [49], i.e.,

$$y(k) = \sum_{m=0}^{M} b_m \left\{ \sum_{n=1}^{N} a_n x(k-m) \right\}^n \tag{5.146}$$

Other models such as parallel Hammerstein models, parallel Wiener models, and Wiener–Hammerstein models are nonlinear memory effect behavioral models.

The memory polynomial model [50] consists of several delay taps and nonlinear static functions. This model describes the amplifier's unbalanced and dynamic AM–AM and AM–PM characteristics. The amplifier behavioral model of discrete-time complex baseband is

$$y(k) = \sum_{m=0}^{M-1} \sum_{n=1}^{N} a_{2n-1,m} |x(k-m)|^{2(n-1)} x(k-m) \tag{5.147}$$

In fact, the memory polynomial model and the Hammerstein are same models with different structures. The polynomial model is also a Volterra series model that takes only the diagonal Volterra kernel. It is suitable for weakly nonlinear systems with memory effects.

The RBF neural network model is a behavioral model describing the nonlinear characteristics of amplifiers in recent years. In this model, some input and output data points are used as training samples to calculate the model parameters, such that the dynamic nonlinear behavior of the amplifier can be reproduced. Neural network model has a rather high accuracy, and the computational complexity is also high; consequently, the complexity for modeling and simulating the EMC of the neural network model as a sub-module in the system increases. This book will not elaborate on the neural network model.

Volterra model is a generalization of the Taylor series model. Theoretically, the Volterra series can approximate a nonlinear continuous function with arbitrary precision. However, in fact, because the number of model parameters increases exponentially with the increase of model order and memory effect, the curse of dimensionality is likely to happen in function identification [51]. Therefore, Volterra model is usually only applicable to weak nonlinear time-invariant systems with attenuated memory. This book will give a specific description of the Volterra model and discuss how to simplify the model so that it can be applied to strong nonlinear systems.

(2) **Simplification of Volterra series behavioral model**

Volterra series is usually used to describe the behavioral characteristics of nonlinear circuits or systems. Compared with simulation calculations that include specific physical structures, Volterra series can greatly reduce the computational complexity and can be easily applied to system behavioral simulation. Modern wireless communication systems usually have the characteristics of high frequency, high speed, and wide frequency band. In many nonlinear modules such as the low-noise amplifiers and mixers, RF power amplifiers are the main source of nonlinear distortion. The distortion and out-of-band interference result from the nonlinear modules will be more serious because of the existence of memory effect, so it is necessary to accurately establish the behavioral model of the power amplifier with nonlinear characteristics of the memory effect. This is also an important topic in the research of RF power amplifier modeling. This book will use a typical RF power amplifier as an example to describe the nonlinear behavior of the RF power amplifiers using Volterra series.

(1) Volterra series model. The Volterra series was first proposed by the Italian mathematician *Vito Volterra* in 1889. It is a generalization of the convolution operation in linear system theory to the analysis of nonlinear systems.

① Volterra series of continuous and discrete systems. In a continuous nonlinear time-invariant system, $x(t)$ and $y(t)$ denote the input and output signals of the system, respectively, and the Volterra series expression is

$$y(t) = \sum_{n=0}^{\infty} y_n[x(t)] \tag{5.148}$$

where $y_n[x(t)]$ is the n-th order component of the system response. The expression is

$$y_n[x(t)] = \int_{-\infty}^{\infty} \cdots \int_{-\infty}^{\infty} h_n(\tau_1, \tau_2, \cdots, \tau_n) x(t-\tau_1) x(t-\tau_2) \cdots x(t-\tau_n) d\tau_1 d\tau_2 \cdots d\tau_n \quad (5.149)$$

where $h_n(\tau_1, \tau_2, \cdots, \tau_n)$ is the n-th order Volterra kernel.

When all of the Volterra kernels higher than second order are zero, the nonlinear system degenerates into a linear system. In addition, the Volterra kernel function is symmetric [52]. For example, the third-order kernel satisfies: $h_3(\tau_1, \tau_2, \tau_3) = h_3(\tau_2, \tau_1, \tau_3) = \cdots = h_3(\tau_3, \tau_2, \tau_1)$.

In general simulation process, we process the digital signal after signal sampling. The discrete form of Volterra series is

$$y(k) = \sum_{n=1}^{\infty} \sum_{i_1=0}^{\infty} \cdots \sum_{i_n=0}^{\infty} h_n(i_1, i_2, \cdots, i_n) \prod_{j=1}^{n} x(k - i_j) \quad (5.150)$$

where $x(k)$ and $y(k)$ are the input and output signals, respectively; $h_n(i_1, i_2, \cdots, i_n)$ is the n-th order Volterra kernel.

In practice, sufficient accuracy can be obtained using a finite number of orders and a finite memory length, and the Volterra series is truncated to the sum of the finite terms. In the communication system, the input and the output are generally modulated signals. Assume the input signal of the RF power amplifier is $x(t) = \mathrm{Re}[\tilde{x}(t) \cdot e^{j\omega_0 t}]$ and the output signal is $y(t) = \mathrm{Re}[\tilde{y}(t) \cdot e^{j\omega_0 t}]$, where ω_0 is the carrier angle frequency, and $\tilde{x}(t)$ and $\tilde{y}(t)$ are the complex envelope of the input and the output signals, respectively. The Volterra model of discrete-time complex baseband is

$$\begin{aligned}\tilde{y}(k) = & \sum_{i=0}^{M-1} \tilde{h}_1(i) \times \tilde{x}(k-i) + \sum_{i_1=0}^{M-1} \sum_{i_2=i_1}^{M-1} \sum_{i_3=0}^{M-1} \tilde{h}_3(i_1, i_2, i_3) \times \tilde{x}(k-i_1)\tilde{x}(k-i_2)\tilde{x}^*(k-i_3) \\ & + \sum_{i_1=0}^{M-1} \sum_{i_2=i_1}^{M-1} \sum_{i_3=i_2}^{M-1} \sum_{i_4=0}^{M-1} \sum_{i_5=i_4}^{M-1} \tilde{h}_5(i_1, i_2, i_3, i_4, i_5) \times \tilde{x}(k-i_1)\tilde{x}(k-i_2)\tilde{x}(k-i_3)\tilde{x}^*(k-i_4)\tilde{x}^*(k-i_5) \\ & + \cdots \end{aligned} \quad (5.151)$$

where $\tilde{h}_n(i_1, i_2, \cdots, i_n)$ is the n-th order complex Volterra kernel of the system; M is the length of the memory; $(\cdot)^*$ is the complex conjugate of the signal; $\tilde{x}(k)$ and $\tilde{y}(k)$ are the input and the output complex envelope discrete signals, respectively. The symmetry of the kernel has been considered in Eq. (5.151). At the same time, using the band-limited modulation characteristic of the communication system, even-order terms are removed, and a limited Volterra series model with the order N and memory length M is truncated to use in the analysis.

② Model parameter identification. The Volterra series model parameter identification is a process to find the Volterra kernel using the input and the output data. Assuming the sampling process starts from time k and a set of sampling data of length L is obtained, the input and the output matrixes of the system are

$$X = [X(k), X(k+1), \cdots, X(k+L-1)]^T \tag{5.152}$$

$$Y = [\tilde{y}(k), \tilde{y}(k+1), \cdots, \tilde{y}(k+L-1)]^T \tag{5.153}$$

where

$$\begin{aligned} X(k) = [&\tilde{x}(k), \cdots, \tilde{x}(k-M+1), |\tilde{x}(k)|^2\tilde{x}(k), \tilde{x}^2(k)\tilde{x}^*(k-1), \cdots, \\ &|\tilde{x}(k)|^{N-1}\tilde{x}(k), \cdots, |\tilde{x}(k-M+1)|^{N-1}\tilde{x}(k-M+1)] \end{aligned} \tag{5.154}$$

The kernel vector of the system is defined as

$$\begin{aligned} H =&[\tilde{h}_1(0), \cdots, \tilde{h}_1(M-1), \tilde{h}_3(0,0,0), \tilde{h}_3(0,0,1), \cdots, \tilde{h}_3(M-1, M-1, M-1), \cdots, \\ &\tilde{h}_N(0, \cdots, 0, 0), \tilde{h}_N(0, \cdots, 0, 1), \cdots, \tilde{h}_N(M-1, \cdots, M-1)]^T \end{aligned} \tag{5.155}$$

The Volterra model in Eq. (5.151) can be written as

$$Y = XH \tag{5.156}$$

The solution of the parameters in Eq. (5.156) is an estimation problem. The collected data is far more than the number of parameters to be estimated. We can use the least square method to solve this pure overdetermined problem. The estimated kernel vector $\hat{H}$ is [53]

$$\hat{H} = (X'X)^{-1}X'Y \tag{5.157}$$

However, when the order and memory length of the nonlinear system increase, the number of the Volterra kernel to be identified grows exponentially, and the elements in the matrix $X'X$ are easily correlated. That is, the matrix appears ill-conditioned so that the singular values may appear during the solution. When the number of the Volterra kernels to be identified is larger than the number of the mutually unrelated elements, the solution process becomes a pure underdetermined problem (the least L_2 norm problem), and the estimated kernel vector is $\hat{H}$ [53]:

$$\hat{H} = X'(XX')^{-1}Y \tag{5.158}$$

After Volterra behavioral model is identified, the model needs to be verified. Many system performance indicators can be used to evaluate the accuracy of the model, such as the bit error rate (BER) and the adjacent channel power ratio. The normalized mean square error (NMSE) [54] is used here as

$$NMSE = 10\log\left(\frac{\sum_{s=1}^{P}\left(\left|\tilde{y}_s - \tilde{y}_s^{\text{mod}}\right|^2\right)}{\sum_{s=1}^{S}|\tilde{y}_s|^2}\right) \tag{5.159}$$

where P is the number of sampling points; $\tilde{y}_s^{\text{mod}}$ is the output complex envelope of the model; $\tilde{y}_s$ is the simulation output complex envelope.

The smaller the NMSE, the higher the accuracy of the model is.

(2) Method for Volterra behavioral model simplification. Theoretically, the Volterra series can approximate the nonlinear continuous function with arbitrary precision. However, the number of identification parameters exponentially increases with the increase of model order and memory length, which easily leads to the curse of dimensionality in function identification. Although the number of parameters in the calculation of the power amplifier model can be reduced using the symmetry of the Volterra kernel and the band-limited characteristics of the communication system, the computational cost of the parameter identification is still large. Therefore, the Volterra series is usually used to handle weak nonlinear systems. The simplification method is to set the identification kernel to zero one by one and then estimate the variation of the model error, leaving only the kernels that have a large impact on the model. This method is complicated and time consuming. In order to reduce the complexity of the Volterra model, many special models have emerged, such as the Wiener model and the Hammerstein model. Both models are adapted to power amplifiers with linear memory effects, but the extraction of model parameters is still difficult [55]. The fast calculation of parallel behavioral model can effectively improve the data processing speed and reduce the calculation time, but the problem of reducing the number of model parameters remains [56]. The memory polynomial model only takes into account the diagonal Volterra kernel. When the nondiagonal Volterra kernel contributes more to the output of the Volterra model than the diagonal Volterra kernel, the accuracy of the memory polynomial model decreases.

In fact, in most RF power amplifiers, the memory effect becomes weaker over time. Additionally, the higher the order of the input components with memory effects, the smaller the effect they have on the output. In order to simplify the Volterra series model, two Volterra kernel parameter control factors are introduced with fixed order and memory length: One is an adjacent diagonal kernel control factor to remove the long memory delay component that has less influence on the output from input signal vector X in Eq. (5.152); the other is a dynamic deviation order control factor that removes the high-order components with memory from the input signal vector X in Eq. (5.152) and keeps the parameters that have a large impact on the output in the Volterra series model. Therefore, the parameters of the model can be greatly reduced under the premise of retaining the accuracy.

Adjacent diagonal kernel control factor: Only the diagonal Volterra kernel is retained in the memory polynomial model, i.e., when $i_1 = i_2 = \cdots = i_n$, $\tilde{h}_n(i_1, i_2, \cdots, i_n) \neq 0$. In the case where the nondiagonal Volterra kernel's contribution to the output cannot be ignored, a looser restriction condition is needed. The adjacent diagonal Volterra kernel control factor S is introduced to retain the kernels near the diagonal Volterra kernel [52]. Taking the n-th order kernel as an example, by setting the integer value for S $(0 \leq S \leq M-1)$ to lose the constraint condition $(\left|i_p - i_q\right| \leq S)$, we can gradually retain more kernels near the diagonal kernel: If $\max\{\left|i_p - i_q\right|\} > S$, then $\tilde{h}_n(i_1, i_2, \cdots, i_n) = 0$; otherwise, the kernel is

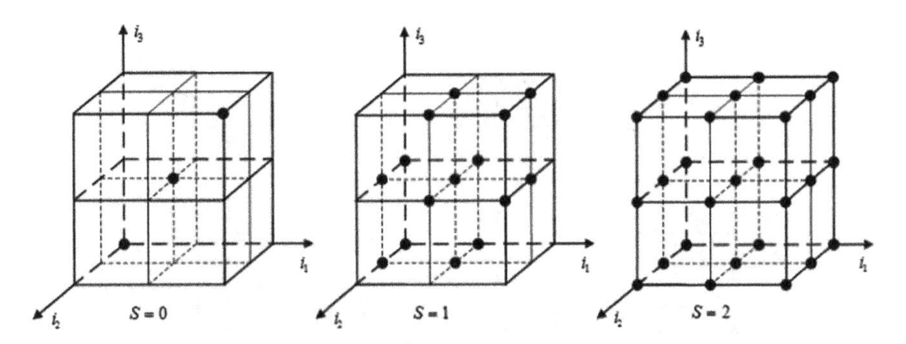

Fig. 5.46 Quantity variation of the third-order Volterra kernel with the control factor S

kept. Figure 5.46 uses the third-order term (i.e., memory length $M = 3$) as an example to illustrate the relationship between the control factor and the parameters to be identified. The origin of the coordinates is the Volterra kernel $\tilde{h}_3(0, 0, 0)$, and i_1, i_2, and i_3 on the axes are all integers. For simplicity, we do not consider the symmetry of the Volterra kernel. It can be seen that: When $S = 0$, Eq. (5.151) becomes the memory polynomial model; when S gradually increases, the identification parameter increases; when $S = M - 1$, a complete Volterra series model can be achieved.

Dynamic deviation order control factor: When the order of the input signal with the memory in the power amplifier increases, its influence on the output will gradually reduce. The distortion of most power amplifiers is mainly caused by the memoryless nonlinearity and low-order memory nonlinearity. The previously discussed Taylor series model is actually a special form of the Volterra series, in which only the memoryless input is retained, i.e., when $i_1 = i_2 = \cdots = i_n = 0, \tilde{h}_n(i_1, i_2, \cdots, i_n) \neq 0$. Another control factor R is introduced here to control the order of the memory input component of the input vector X in Eq. (5.152), which is called the dynamic deviation order control factor. R can also be interpreted as the maximum number of the nonzero elements in the set $\{i_1, i_2, \cdots, i_n\}$, and $0 \leq R \leq N$. For example, for the n-th order term, when $R = 1$, there can only be one input component with memory, i.e., $\tilde{x}^{n-1}(k)\tilde{x}(k - i)$. Now taking the third-order term (i.e., memory length $M = 3$) in Fig. 5.47 as an example, we will explain the relationship between the control factor and the parameters to be identified. The origin of the coordinates is the Volterra kernel $\tilde{h}_3(0, 0, 0)$, and i_1, i_2, i_3 are rounded on the axis. It can be seen that: When $R = 0$, Eq. (5.151) becomes a memoryless Taylor series model; when R gradually increases, the number of the identification parameters increases; when $R = N$, the complete Volterra series model can be achieved.

Under the control of S or R, the number of Volterra kernel increases almost linearly with the increase of order and memory length.

However, both S and R are integers and have their own value range: $0 \leq S \leq M-1$ and $0 \leq R \leq N$. When the memory length M is large and order N is small, the value of S is large and the parameter changes can be controlled easily, but the value of R is small which results in fast parameter growth. Therefore, S is preferred as the control

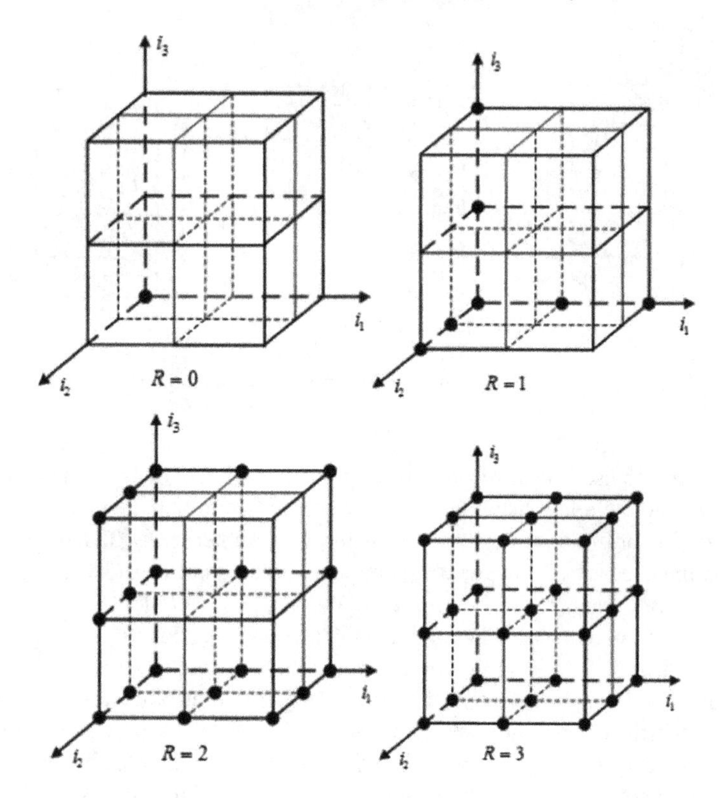

Fig. 5.47 Quantity variation of the third-order Volterra kernel with the control factor R

factor. When the memory length M is small and order N is large, R is preferred as the control factor. When the memory length M and the order N are equal, the combination of the two factors shall be used.

In the actual power amplifier behavioral modeling process, combining the two control factors S and R, there will be four variables to control the model accuracy, which are the order N, memory length M, and the control factors S and R. Because the Volterra kernels generated using different S and R will partially overlap, here we use "or" to preserve the corresponding Volterra kernel parameters to be identified. By gradually increasing S and R, the number of parameters to be identified is controlled, such that the memory nonlinearity of the power amplifier can be described using less Volterra kernel.

(3) Verification of power amplifier behavioral model. In order to verify the simplified method of the power amplifier behavioral model, a 10 W power amplifier has been designed using the MRF6S21140H from Freescale. A commercial circuit software was used for simulation and raw data collection. When extracting parameters, in order to avoid the ill-conditioned matrix, and to fully reflect the nonlinearity of the power amplifier, Gaussian white noise signal is usually used as the input signal [54]. In

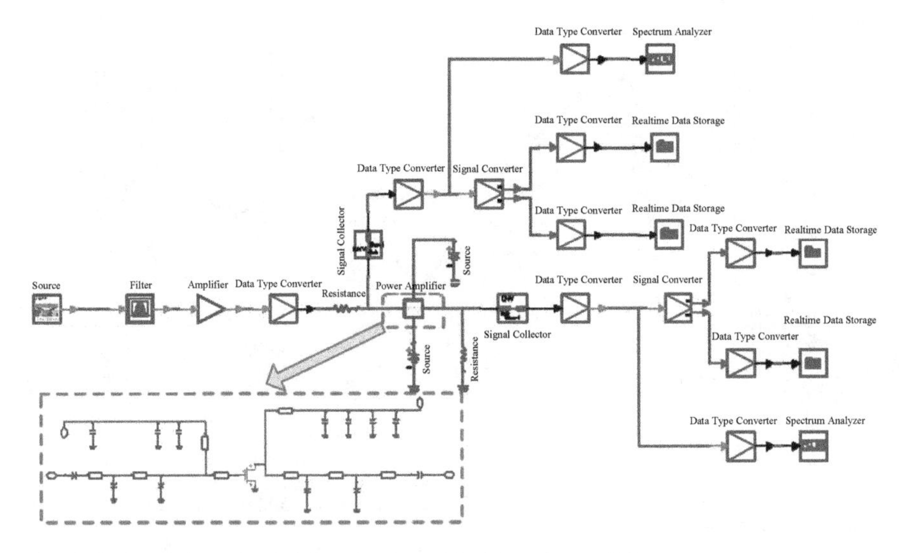

Fig. 5.48 Simulation of underlying circuit model with power amplifier

fact, the Gaussian white noise method is difficult in both simulation and experimental measurements, and the power amplifier is not used to amplify the Gaussian white noise signal. A WCDMA signal with an input code rate of 3.84 Mb/s was used as the input stimulus for the power amplifier with carrier frequency of 2.14 GHz. Figure 5.48 shows the simulation model that includes the power amplifier's underlying circuits. The matching and biasing circuits in the amplifier's underlying circuit were the main source of memory effects. This book regards the output of the simulation results of the circuit model as experimental data. About 4,000 input and output data points have been used for model parameter identification and validity verification.

The power amplifier circuit model is replaced with a Volterra series, of which the order N is 5 and the memory length M is 5.

First, the value of the control factors S and R was changed individually to analyze the goodness of fit of the model. Figures 5.49 and 5.50 show the partial time-domain complex envelope output waveforms for different behavioral models corresponding to different values of S and R, respectively. For the sake of clarity, only the real output waveform when both S and R were at the maximum value of 2 was considered. As the value increased, the output waveform of the behavioral model got closer and closer to the original data waveform.

For the different values of control factors, the number of parameters to be identified and the corresponding NMSE values are shown in Table 5.9.

When the NMSE is required to be no higher than -40 dB, the relative error between the model output and the simulated acquisition signal cannot be higher than 0.01%. As can be seen from Table 5.9, the requirement will be met with $S = 2$ (203 identification parameters) or $R = 3$ (225 identification parameters).

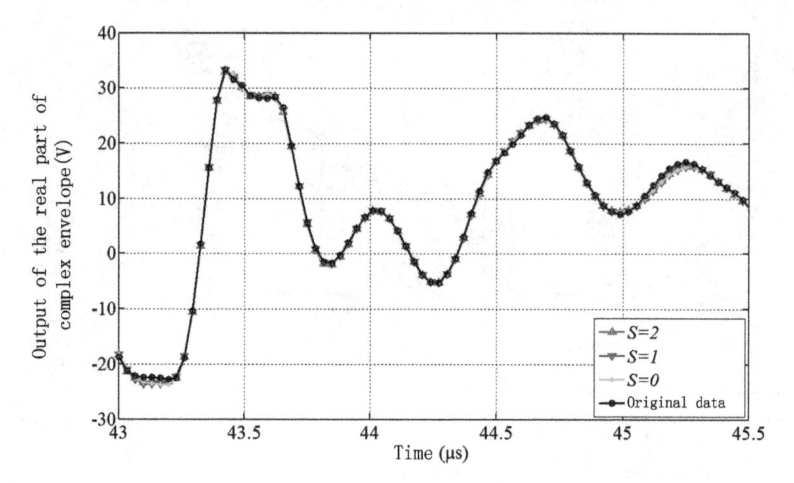

Fig. 5.49 Partial real part output waveforms of the time-domain complex envelope for behavioral models with different values of S

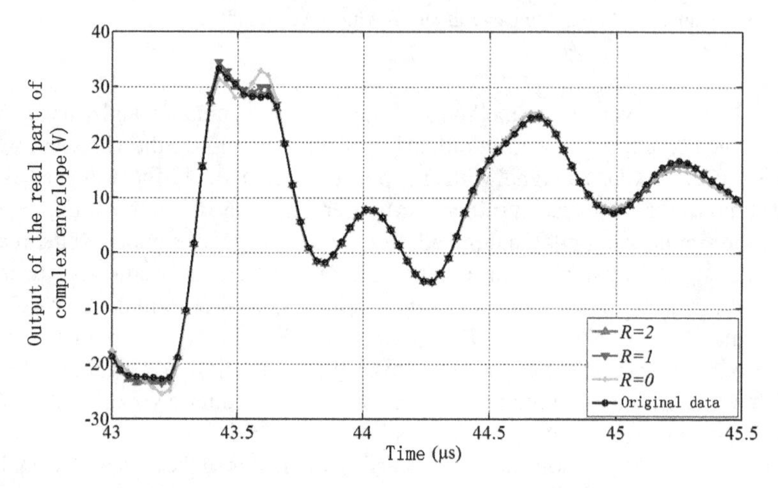

Fig. 5.50 Partial real part output waveforms of the time-domain complex envelope for behavioral models with different values of R

The memory effect of most power amplifiers gradually weakens as time goes by. At the same time, the influence from the input components with memory effect to the output gradually decreases as the order increases. Therefore, two control factors with large contribution to the output can be used to retain the input components, such that the number of parameters to be identified is further reduced. According to the data in Table 5.9, let $S = 1$, $R = 2$, and all Volterra kernels satisfying $S = 1$ or $R = 2$ are listed as parameters to be identified. The total number of identification parameters is 140 after calculation, and the corresponding NMSE is -40.2 dB. Figure 5.51 is the partial time-domain complex envelope waveform output from the power amplifier.

Table 5.9 Number of parameters to be identified and the corresponding NMSE under different control factor values

Control factor		The number of parameters to be identified	NMSE (dB)
S	0	15	−37.2
	1	71	−39.0
	2	203	−40.2
	3	407	−41.4
	4	605	−42.2
R	0	3	−30.3
	1	23	−37.5
	2	85	−39.5
	3	225	−40.8
	4	405	−41.7
	5	605	−42.2

When $S = 1$ and $R = 2$, the output signal obtained from the simplification of the behavioral model can fit the actual output waveform well. In this case, the use of two control factors can better simplify the Volterra series model. The actual parameters to be identified are reduced from 605 to 140. Most of the extraneous parameters are eliminated, which greatly reduces the complexity of the solution matrix. Although the memoryless Taylor series model has a relatively small number of parameters, the deviation from the test data curve is still large (this is because the Taylor series model does not contain the term describing the memory effect), which can be seen from the results of the real part. It can also be seen that the power amplifier circuit has obvious memory characteristics, such that some memory items need to be reserved.

The signal output spectrum is further discussed based on the solved model. The behavioral model of the mathematical expression can be easily embedded in commercial circuit simulation software to build the digital circuit behavioral model containing the behavioral model of power amplifier, as shown in Fig. 5.52. Compared with Fig. 5.51, we suggest that time-domain simulator (envelope simulator) be removed from the model to reduce the transient time-domain calculations for the analog circuits.

Figure 5.53 shows the output spectrum of the input signal after passing through the power amplifier circuit. From the figure, we see that the output of the simplified Volterra model is basically the same as the output spectrum of the original circuit model and the out-of-band signal spectrum is exactly what is needed for the analysis of the abnormal EMC signals; since the memoryless Taylor model does not consider the memory effect, the out-of-band output spectrum power value is significantly reduced and does not accurately describe the out-of-band power spectrum.

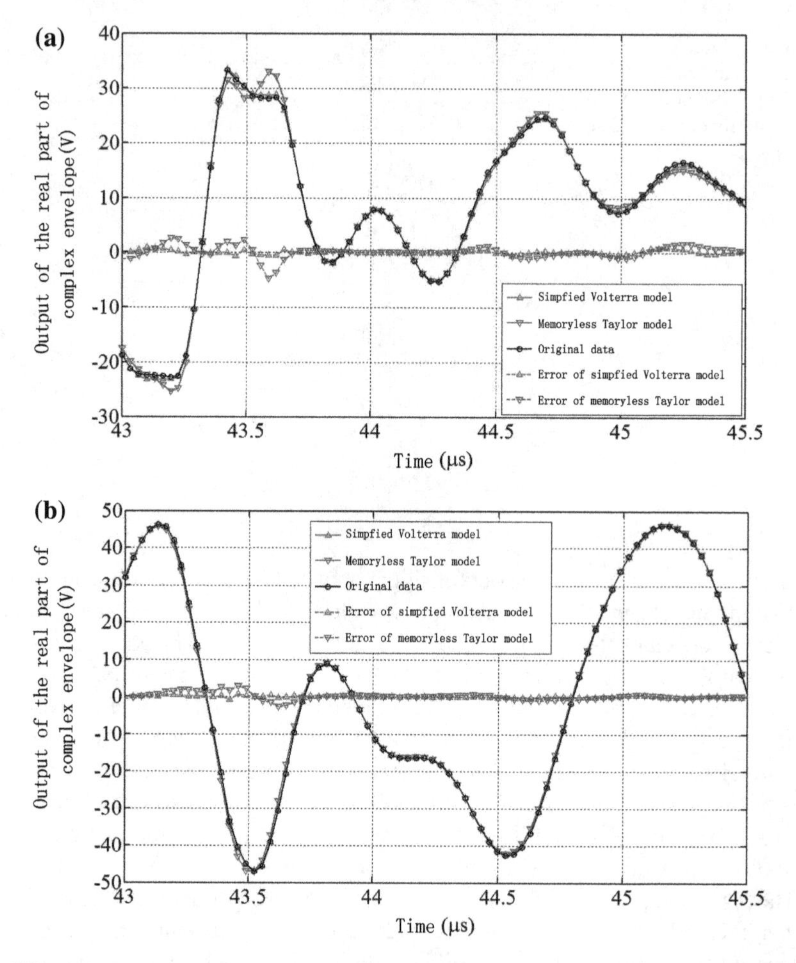

Fig. 5.51 Partial time-domain complex envelope output waveform for simplified behavioral model, **a** real part and **b** imaginary part

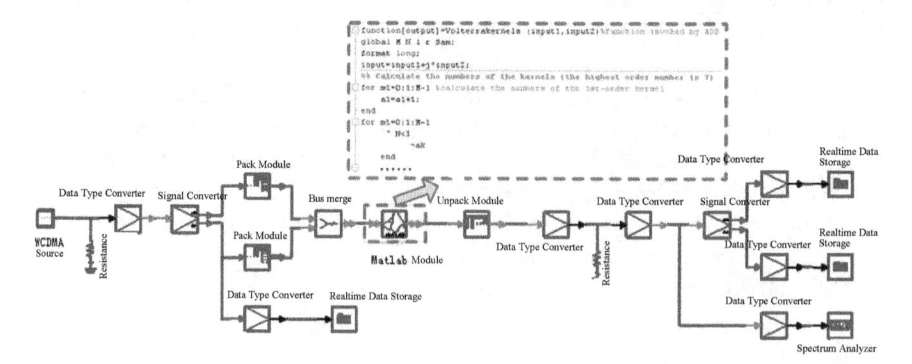

Fig. 5.52 Simulation of behavioral model including power amplifier

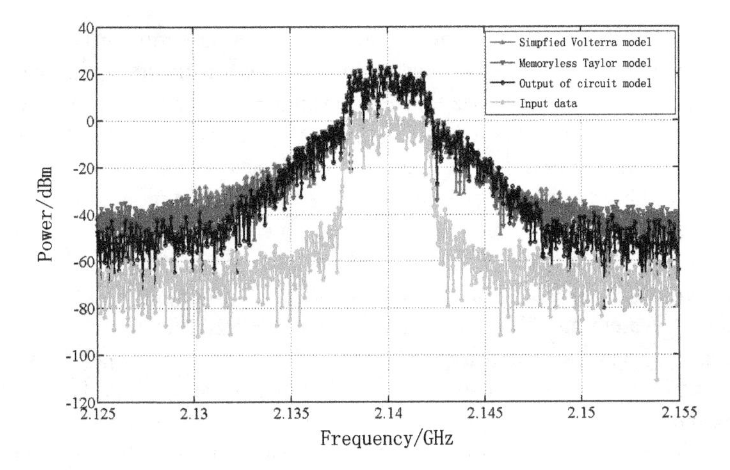

Fig. 5.53 Output spectrum of power amplifier behavioral model

5.3.3 EMC Modeling Method Based on Gray System Theory

Equipment's EMC analysis model is the basis for EMC behavioral modeling. It is often required to specify the external electrical characteristics of the equipment and detailed design parameters for the EMC analysis model. However, in many applications, the electrical design parameters of the equipment are difficult to obtain, and even the key parameters including power and operating frequency cannot be accurately obtained. It also happens that the EMC behavioral model cannot be established due to the complex structure of the equipment. In order to obtain the main parameters of the equipment in the system-level EMC analysis, we propose a method which extracts the EMC characteristics of the equipment, such as the harmonic interference characteristics and the noise spectrum of the power supply through testing, generalize regular expressions for use in design, and further build EMC analysis model of the equipment/module under the condition of incomplete parameter information. This method is called the EMC modeling based on gray system theory method.

The incomplete information of the electronic system is categorized into four types: incomplete design information, incomplete structure information, incomplete boundary information, and incomplete system operation behavior information (response characteristics).

A system with incomplete information is called a gray system, where "gray" means "incomplete information." In different scenarios and from different perspectives, the meaning of "gray" can also be extended. A gray system can be an intrinsic gray system or an extrinsic gray system. The basic characteristics of the intrinsic gray system are: There is no physical prototype, and there is no information to establish a definite relationship. The basic characteristics of the system are multiple parts that depend on each other, restrict each other, and are combined in a certain order. The intrinsic gray system has one or more kinds of functions. An extrinsic gray system refers to a physical system whose information is temporarily unknown or not yet obtained. The research object of the gray system theory is the uncertainty system with "insufficient information." The system has "part of information known and part of information unknown." With this theory, the real world is described and understood through generation and developing of the "partial" known information. According to the gray system classification, the gray system modeling method also comes in categories: One is the hybrid method, which uses the deductive method for the white part with known information and the inductive method for the black part with unknown information. This method is applicable to extrinsic gray system; the other is gray system modeling method applied to intrinsic gray system.

The gray system modeling method reveals the internal movement law by processing gray information. It uses the system information to quantify the abstract concept, build model for the quantified concept, and finally optimize the model. It not only tries to isomorphize the system model using output information, but also attaches great importance to correlation analysis, so as to make full use of the system information and transform the disordered data into an ordered sequence suitable for modeling of differential equations. The gray system modeling method adopts gray number processing represented by interval and interval operations, which is a simple and practical method.

The idea for gray modeling: Through gray generation or sequence operators, the randomness is weakened and the potential laws can be discovered; through the interchange of gray difference equations and gray differential equations, continuous dynamics differential equations can be constructed using discrete data sequences.

The basic ideas of gray system modeling are: (1) qualitative analysis, quantitative modeling, and close combination of qualitative and quantitative analysis; (2) clarification of system factors, relationship between factors, and the relationship between the factors and the system (the relationship between factors and the relationship between the factors and the system are not absolute); (3) understanding the basic performance of the system through modeling; (4) system diagnosis through the modeling to reveal potential problems; (5) obtaining as much information as possible from the model, especially the changing information; the commonly used data for building models includes scientific experimental data, empirical data, and production data; the sequence generation data is the basis for establishing a gray model; (6) constructing gray differential equation for the data which might follows implicit rules; (7) the model accuracy is set by gray numbers or gray sequences; (8) the accuracy of the model can be improved by different generation methods of gray data or gray sequence, data selection, sequence adjustment and correction, gray function form, and supplementation methods for different levels of residual gray model.

The gray system theory uses three methods to test and verify the accuracy of the model: ① residue size test, which is the point-by-point test of the error between the model value and the actual value; ② correlation test, which is a test by examining similarity between the model value curve and the modeling sequence curve; and ③ posterior difference test, which tests the statistical characteristics of the residual distribution.

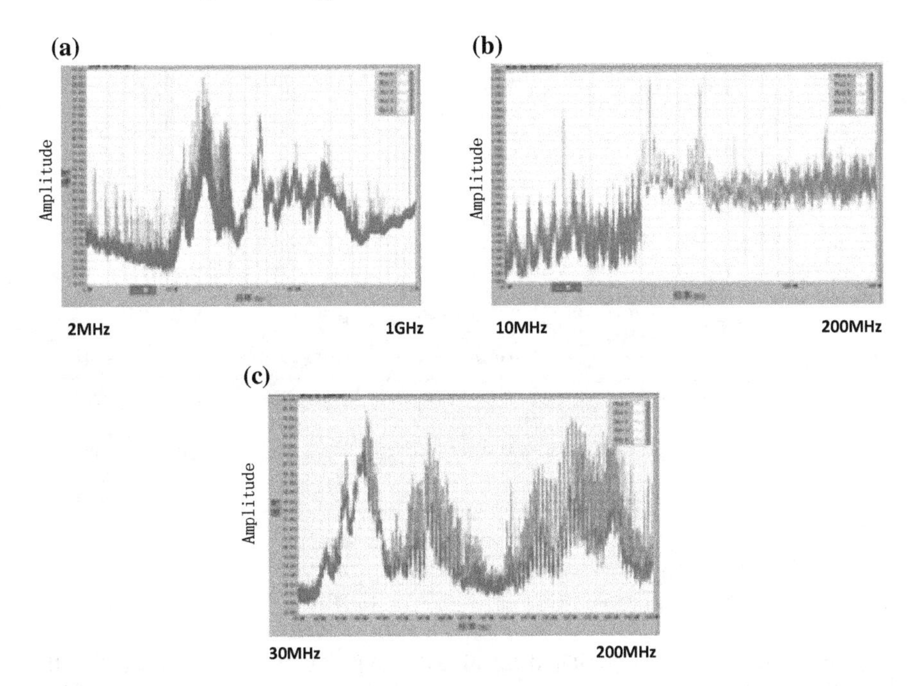

Fig. 5.54 Electromagnetic emission test data of three pieces of airborne equipment. **a** Equipment 1; **b** Equipment 2; and **c** Equipment 3

With the method of EMC modeling based on gray system theory, we can analyze the interference factors in the interference spectrum of the equipment, extract the harmonic interference and broadband interference information in the equipment, obtain the electromagnetic emission law of the equipment, and establish the EMC gray box model of the equipment based on the test data. The modeling accuracy is determined according to the least squares criterion.

One of the most important applications of the EMC modeling based on gray system theory is to investigate the potential interference sources and to design the corresponding EMC improvement schemes in the subsystem-level joint test and system-level precompliance test. Figure 5.54 is an example to illustrate the unique role of gray system method. The figure provides the electromagnetic emission test data for the three pieces of airborne equipment, respectively. As can be seen from Fig. 5.54, even for a single piece of onboard equipment, its electromagnetic emission data covers a wide range of frequencies which is usually hundreds of frequency octaves. When there are several pieces of such equipment in a system or a subsystem, the electromagnetic emission spectrum that they jointly produce will be more complicated. If a piece of equipment that works with these three pieces of equipment is susceptive, it is necessary to determine the cause of the susceptibility and design the interference suppression device.

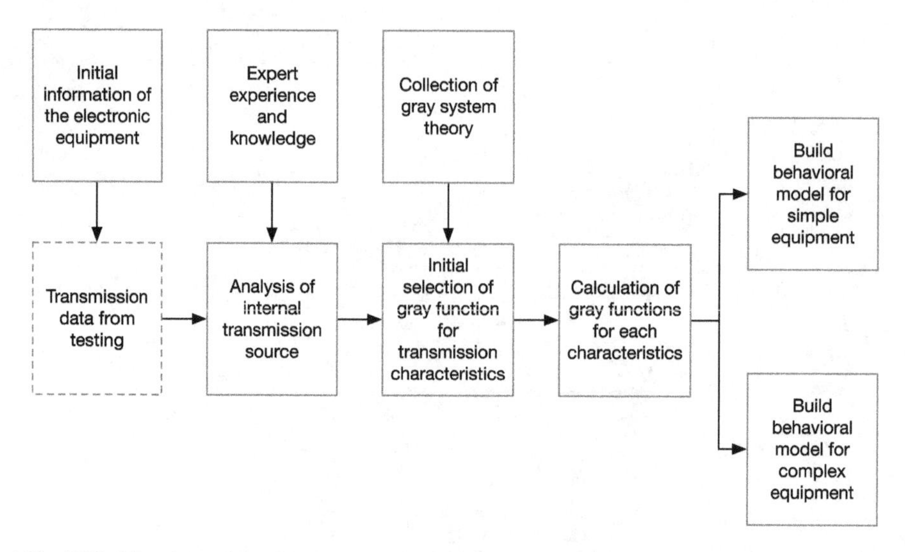

Fig. 5.55 Flowchart of the gray behavioral modeling and simulation of electronic system

In this situation, it is difficult to determine the main source of interference directly based on the electromagnetic emission test data. It is also difficult to design an effective interference suppression scheme (because the usual interference suppression program—filtering, shielding, etc.—is narrowband technology, while the frequency band of the emission spectrum as shown in Fig. 5.54 is very wide). For large systems (e.g., the system contains dozens of equipment), it is more difficult to identify the sources of interference and design effective suppression schemes.

The EMC modeling based on gray system theory method proposed in this book is very effective in solving this kind of problem. We will introduce the EMC modeling based on gray system theory method and provide examples of applications in the following paragraphs.

In general, electronic equipment includes units such as power supply, analog circuits, digital circuits, power amplifiers, and filtering. The communication system also includes nonlinear devices such as power amplifiers, mixers, and antenna matching networks. The external emission characteristics of the equipment are the combined emission characteristics of the various units above. In order to obtain the internal emission characteristics of the equipment from the external characteristics, it is necessary to test the emission characteristics of the equipment based on the internal potential interference module of the equipment. The test data is used to decompose and extract the external emission characteristics of each interference module to establish the behavioral model for different modules. Finally, we can build a behavioral model of the equipment based on the behavioral models of each module according to the equipment characteristics. The electronic system gray behavioral modeling simulation flow is shown in Fig. 5.55.

From the perspective of EMC, the radiation interference from most systems can be categorized into harmonic interference and broadband interference. According to the gray behavioral modeling and simulation method and the electrical characteristics of the system, we can decompose the emission spectrum characteristics of the known system according to the different spectral distribution characteristics and use the test data to deduce the gray function of the spectral distribution of various characteristics to establish a gray behavioral model of external emission characteristics of the entire electronic system. In order to fully explain how to extract the gray function from the test data and establish a gray behavioral model of electronic system, we divide the modeling method into two steps: The first step is to establish a gray behavioral model for the single emission law of electronic systems; the second step is to study the gray behavioral model of the complex electronic systems with multiple emission characteristics.

1. Establish a simple gray behavioral model for single emission interference characteristics of the electronic system

For a simple emission interference system, its emission test data contains a single regular emission spectrum characteristic. According to the gray system theory, a gray function can be used to quantify the spectral distribution, and then a mathematical function can be used to establish a behavioral model under a single characteristic of a simple emission system. Assume that according to the preliminary judgment of the characteristics of the electronic system, the test data $X = \{x_i(k)(k = 1, 2, \ldots, m, i = 1, 2, \ldots, n)\}$ is obtained, where m indicates the data length and n is the time of repeated test to avoid the influence of the environment on the test data. The test data can be abbreviated as $X = \{x_i(k)\}$. The method for establishing a gray behavioral model for a single emission law of an electronic system is as follows:

(1) Extract test data and construct a reference data sequence. In order to reduce the test error in the test data, data at the k-th point is used as the center to calculate the arithmetic average $x_0(k) = \frac{1}{n}\sum_{i=1}^{n} x_i(k)$, of the test data of the n group. $X = \{x_0(k)\}$ is used as the reference data sequence.
(2) Extract the gray sequence according to the maximum correlation. The gray correlation of the gray data sequence is defined as

$$r(x, x_0) = \frac{\xi \|d(x, x_0)\|}{|x - x_0| + \xi \|d(x, x_0)\|} \tag{5.160}$$

where $r(x, x_0)$ is the gray correlation between x and x_0. The larger the value of $r(x, x_0)$, the closer the correlation; $\|d(x, x_0)\|$ is the correlation distance between x and x_0; ζ is the resolution coefficient and $\xi \in (0, 1]$. We take $\xi = 0.5$.

n sequences can be substituted to calculate the gray correlation. The data sequence $\hat{X} = \left\{\hat{x}(k)(k = 1, 2, \ldots, m)\right\}$ with the greatest gray correlation with the reference data sequence can be extracted, namely gray sequence.
(3) Create a gray function. According to the regularity trend of gray data series, we can assume the corresponding gray function form. For example, for a series

that approximates an exponential trend, it can be assumed that the data satisfies the exponential equation. We can then calculate and build a gray function

$$\hat{x}^{(0)}(k) + a\hat{x}^{(1)}(k) = b \tag{5.161}$$

where a, b are parameters.

Thus, the gray model of the exponential series is

$$\frac{d\hat{x}}{dt} + a\hat{x} = b \tag{5.162}$$

For the common form of mathematics law, there are well-defined gray function forms and function parameter calculation methods, which generally meet the requirements of extracting the corresponding law functions and establish the gray model.

(4) Solve the parameters of the gray function. The gray function can be written as

$$Y = \phi\theta \tag{5.163}$$

where

$$Y = \begin{bmatrix} \hat{x}^{(0)}(2) \\ \hat{x}^{(0)}(3) \\ \vdots \\ \hat{x}^{(0)}(m) \end{bmatrix}, \phi = \begin{bmatrix} -\hat{x}^{(1)}(2) & 1 \\ -\hat{x}^{(1)}(3) & 1 \\ \vdots & \vdots \\ -\hat{x}^{(1)}(m) & 1 \end{bmatrix}, \theta = \begin{bmatrix} a \\ b \end{bmatrix}$$

Then, the least square estimation parameter of the gray function is

$$[a, b]^T = \left(\phi^T\phi\right)^{-1}\phi^T Y \tag{5.164}$$

(5) Create a gray expression. Substitute each parameter into a gray function, and the gray function expression of the test data is

$$\begin{aligned} \hat{x}(f) &= (\hat{x}(f_0) - b/a)e^{-a\frac{f}{f_0}} + a/b \\ &= Ce^{-Af} + B, f \in [f_{\min}, f\max] \end{aligned} \tag{5.165}$$

where the independent variable f represents the frequency of the test data; A, B, and C are

$$A = \tfrac{a}{f_0}, B = a/b, C = \hat{x}(f_0) - b/a \tag{5.166}$$

This way, a large amount of data is simplified into a simple mathematical expression that facilitates modeling and simulation.

(6) Create a gray behavioral model based on the expression. Substitute the gray expressions into the behavioral simulation system; thus, we can establish a gray behavioral model of a simple system with single law.

2. Establish a gray behavioral model of a complex emission system under multiple radiation conditions.

For complex electronic systems, there are many factors that affect the emission characteristics of the system. When electromagnetic interference analysis is performed, the original test data must include a variety of implicit laws. Various regular gray data sequences can be extracted from the original test data. According to the gray system theory, the regularity of one can be improved by transformation, and the influence of the other laws can be reduced. Thus, a regular gray sequence analysis formula can be obtained first, and a single characteristic behavioral model can be established. Then, we can identify the factors that make up the system and the correlation between the factors, and gradually obtain other regular gray sequences, so as to establish a complex system gray behavioral model containing multiple emission characteristics. The modeling method is as follows:

(1) Determine the gray characteristics of the system. As mentioned before, the type of externally emitted interference from the electronic system can be generally categorized into harmonics with periodic distribution of frequencies and broadband interference with continuous distribution; e.g., the electronic systems have different forms of power supply which have different emission characteristics. A common switching power supply exhibits harmonic characteristics externally. The AC power supply exhibits broadband characteristics externally. Different interference components have different frequency distributions, different amplitudes, and different influences. At the same time, different functional modules within the system will also show a variety of distribution forms; for example, the crystal oscillator and the switching power supply will exhibit harmonic interference characteristics, but the general harmonic frequency of the crystal is higher. Therefore, according to the existing EMC knowledge, we may analyze the possible functional modules and interference components in the system from the external output spectrum test data, infer the internal characteristics of the system from the external test data, and establish a behavioral simulation model.

Figure 5.56 shows an example of the decomposition of interference emitted by an electronic system. The gray function and gray model of each module are clearly defined according to the type of spectrum to establish a gray behavioral model of the entire electronic system. Suppose the system has g number of gray modules, gray characteristics can be described using the g gray model functions $f_i (i = 1, 2, \ldots, g)$ such that the number and type of models can be evaluated.

(2) Determine the functional form of each gray characteristic. In the system behavioral model, the exact function expressions of each gray module are given according to the harmonic model and the broadband model, and then the gray behavioral description function corresponding to each gray module is obtained. If the form of

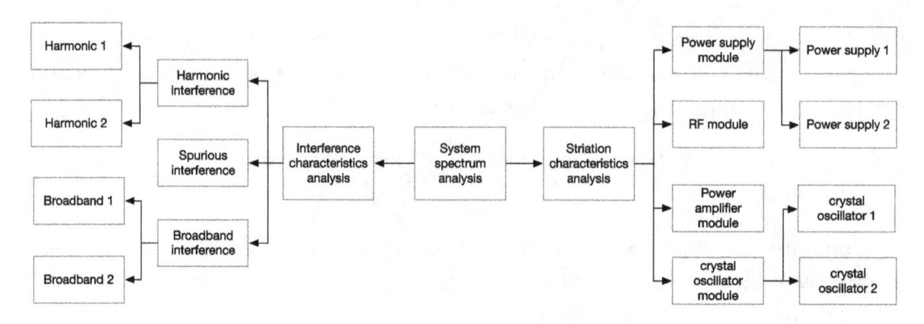

Fig. 5.56 Decomposition of the emission type of an electronic system

the module is relatively complex, we can use a combination of multiple piecewise functions to create the exact function of the behavioral model.

(3) Establish the gray behavioral model of each gray module according to the system emission characteristics. For the same gray module, the gray function has the same form, and the variable forms of the function should also be the same. Assuming that the i-th module has n functions and each function has g parameters, we can create a parameter matrix in the form of the same type of function in the module.

$$R_i = \begin{vmatrix} R_1 \\ R_2 \\ \vdots \\ R_n \end{vmatrix} = \begin{vmatrix} r_{i11} & r_{i12} & \cdots & r_{i1g} \\ r_{i21} & r_{i22} & \cdots & r_{i2g} \\ \vdots & \vdots & \vdots & \vdots \\ r_{in1} & r_{in2} & \cdots & r_{ing} \end{vmatrix} \tag{5.167}$$

(4) Calculate the error of each model and modify the models. Compare the gray sequence of each module with the output of the model, and ensure that the model's error is less than 3 dB when the output data converges; otherwise, adjust the gray function form or raw data interval and recalculate until the model's error is less than 3 dB. The error calculation formula is

$$e = \max\left|\hat{x}(k) - x(k)\right|, (k = 1, 2, \ldots, m) \tag{5.168}$$

(5) Establish a gray behavioral model of the entire complex electronic system. On the basis of the behavioral model of each module, according to the overall emission of the system and the cross-linking relationship between the modules, the gray behavioral model of the system is established. The inter-system EMC analysis focuses on the external emission characteristics of the system. Although the current distribution and transmission path within the system are not clear, a behavioral model in which the transmission output is consistent with the actual system can still be established on the simulation platform.

3. An example of gray behavioral modeling

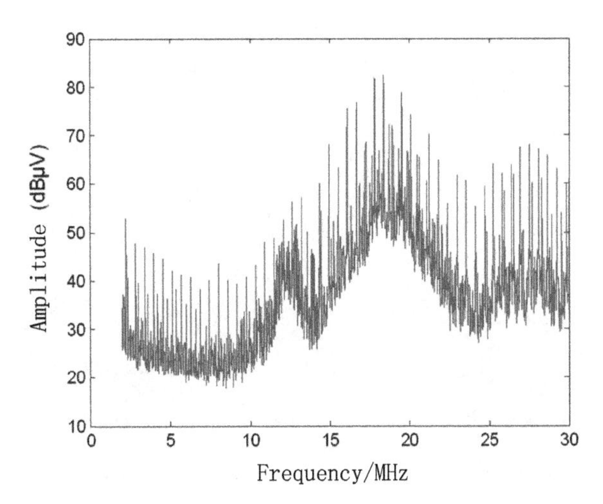

Fig. 5.57 Test spectrogram of an electronic system

We will use an example to describe the entire process of the gray behavioral modeling method of the system emission characteristics in detail. The tested external emission spectrum of an electronic system is shown in Fig. 5.57. The test spectrum is 2–30 MHz, and the frequency interval is 10 kHz.

Through the analysis of spectrum, there are multiple interference sources with different principles and functions in the system. It can be identified that the harmonic interference and broadband interference coexist. It can be initially determined that there are three interference sources in the system (two with harmonic characteristic and one with broadband characteristic), and the interference energy is mainly concentrated in the broadband interference. Assume that the original test sequence is $X = \{x(k)\}$, and the broadband gray sequence $X_W = \{x(k)\}$ (the subscript W denotes the broadband interference component) is extracted according to the principle of maximum correlation. The gray sequence of the first and second harmonics is $X_{H1} = \{x_{H1}(k)\}$ and $X_{H2} = \{x_{H2}(k)\}$.

Assuming that both the harmonic component and the broadband component satisfy the exponential function, then the broadband component and the harmonic component can be represented by a set of exponential mathematical models

$$\begin{cases} \dfrac{dx_{w1}}{dt} + a_{w1}x = b_{w1} \\ \dfrac{dx_{w2}}{dt} + a_{w2}x = b_{w2} \\ \qquad \qquad \cdots\cdots \\ \frac{dx_{wg}}{dt} + a_{wg}x = b_{wg} \end{cases} \tag{5.169}$$

where g is the number of exponential model in each interference component.

The mathematical model of the broadband interference is

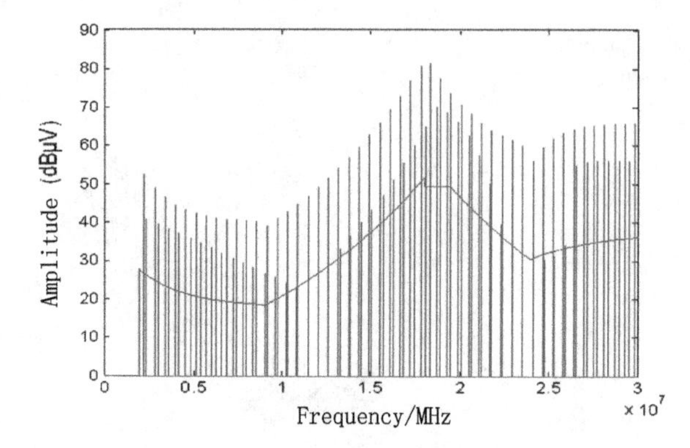

Fig. 5.58 Output of the gray system simulation model

$$\begin{cases} \hat{x}_{w1}(f) = C_1 e^{-A_1 f} + B_1,\ f_1 \in [2M, 15M) \\ \hat{x}_{w2}(f) = C_2 e^{-A_2 f} + B_2,\ f_2 \in [15M, 22M) \\ \hat{x}_{w3}(f) = C_3 e^{-A_3 f} + B_3,\ f_3 \in [22M, 30M] \end{cases} \tag{5.170}$$

After the decomposition of the emission characteristics of each interference module in the system, gray sequence of each component is extracted. Then, we can calculate the function parameters of each interference component and define the function of analytical expression. The parameter transposed matrix for the broadband interference is

$$\begin{bmatrix} 5.014 \times 10^{-5} & 1.11453 \times 10^{-4} & 5.98319 \times 10^{-4} \\ 4.5971 & -0.1092 & 0.7430 \\ -4.187 & 0.3624 & -0.0570 \end{bmatrix} \tag{5.171}$$

After getting the gray function of each interference component, the gray behavioral model of each module is established, and the behavioral simulation results of each module are compared with the actual characteristics to ensure that the EMC simulation output error is less than 3 dB. Figure 5.58 shows the gray system model simulation output of this electronic system. Based on the gray model of the above modules, the overall behavioral simulation model of the system is established, as shown in Fig. 5.59.

Due to the effective differentiation of each module, the EMC design method to reduce the external emission of the system is predicted using the behavioral simulation method. For different interference modules, the corresponding quantitative suppression requirements and improvement measures are clearly defined, which can effectively solve problems such as harmonic interference, broadband interference, and environmental noise interference. In practical applications, we can take specific measures: Add port filters for the two types of harmonic interference sources; add power filters that suppress broadband interference to the power supply.

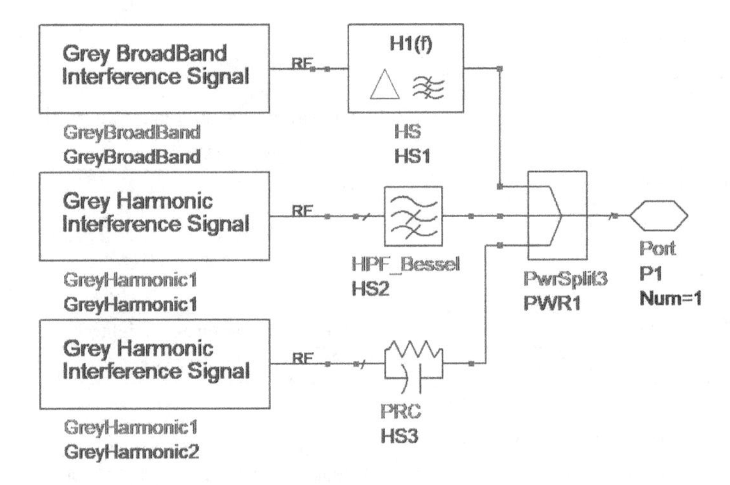

Fig. 5.59 Gray system behavioral model of the electronic system

Using the gray behavioral model, in the electronic system design stage, based on the precompliance test results, we can investigate the EMC problems existing in the equipment in a timely manner and formulate quantitative solutions, which effectively reduces the problem of excessive emission from the equipment in a later phase and reduces the interference between equipment. Thus, the product's EMC quality can be improved.

5.4 EMC Simulation Method

The process of EMC simulation method is proposed as follows:

(1) EMC concept description based on equipment characteristics, which include: ① Sort the main radiation sources and susceptive equipment in the system, and mark the type of the equipment (transmitter, receiver, transceiver); ② sort the energy transmission relationships from the radiation source to the susceptive point through field, field–circuit, and circuit, and construct the system interference correlation relationship; ③ list the factors affecting EMC in the interference correlation relationship; i.e., list the main EMC indicators for each equipment.
(2) Build the equipment's EMC behavioral model. The modeling process mainly includes: using a formal modeling language and various modeling tools to quantify the conceptual model and establish an executable behavioral model covering the functions of the system. Two methods are used to create the behavioral model: (1) behavioral modeling based on equipment characteristics and (2) behavioral modeling based on interference correlation matrix. The former adopts a graphical circuit language for simulation, and the latter uses an analytical mathematical expression for simulation.

Table 5.10 Antenna parameters

Antenna mode	Affiliated system and function
Polarization mode	Operation frequency and bandwidth
Input impedance	Standing wave ratio
Installation position	Having a radome or not
Transmitting power	Working principle
Receiver sensitivity	Dynamic range of receiver
Feed mode	Feed loss
Antenna efficiency	Equivalent aperture or height
Phase center	
Structure size and characteristic parameters	
Out-of-band characteristics of antenna	
Beam characteristics of the E surface	Beam characteristics of the H surface
Half power beam width	Half power beam width
Main lobe gain	Main lobe gain
Side lobe gain	Side lobe gain
Depth at zero point	Depth at zero point
Beam width of the first zero point	Beam width of the first zero point

(3) Conduct EMC evaluation and performance optimization so that the system meets the EMC requirements.

We illustrate the EMC simulation process in detail using several cases.

5.4.1 System-Level EMC Simulation Method for an Aircraft (Case 1)

When performing system-level EMC simulation, we often face problems of incomplete system/subsystem parameters or unknown operation mechanisms. Therefore, we need to collect equipment parameters, build basic models, and perform precompliance testing of design phase simultaneously.

1. Basic information collection

(1) Sort the system working principle and cross-linking relationship. Understand the use of the equipment, working principle, installation location, and input and output interfaces with other systems.
(2) Sort the cross-linked signal characteristics. Understand the input and output signal characteristics of the equipment and other systems, including: type, voltage, frequency, filtering mode of the power supply; operating mode, rotational speed of the motor; frequency, whether there is frequency division or frequency

Table 5.11 Receiver parameters

Communication equipment name	Communication equipment code
First LO frequency	First LO bandwidth
Second LO frequency	Second LO bandwidth
First IF frequency	First IF bandwidth
Second IF frequency	Second IF bandwidth
Working mode	
Selectivity (type, parameters specification, passband, and stopband of the filter)	
Gain and noise figure of the system amplifier	
Gain and noise figure of the system mixer	
Local oscillator crystal frequency	
Receiver sensitivity	
Demodulation method	
Isolation, out-of-band rejection, spurious, harmonics, etc.	

multiplication, frequency stabilization mode, out-of-band rejection, spurious, attenuation of each harmonic, etc., of the crystal oscillator; type, amplitude, waveform, bandwidth, modulation included or not, modulation characteristics, conduction mode, harmonic characteristics of the analog signal; waveform, amplitude, pulse width, frequency, bandwidth, duty cycle, repetition frequency of the digital signal; type, model, impedance, and other performance parameters of the transmission cable; type, location, and performance parameters of connectors.

(3) Analyze EMI threats. Understand the EMI threats of the equipment, its past and potential failures, and test items that are difficult to pass.

(4) Sort the conventional EMI prevention measures. Understand the EMC design measures taken and their effects, including: cable layout; layout of PCB board and board devices, circuits, routing; grounding and grounding resistance (grounding position and mode of the analog ground, digital ground, power ground, shielding ground, safety ground, etc.); filtering (including threats, filtering measures, expected results, and test results); shielding (including threats, shielding measures taken, expected results, and test results); lap joints (including lap handling and lap joint resistance); antenna parameter table (Table 5.10), receiver parameter table (Table 5.11), and transmitter parameter table (Table 5.12).

2. Basic model library

Establishing a behavioral simulation model library of commonly used airborne equipment is very useful for EMC simulation. The EMC models of common airborne equipment are shown in Fig. 5.60.

Table 5.12 Transmitter parameters

Communication equipment name	Communication equipment code
First LO frequency	First LO bandwidth
Second LO frequency	Second LO bandwidth
First IF frequency	First IF bandwidth
Second IF frequency	Second IF bandwidth
Working mode, transmitting power, dynamic gain	
Modulation method	
Frequency points and the suppression of harmonics	
Filter (types, parameters, passband, and stopband)	
Gain and noise figure of the system amplifier	
Gain and noise figure of the system mixer	
The suppression of out-of-band, spurious, harmonics, etc.	

The equipment model library is mainly based on the design principle diagrams, working principle diagrams, and EMC test data of the subsystems and equipment. The equipment model is a functional description of the object. The model library parameter values can be filled and modified according to the test results of a different equipment. In order to ensure the accuracy and validity of the model, and to ensure that the model effectively includes all of the EMC failure factors, it is necessary to verify the model iteratively based on the test data.

3. Design precompliance test

In order to determine the EMC requirements and provide basis for EMC design, an overall design EMC test is required, including: analysis and test of electromagnetic environment, simulation test of equipment, subsystem and antenna optimal layout, and test for overall interference control.

(a) Shortwave radio;
(b) Microwave landing system;
(c) Ultrashort wave radio;
(d) Fuel gauge;
(e) TACAN;
(f) Airborne velocity radar;
(g) Airborne audio communication system;
(h) DSSS–QPSK communication system;
(i) Airborne GSM mobile communication system.

4. Conceptual description of EMC based on equipment characteristics

By establishing a digital aircraft model, it is possible to analyze the interference relationships among the subsystems at the whole aircraft level, including the circuit–circuit, circuit–field, and field-level interference/susceptive interference relationships of the interference/susceptive equipment. Then, we can build analysis model for subsystems, create the design signal flow and the interference signal flow, and predict

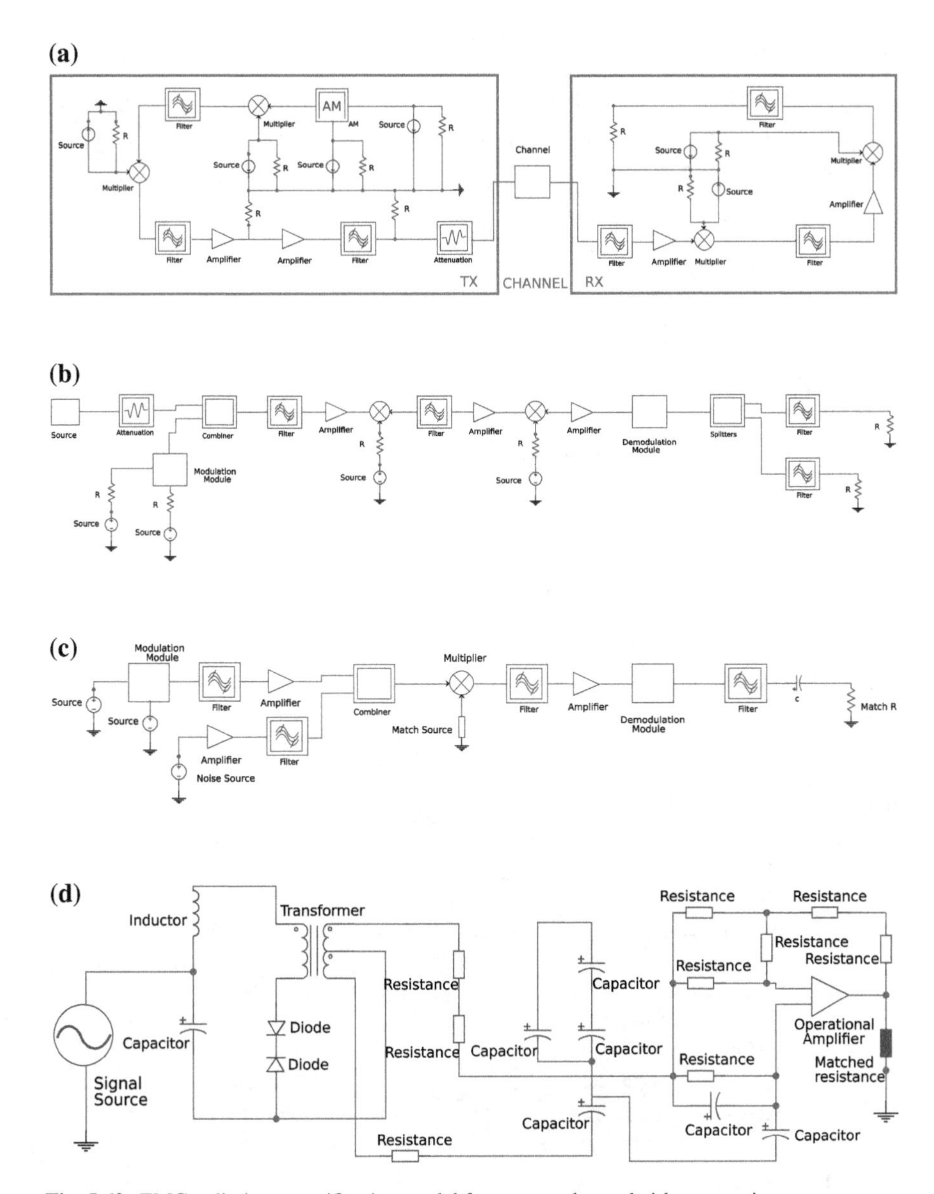

Fig. 5.60 EMC radiation quantification model for commonly used airborne equipment

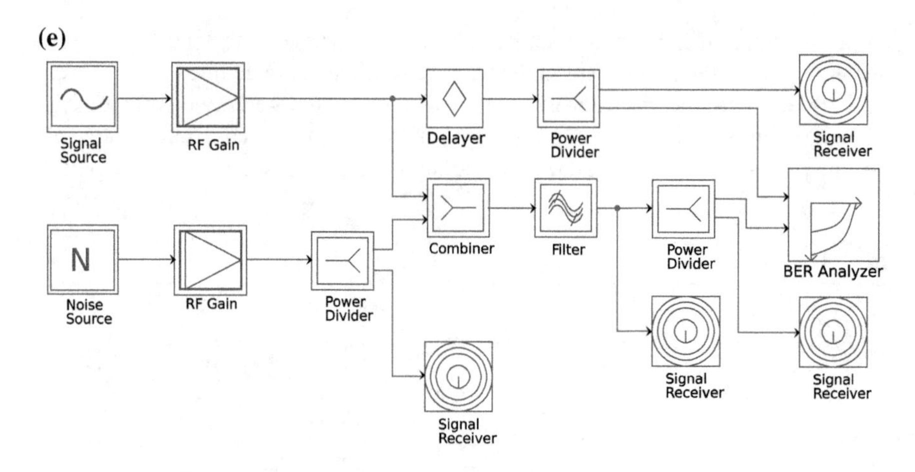

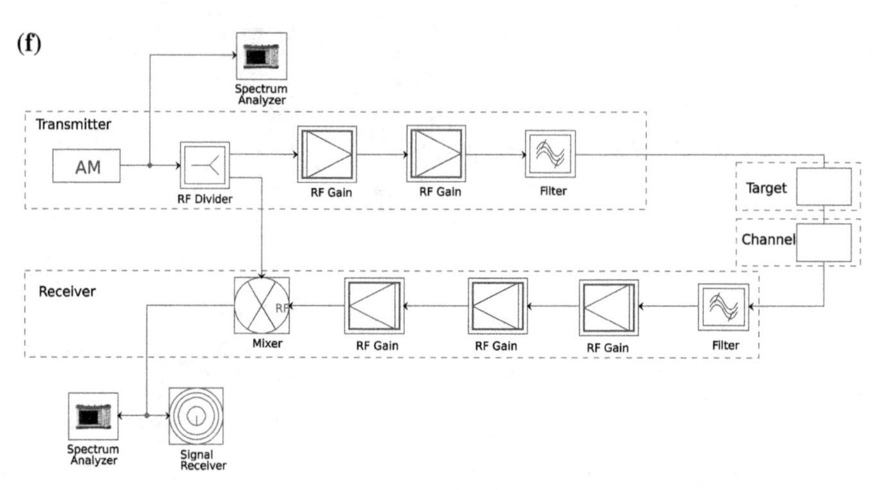

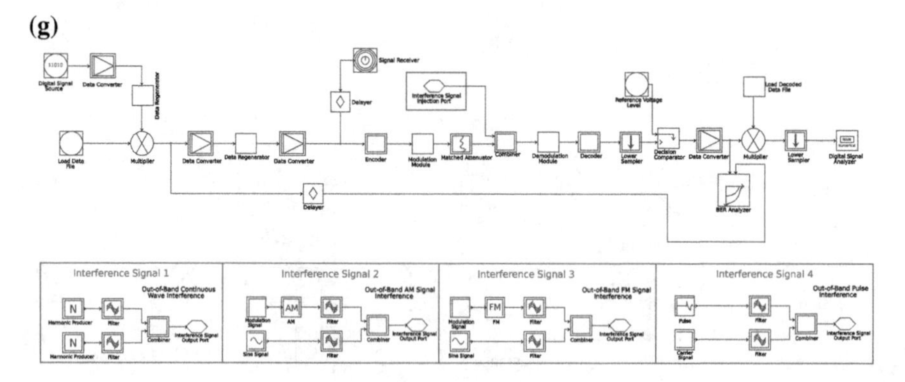

Fig. 5.60 (continued)

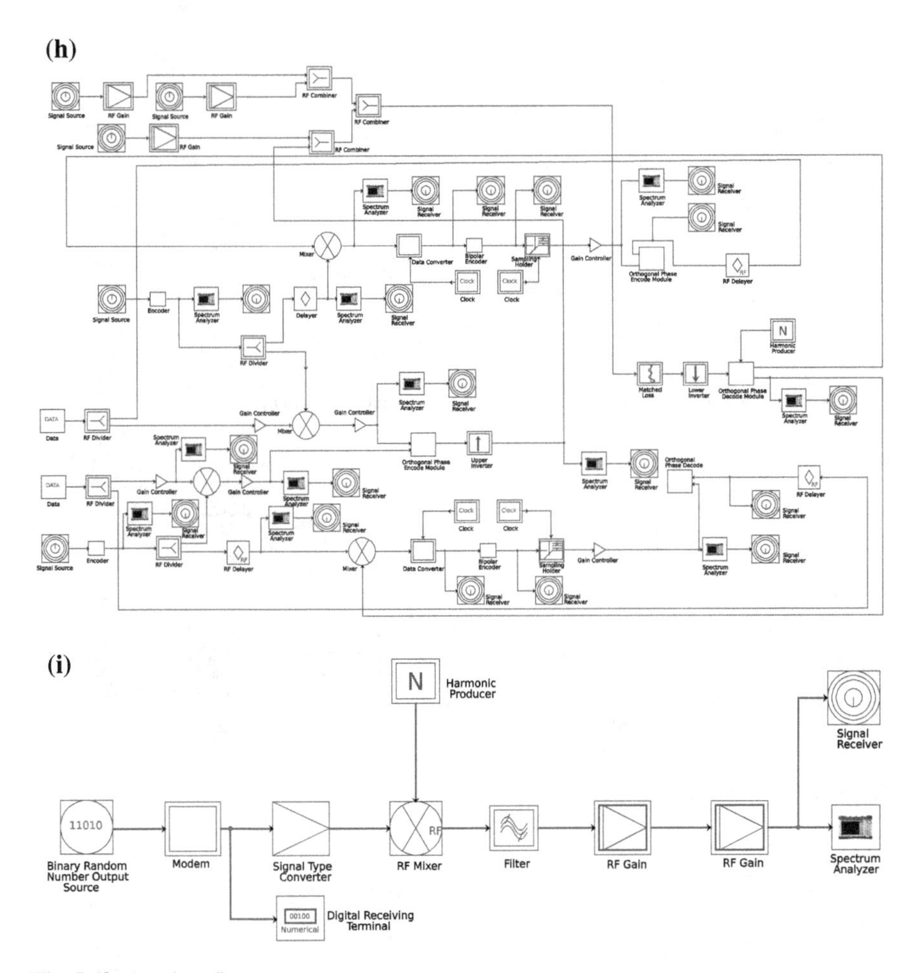

Fig. 5.60 (continued)

the radiation emission quantitatively to determine the interference the equipment suffers from by using the time-domain, frequency-domain, hybrid field, and field—circuit coupling coordination methods. Next, we can build a behavioral simulation model to analyze the factors affecting the EMC design and decompose the EMC indicators, such as shielding performance of transmitters and susceptive equipment, cable layout, distribution of electromagnetic environment, resonance characteristics of the cabin, equipment layout, tolerable degradation, equipment safety priorities, and safety margins of susceptive equipment. This way, we can complete the quantitative collaborative EMC design for the aircraft. Figure 5.61 shows the interference relationship of the electronic equipment of an aircraft, including over ten airborne equipment including shortwave radios, ultrashort wave radios, servo systems, fire control systems, integrated navigation, fuel gauges, and radio compasses.

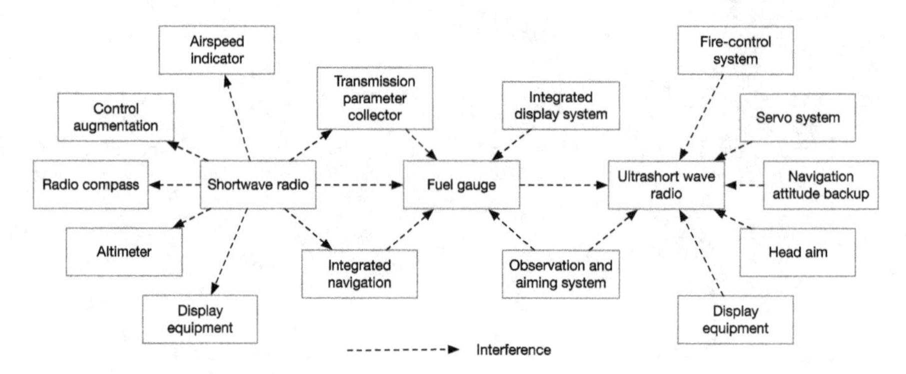

Fig. 5.61 Interference relationship of electronic equipment of an aircraft

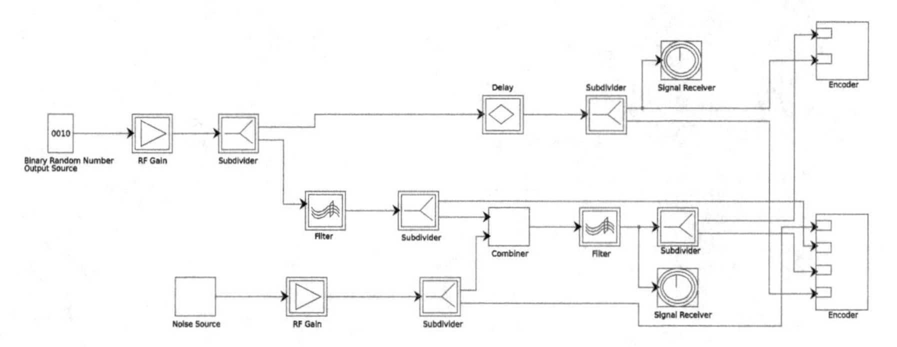

Fig. 5.62 EMC digital model of an aircraft

Figure 5.62 shows the EMC digital model of an aircraft.

5. EMC simulation analysis of equipment

The equipment model can be modified, and the equipment's EMC can be simulated according to the basic model and characteristics of the equipment. For example, the harmonics and broadband noise generated by the fundamental frequency of a shortwave radio can interfere with the highly sensitive receiving equipment (such as ultrashort wave radios) on board. For any frequency in the broadband, the final-stage transistor load of the shortwave radio power amplifier may not be in the best matching state and may work in the saturation region or the cutoff region, thereby causing a large nonlinear distortion and resulting in a series of harmonics and broadband noise. In order to formulate radio design requirement that meets the requirement of CE106, behavioral simulation methods are used to analyze equipment performance and reduce the harmonic rejection ratios. This way, the design requirements can be proposed.

The behavioral model of shortwave radio emission is shown in Fig. 5.63. The characteristics of the output spectrum of the shortwave radio in the modeling and simulation are shown in Fig. 5.64.

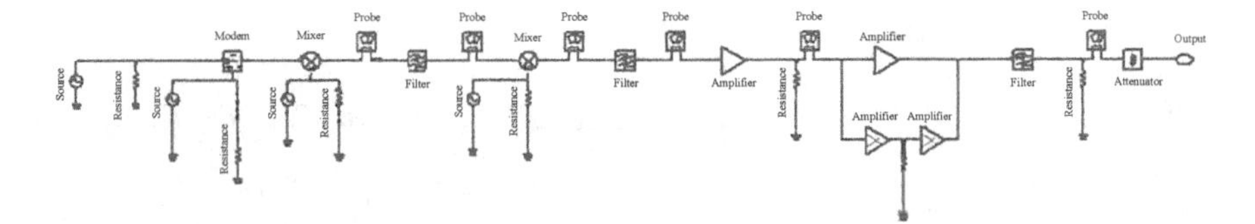

Fig. 5.63 Shortwave radio behavioral emission model

6. EMI simulation analysis between subsystems

Shortwave radios and ultrashort wave radios use adjacent frequencies, and out-of-band harmonics and broadband spurious interferences generated by shortwave radios may cause interference to ultrashort wave radios. The following section analyzes the interference of shortwave radios to ultrashort wave radios.

Based on the behavioral models of shortwave radios and ultrashort wave radios, a detailed model is established for single-frequency noise interference and broadband noise interference, and the interference characteristics of the two radios under various modes are analyzed. Ultrashort wave radio in the system uses two antennas and has two operating modes, AM and FM as shown in Fig. 5.65.

According to the above analysis, the network models that need to be established in the simulation analysis include shortwave models (including nonlinear power amplifier modules), ultrashort wave radio susceptibility models (AM and FM thresholds), and two antenna coupling modules. The specific simulation analysis content is shown in Table 5.13.

Table 5.14 provides the results of "interference on the ultrashort wave FM receiving system from the harmonic emission of the shortwave radio through ultrashort wave antenna 1" to determine if the design requirements are met based on whether the susceptibility threshold is exceeded.

5.4.2 Out-of-Band Nonlinear Interference Simulation Method for Transmitting and Receiving Systems (Case 2)

The traditional development of transmitting and receiving systems only takes into account the in-band parameters such as power gain, pattern, impedance, and radiation efficiency. Each piece of equipment is provided by a different vendor and has its own design specifications. It is rare if there is any systematic analysis on the influence other than the functional specifications, such as the out-of-band spurious and out-of-band susceptibility. Many problems have been found in the implementation of EMC engineering in recent years, and engineers have tried to systematically study the out-of-band characteristics of the equipment. Generally, multiple transmitters and receivers are distributed on the same platform. Although each transmitter or receiver

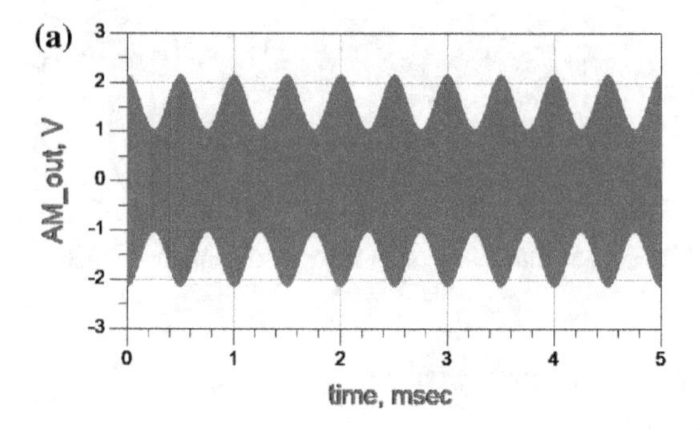

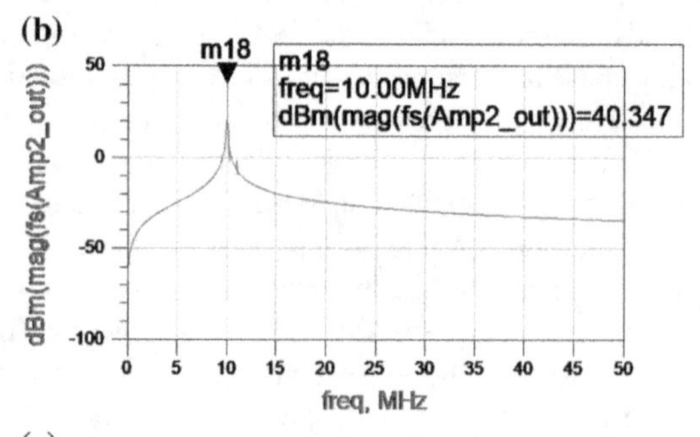

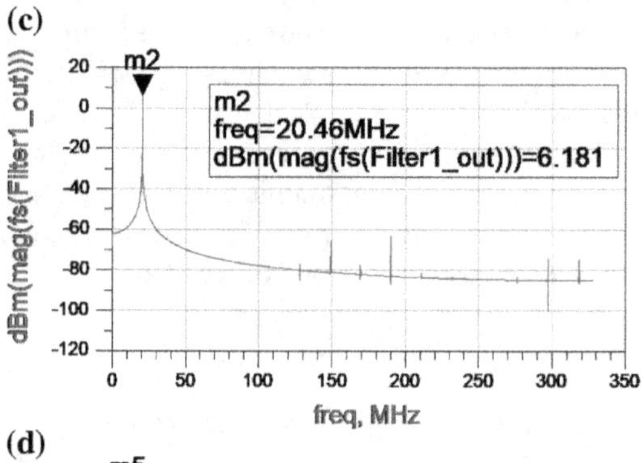

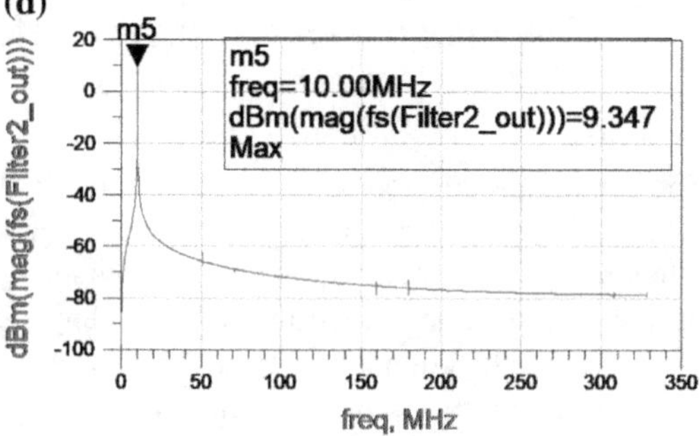

Fig. 5.64 Analysis of output spectrum characteristics of shortwave radio behavioral modeling and simulation, **a** time domain output of the ideal model; **b** frequency domain output of the ideal model; **c** the first IF output; **d** the second IF output; and **e** nonlinear output through the nonlinear amplifier

Table 5.13 List of interference simulation decomposition between subsystems

Interference equipment	Interfering mode	Coupling path	Susceptive equipment	Susceptive mode	Content for simulation analysis
Shortwave radio	Harmonic transmission	Ultrashort wave antenna 1	Ultrashort wave radio	AM susceptive threshold	Interference on the ultrashort wave AM receiving system from the harmonic emission of the shortwave radio through the ultrashort wave antenna 1
				FM susceptive threshold	Interference on the ultrashort wave FM receiving system from the harmonic emission of the shortwave radio through the ultrashort wave antenna 1
		Ultrashort wave antenna 2		AM susceptive threshold	Interference on the ultrashort wave AM receiving system from the harmonic emission of the shortwave radio through the ultrashort wave antenna 2
				FM susceptive threshold	Interference on the ultrashort wave FM receiving system from the harmonic emission of the shortwave radio through the ultrashort wave antenna 2
	Broadband spurious emission	Ultrashort wave antenna 1		AM susceptive threshold	Interference on the ultrashort wave AM receiving system from the broadband spurious emission of the shortwave radio through the ultrashort wave antenna 1
				FM susceptive threshold	Interference on the ultrashort wave FM receiving system from the broadband spurious emission of the shortwave radio through the ultrashort wave antenna 1
		Ultrashort wave antenna 2		AM susceptive threshold	Interference on the ultrashort wave AM receiving system from the broadband spurious emission of the shortwave radio through the ultrashort wave antenna 2
				FM susceptive threshold	Interference on the ultrashort wave FM receiving system from the broadband spurious emission of the shortwave radio through ultrashort wave antenna 2

Table 5.14 Harmonic interference to ultrashort wave radio (125 MHz) from shortwave radio (25 MHz)

Shortwave harmonic emission frequency (MHz)	Harmonics strength (dBm)	Cable loss (dB)	Isolation (dB)	Noise of the input end of the ultrashort wave (dBm)	Receiving out-of-band suppression (dB)	Interference threshold (dBm)	Exceeding the threshold (dB)	Interference or not
f_0	50.9	3	−58.6	−10.7	80	−113	23.3	Yes
$2f_0$	−11.2	3	−35.7	−49.9	80	−113	−16.9	No
$3f_0$	−14.5	3	−46.8	−64.6	80	−113	−31.6	No
$4f_0$	−20.6	3	−48.7	−72.3	80	−113	−38.3	No
$5f_0$	−21.5	3	−50.3	−74.8	0	−113	38.2	Yes

Fig. 5.64 (continued)

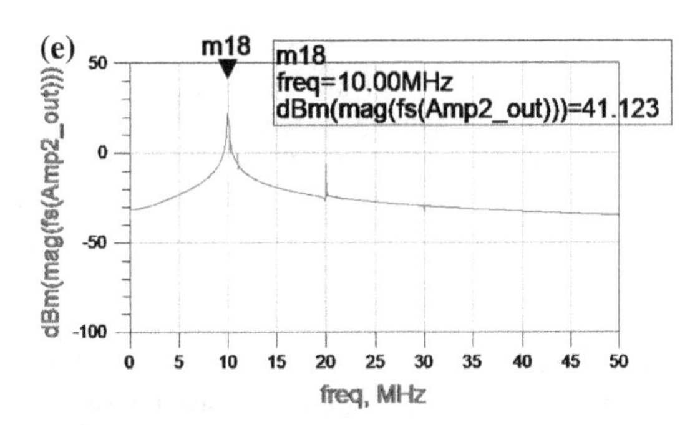

can meet the requirements of the EMC specification, the strong out-of-band signals of some devices in a limited space may still fall within the band of other devices. The out-of-band radiation characteristics may cause mutual interference between transmitters and receivers. In particular, some transmitters have large transmitting power and weak antenna directivity, and the omnidirectional power is easily received by other receivers on the same platform. Therefore, the out-of-band nonlinear interference characteristics have an important influence on the EMC of the entire platform.

In the simulation and analysis model of system-level EMC, in addition to the analysis of the functional signals, other signals with potential interference also need to be analyzed using the behavioral simulation methods. Therefore, it is necessary to establish a behavioral simulation model of the transmitter and the receiver, and fully consider the out-of-band performance of the system. Additionally, we need to focus on the sub-circuit modules with large influence, such as amplifiers and mixers, because they work in the nonlinear regions and their spurious products are likely to cause serious interference.

According to the active nonlinear out-of-band interference characteristics of the transmitters and receivers, the behavioral simulation method is used to establish the behavioral model of the transmitting and the receiving equipment. The spectral distribution of the transmitting equipment and the frequency response characteristics of the receiving equipment are analyzed to investigate the effects of nonlinear inter-

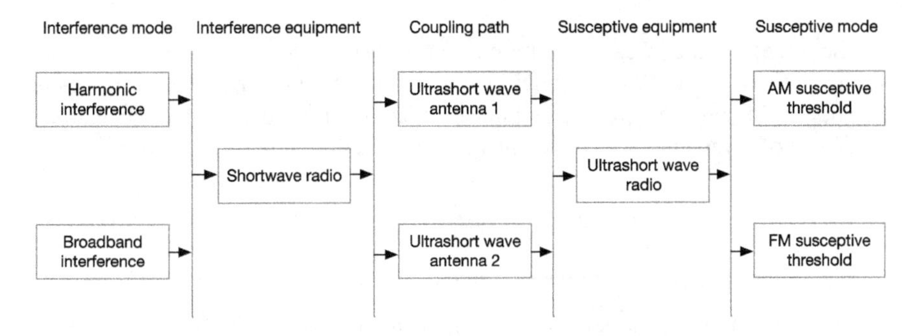

Fig. 5.65 Behavioral modeling, simulation, and analysis of interference between subsystems

ference signals from the transmitting system on other receiving equipment on the same platform. At the same time, we propose a method for the allocation of the third-order intermodulation distortion parameters with a large influence on system modeling and provide a modified cascaded model of the third-order intermodulation distortion parameters. Finally, we discuss a practical analog predistortion circuit that suppresses the third-order intermodulation distortion.

1. Out-of-band nonlinear interference of transmitters and receivers

An actual RF system needs nonlinear and active components including crystal diodes and bipolar transistors. It can be used for signal detection, mixing, amplification, frequency multiplication, switching, and so on. Nonlinear interference signals are also the result of signal modulation of amplifiers or mixers in nonlinear regions. It is a system-level interference phenomenon which usually occurs when the signal groups are large and the signals are easily coupled. In the process of EMC analysis, it is often found that each device in the system satisfies its own requirements, but problems still occur when integrated into the platform. Under certain circumstances, the radiation intensity of the transmitting equipment is still relatively strong and the energy of out-of-band radiation is likely to cause interference to susceptive equipment within a short distance. Additionally, the susceptive equipment itself may also cause unnecessary response due to the nonlinear characteristics.

(1) EMC interference prediction principle

The occurrence of any EMI follows the three aspects of EMI, and the interference prediction problem can be described by a general mathematical model. The interference model can be used to describe any interfering equipment with radiation emission characteristics in the system. Assuming the interference power output by the interference source is $P_T(t, f)$, the influence of the propagation path on the interference signal is represented by the transfer function by $T(t, f, r, \theta)$ (where t is time, f is frequency, r is distance, and θ is the propagation direction). Then, the effective power of the interference generated by the interference source at the susceptive equipment is $P_I(t, f, r, \theta)$, which can be expressed in decibels as [57]

$$P_I(t, f, r, \theta) = P_T(t, f) - T(t, f, r, \theta) \tag{5.172}$$

The susceptive model can be represented as any equipment with radiation susceptibility in the system. $P_S(t, f)$ describes the radiation susceptive characteristics of the equipment coupled with external interference. In order to quantitatively express the degree of compatibility and incompatibility, the interference margin is defined as

$$M(t, f) = P_I(t, f, r, \theta) - P_S(t, f) \tag{5.173}$$

The performance of susceptive equipment can be evaluated by $M(t, f)$:

(1) If $M(t, f) > 0$, the susceptive equipment will be interfered. The value of $M(t, f)$ indicates the intensity of the interference.

(2) If $M(t, f) = 0$, the susceptive equipment is in a critical state of interference; that is, the equipment may be subject to interference, and the safety margin is zero.
(3) If $-6 < M(t, f) < 0$, the susceptive equipment will not be interfered, but it is likely to be interfered.
(4) If $M(t, f) < -6$, the susceptive equipment will not be interfered and can work safely and stably.

In practical applications, when the susceptive equipment is affected by n interference sources, (5.173) can be generalized to

$$M(t, f) = \sum_{i=1}^{n} P_{I_i}(t, f, r, \theta) - P_S(t, f) \tag{5.174}$$

Equation (5.174) is usually called the interference prediction equation, which is universal.

(2) Model of radiation characteristics of the transmitting equipment

In EMC analysis and design, besides the spectrum power distribution within the band of the transmitting equipment, we also need to understand the spectral power distribution of the out-of-band spurious and harmonics. Since the output powers of a different transmitting equipment are different, circuits with the same function in different devices will also be different. In this case, the spectrum allocation rule is generally described by a statistical method. Ideally, the spectral power distribution of the transmitting equipment can be expressed as [58]

$$P_t(f) = \begin{cases} \bar{P}_B + a\delta_B, & f_L \le f \le f_H \\ \bar{P}_N, & f < f_L, f > f_H \end{cases} \tag{5.175}$$

where the parameters have the same explanation as in Eq. (5.71).

The expressions of the statistical mean of the fundamental radiation power and the standard deviation are

$$\bar{P}_B = \frac{1}{m} \sum_{i=1}^{m} P_i \tag{5.176}$$

$$\delta_B = \sqrt{\frac{1}{m-1} \sum_{i=1}^{m} (\bar{P}_B - P_i)^2} \tag{5.177}$$

where P_i is the measured value of a single transmitter's output power of the fundamental wave; m is the sampling number of the transmitter.

However, in engineering practice, due to the nonlinear effects of signal sources, power amplifiers, filters, and other electronic components in the transmitting equipment, the output power of the transmitting equipment is not completely limited to a certain frequency or within a narrow band. Harmonic and parasitic radiation components also exist. The general mathematical expression of the harmonic radiation of the transmitting equipment is

$$\overline{P_h}(N_h f_0) = \bar{P}_B + A \lg N_h + B \tag{5.178}$$

where f_B is the fundamental frequency; N_h is the number of harmonics; A and B are the harmonic suppression constant of the transmitting equipment.

Similarly, the average power of parasitic radiation is

$$\overline{P_p}(f) = \bar{P}_B + A' \lg\left|f / f_B\right| + B' \tag{5.179}$$

The radiation signal of a common transmission equipment is a signal modulated by different types, and the energy is mainly distributed in a frequency band near the fundamental wave. Therefore, the transmission power model is

$$P(f) = \bar{P}_{\max} + W(f) \tag{5.180}$$

where $\bar{P}_{\max}$ is the statistical mean of the maximum radiation power of the transmitting equipment, $W(f)$ represents the relative value (dBc) of a certain spectrum component and the maximum transmitting power of the fundamental wave, and its expression is

$$W(f) = W_i + N_i \lg(\frac{|f - f_B|}{\Delta f_i}), \quad (f \in \left[\Delta f_{i-1}, \Delta f_i\right]; i \in [1, n]) \tag{5.181}$$

where the parameters have the same explanation as in Eq. (5.74).

Similarly, for the harmonics and spurious components of other carrier frequencies, the modulation envelope can be taken into account in a similar manner of Eq. (5.181).

The model built based on priori data is generally more universal, but in the actual analysis process, due to the lack of data, it is not easy to derive an accurate radiation characteristics model for the emission equipment. Therefore, we need to use the behavioral simulation method of the circuit to build a behavioral simulation model of the transmitting equipment. For developed equipment, we can also obtain accurate radiation characteristics through testing.

Figure 5.66 is a principle diagram of the RF end of a typical second-order superheterodyne transmitter. After the modulation signal and the carrier signal enter the modulation module, they go through the first-stage filtering and then enter the first mixer for mixing. After filtering and amplification, the modulation signal with higher frequency is obtained. Then, the signal enters the second mixer for frequency conversion. Next, the signal is filtered and amplified again. Finally, the signal is amplified by the power amplifier to obtain the radio frequency (RF) signal output.

The power of each frequency point calculated by the model can be expressed by the following expression (in decibels):

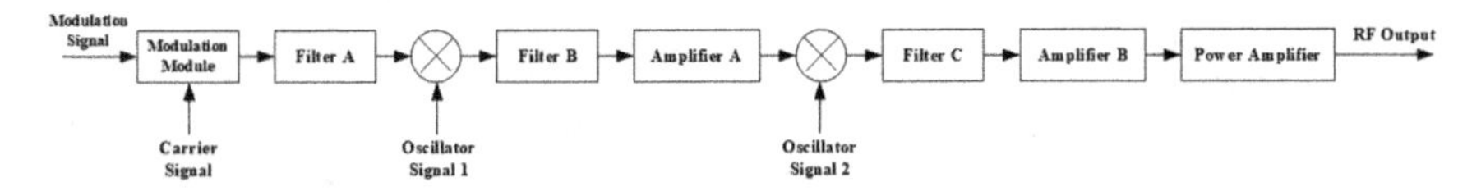

Fig. 5.66 Principle diagram of the RF end of a typical second-order superheterodyne transmitter

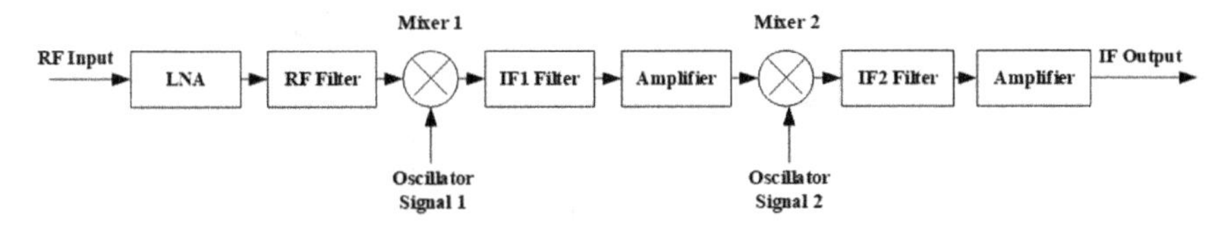

Fig. 5.67 Principle diagram of the RF side of a typical second-order superheterodyne receiving equipment

$$P_t(f) = P_{in\_sig}(f) + G_t(f) + L_t(f) \tag{5.182}$$

where $P_{in\_sig}(f)$ is the spectral power distribution of the input signal for the transmitting equipment; $G_t(f)$ is the total power gain of the amplifier and the mixer, which takes into account the nonlinear factor; $L_t(f)$ is the total attenuation of the transmission power of the filter.

(3) Mode of receiving equipment response

At present, the sensitivity of the receiving equipment is getting higher, and the requirement of anti-interference ability of the external electromagnetic energy is becoming higher. When performing system EMC analysis and design, we need to focus on the frequency selection characteristics of the receiving equipment and the prediction of potential out-of-band signal. A block diagram of the RF end of a typical second-order superheterodyne receiver is shown in Fig. 5.67. Radio frequency (RF) signal enters the low-noise amplifier (LNA), passes the first-stage filtering, then enters the first mixer to be filtered and amplified, and then enters the second mixer for filtering and amplification, and finally the IF signal output is obtained.

Similarly, a generic model has also been formed for the receiving equipment based on a large amount of a priori data. The frequency selectivity of the receiving equipment near the operation bandwidth or in the main receiving channel can also be represented by a piecewise straight-line approximation [58].

$$S(\Delta f) = S(\Delta f_i) + S_i \lg(\Delta f/\Delta f_i) \tag{5.183}$$

where the parameters have the same explanation as in Eq. (5.78).

Signals located within the frequency points and frequency range outside the main receiving channel can also enter the receiver due to the system's nonlinear characteristics, affecting the in-band of the receiving system. These frequencies or frequency bands are called additional receiving channels, which mainly include IF receiving

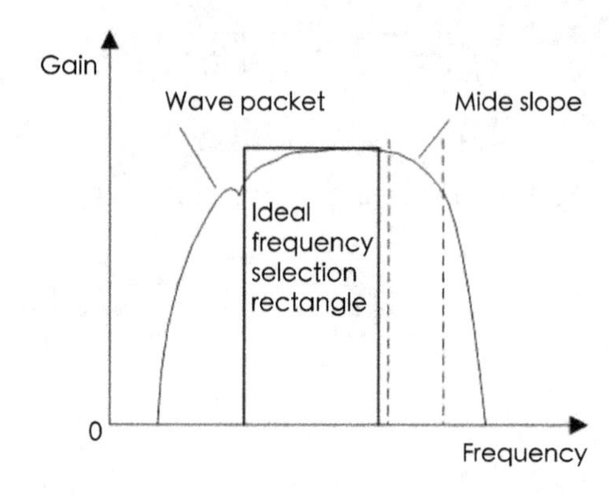

Fig. 5.68 Frequency selectivity of the superheterodyne receiving system

channel, image receiving channel, and harmonic receiving channel. The selective model of the receiving equipment in the main additional receiving channel is

$$S(f_a) = I\,\lg(f_a/\,f_{cen}) + J \tag{5.184}$$

where the parameters have the same explanation as in Eq. (5.79).

For an ideal receiver, the frequency selectivity should exhibit a rectangular characteristic. In reality, however, the selectivity of the receiving equipment is very different from the rectangular characteristics. In addition, due to the lack of sample data, it is difficult to use (5.184) to create an accurate frequency selective model of the receiving equipment. The full-band response characteristics of the receiving equipment can be analyzed using the behavioral simulation method of the circuit. For existing equipment, we can obtain more accurate response characteristics based on testing. The gain of the input RF signal after the entire RF system is

$$G_r(f) = L_{f1}(f) + L_{f2}(f) + L_{f3}(f) + \cdots + L_{fn}(f) + G_{ramp}(f)(\text{dB}) \tag{5.185}$$

where $G_r(f)$ is the gain of the entire system; L_{fn} $(n = 1, 2, \cdots)$ is the attenuation value of the filter at all levels (negative in decibels); G_{ramp} is the total gain of other modules such as the amplifiers and the mixers.

When the attenuation of $|L_{fn}(f)|$ is small in some frequency bands, $G_r(f)$ will increase accordingly, and the fluctuant "wave packets" or the declining gradients in certain frequency bands are slower, as shown in Fig. 5.68. This phenomenon is particularly prominent in broadband receivers, because the interference signals of relevant frequencies are more likely to access receiving systems in complex electromagnetic environments.

Then, we set up a behavioral simulation model of the receiving equipment. The calculated power at each frequency point can be expressed as

$$P_r(f) = P_{RF}(f) + G_r(f) + L_r(f) \tag{5.186}$$

where $P_{RF}(f)$ is the spectral power distribution of the weak signal received by the receiving equipment; $G_r(f)$ is the total power gain of the amplifier and the mixer of the receiving equipment, of which the nonlinear factors need to be considered; $L_r(f)$ is the total attenuation of the transmission power of the filter.

(4) Calculation of out-of-band nonlinear interference between equipment

For transmitting and receiving equipment, the product development regulations have corresponding requirements for their out-of-band characteristics. For example, CS103 specifies the requirements for intermodulation-conducted susceptibility of the antenna terminals, and the equipment cannot have intermodulation products that exceed the tolerance. CE106 specifies that the antenna terminal-conducted emission should not exceed the following values: All harmonic emissions and spurious emissions except the second and third harmonics of the transmitter (transmission state) should be at least 80 dB lower than the fundamental level; the second and third harmonics should be suppressed $50 + 10\lg P$ (where P is the fundamental peak output power (W)) or 80 dB, whichever is less.

As long as the electronic equipment satisfies the national EMC regulations, it can be regarded as qualified. However, the method of tailoring national regulations is not necessarily effective, because when multiple pieces of electronic equipment are integrated into cascade, EMC problem will also appear. Even if the equipment satisfies the emission requirements, when it is positioned on the platform of a complex multi-set communication system, EMC problems may still occur. It is also easy to overlook the radiation interference caused by the nonlinearity outside the communication frequency band of the transmitting equipment. In general, only after the signal interference phenomenon outside the communication band of the transmitting equipment happened, will the engineers start to position the interference source through testing. This approach not only consumes a lot of human and material resources, but may also shorten the life of the equipment due to the large number of tests.

The power of the transmitting equipment is very large. Considering the power consumption, in order to ensure the efficiency of the transmitter output power, the power amplifier in the transmitting equipment is likely to work in the nonlinear area or even the saturation area, which will bring unnecessary harmonics and intermodulation products. Transmitting equipment and receiving equipment usually adopt the superheterodyne-type structure, the shortcoming of which is the production of the unnecessary spurious response during mixing. Therefore, the radiation interference outside the communication band caused by the nonlinearity cannot be ignored. The high-order harmonics and intermodulation products generated by the transmitting equipment are easily received by other highly susceptive receivers, affecting the normal reception of functional signals. The nonlinearity of the amplifier is usually described by gain compression and intermodulation distortion. The nonlinearity of the mixer is described by frequency conversion compression and intermodulation distortion. The nonlinear characteristic parameters have been discussed previously, including the output power $P_{1\text{dB}}$ of the 1 dB gain compression point, the input power P_{IIP3}, and the output power P_{OIP3} of the TOI point.

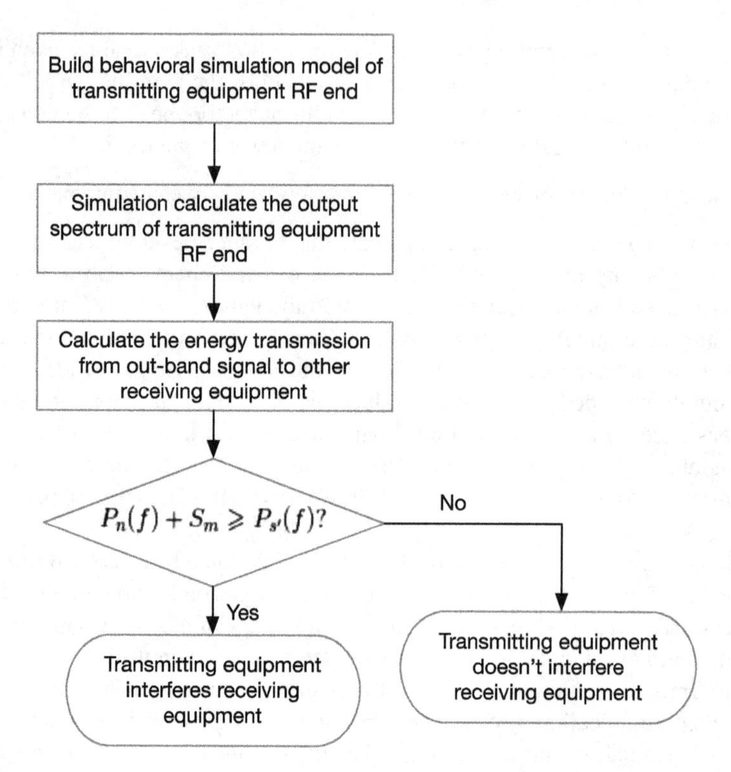

Fig. 5.69 Flowchart of out-of-band nonlinear interference prediction between communication systems

Out-of-band communication signals are the combination of several frequencies in a nonlinear circuit. The combined frequency can be expressed as $mf_1+nf_2+lf_3+\cdots$, where m, n, l, ... are coefficients, and their values are 0,1,2,.... When $m+n+l+\cdots=N$, the N-th order signal is obtained. When there is only a single m or n equal to 2, 3, or larger value, the obtained signal will be second, third, or higher harmonics. After the signal passes through the mixer and amplifier, many harmonic signals and intermodulation signals appear. Especially in the mixing process, the power of the local oscillator required for the operation of the mixer is usually at least 10 dB greater than the power of the signal being processed; as a result, the local oscillator signal may have an energy re-injection effect on the generated out-of-band interference signal.

Large-scale electronic information platforms generally contain multiple sets of communication systems. High-power transmitters and high-sensitivity receivers are installed in a limited space which results in dense electromagnetic signals. For the communication systems on the same platform, EMI problems are easy to expose due to the close placement of the systems. System simulation can be used for prediction analysis from the interference transmitting system to coupled path and then to susceptive system, as shown in Fig. 5.69.

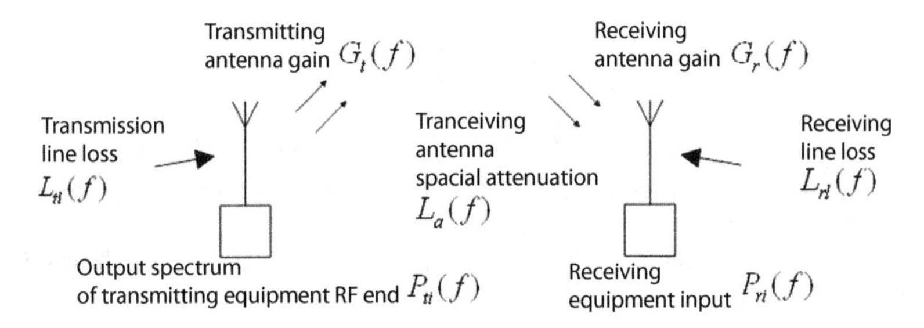

Fig. 5.70 Diagram of the energy transmission between transmitting and receiving equipment

First, we can build the behavioral model of the radio frequency of the transmitting system using circuit behavioral simulation. Considering the nonlinear characteristics of the amplifier and mixer module, we then calculate the out-of-band output spectrum of the transmitting system. The output spectrum includes various spurious components such as harmonics, intermodulation, and frequency conversion, some of which may fall within the communication band of other receiving systems. Then, we can filter out the interference frequency points that may fall into the communication band of other receiving systems and perform energy transmission calculation.

As shown in Fig. 5.70, the energy transmission relationship between the transmitting system and the susceptive system is

$$P_{ri}(f_{in}) = P_{ti}(f_{in})L_{tl}(f_{in})G_t(f_{in})L_a(f_{in})G_r(f_{in})L_{rl}(f_{in}) \tag{5.187}$$

Equation (5.187) can be written in the form of dB as

$$P_{ri}(f_{in}) = P_{ti}(f_{in}) + L_{tl}(f_{in}) + G_t(f_{in}) + L_a(f_{in}) + G_r(f_{in}) + L_{rl}(f_{in}) \tag{5.188}$$

where f_{in} is the operating frequency that can fall within the passband of the receiving system; $P_{ri}(f_{in})$ is the power coupled from the transmitting system to the susceptive system at a certain frequency in the receiving band; $P_{ti}(f_{in})$ is the out-of-band output signal power at the RF end of the transmitting system; $G_t(f_{in})$ is the antenna power gain; $L_a(f_{in})$ is the spatial attenuation; $G_r(f_{in})$ is the receiver antenna power gain; $L_{tl}(f_{in})$ is the transmitting line loss; $L_{rl}(f_{in})$ is the receiving line loss;

In general, transmitting line loss $L_{tl}(f_{in})$ and receiving line loss $L_{rl}(f_{in})$ can be obtained by referring to the cable's technical manual or through testing. The transmitting antenna power gain $G_t(f_{in})$ and receiving antenna power gain $G_r(f_{in})$ can be obtained through simulation software analysis or through testing. The spatial attenuation $L_a(f_{in})$ can be calculated in the following two approaches:

(1) When the receiving antenna is located in the far-field area of the transmitting antenna

$$r > \frac{2D^2}{\lambda} \tag{5.189}$$

where D is the maximum size of the transmitting antenna, r is the distance between the transmitting antenna and the receiving antenna, and λ is the operating wavelength of the transmitted signal.

The spatial attenuation $L_a(f_{in})$ (dB) of the transmitted signal between transmitting equipment and receiving equipment can be calculated by the following formula:

$$L_a = 10\lg\frac{\lambda^2}{(4\pi)^2r^2} = 10\lg\frac{c^2}{(4\pi)^2r^2f^2} \tag{5.190}$$

where f is the operating frequency (MHz) and c is the transmission speed of the transmitted signal ($3 \times 10^8 m/s$).

Substituting π and c in (5.190), we get

$$L_a = -27.56 + 20\lg f + 20\lg r \tag{5.191}$$

(2) When the distance r between the receiving antenna and the transmitting antenna satisfies

$$r \leq \frac{2D^2}{\lambda} \tag{5.192}$$

The spatial attenuation between the antennas can be obtained by testing or using existing calculation software. Generally, according to the actual antenna position, we can build the simulation model of the transmitting and the receiving antennas using electromagnetic field simulation software. Based on the model, we can calculate the spatial attenuation of the transmitted signal when it arrives at the susceptive equipment.

In different communication platforms, the diffraction attenuation, coupling shielding, and polarization mismatch between the transmitting and the receiving antennas vary greatly. Neglecting the above factors, we can calculate the radiation interference signal of the transmitting equipment falling into the passband of the receiving equipment of another communication system without considering the suppression of the received signal by the receiving equipment. Thus, the attenuation of the interference signal will not be affected by the above factors and reaches minimal; then, we can analyze the worst-case interference and predict more potential interfering frequencies. In general, the distance between the transmitting and the receiving antennas is short on the same platform, and the receiving antenna is usually located in the near-field area of the transmitting antenna. The spatial attenuation $L_a(f)$ of the signal between the transmitting equipment and the receiving equipment is calculated using field simulation software. In certain cases, $L_a(f)$ can also be obtained by testing at the same time.

The determinant condition of the transmitting system interfering the receiving system is

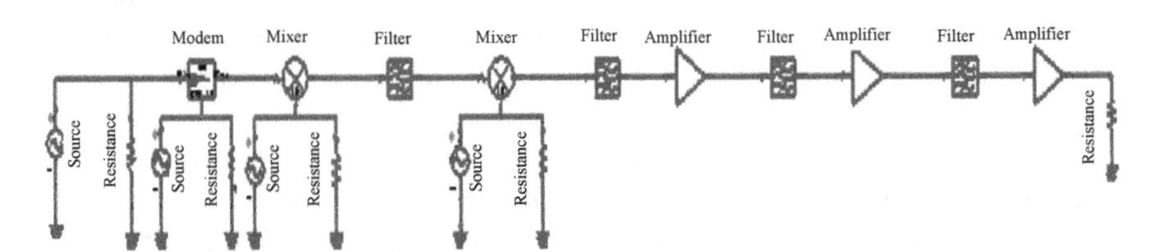

Fig. 5.71 RF end behavioral simulation model for the shortwave radio transmitting system

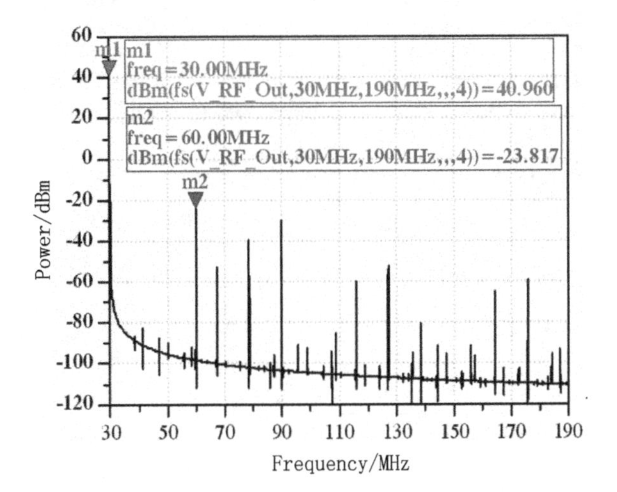

Fig. 5.72 Out-of-band output spectrum of the RF end of the transmitting system

$$P_{ri}(f) + S_m \geq P_{s'}(f) \tag{5.193}$$

where $P_{s'}(f)$ is the sensitivity of the receiving system; S_m is the safety margin of the receiver (S_m is 6 dB in national regulation).

Now, we will analyze the interference of one shortwave radio (as a receiving system) from another ultrashort wave radio (as a transmitting system) in a real platform. The shortwave radio is a second-order superheterodyne structure as shown in Fig. 5.66. The modulated frequency is 512 kHz, the first local oscillator signal frequency is 12.466 MHz, and the second local oscillator signal frequency is 42.978 MHz. Thus, the output frequency is 30 MHz and the transmitting power is 12.5 W (40.96 dBm). Considering the nonlinear parameters of each module in the system, the simulation model of the RF end of the transmitting system is shown in Fig. 5.71.

The calculated spectrum range of the output is 30–190 MHz. Figure 5.72 shows the out-of-band output spectrum of the shortwave radio transmitter. The intensity of the signal at $f = 60$ MHz is -23.817 dBm.

The operating frequency band of another ultrashort wave radio is 30–88 MHz, and the voice receiving sensitivity is not higher than 1.5 μV in the AM mode and 0.5 μV in the FM mode. We can use the signal source to simulate the voice signal, and transmit the FM signal to the RF end of the receiving system. Then, it can be found that when there is apparent whistle sound in the ultrashort wave radio headphone, the susceptibility at 60 MHz is −85 dBm.

The out-of-band radiation at 60 MHz for the shortwave emission system is selected from Fig. 5.72. In a communication platform in a limited space, the receiving antenna is usually located in the near-field area of the transmitting antenna. $L_a(f)$ is provided by testing, and the value is 65.4 dB. In addition, through the technical manual, we get that $L_{ti}(f)$ equals 3 dB, $L_{ri}(f)$ equals 3 dB, $G_t(f)$ equals 4 dB, and $G_r(f)$ equals 3 dB.

It can be seen from Eq. (5.186) that $P_{ri} = -88.217$ dBm. According to (5.193), $P_{ri}(f)+S_m = -82.217$ dBm (more than $P_{s'} = -85$ dBm). Therefore, we can conclude that on multiple communication platforms in a narrow space, ultrashort wave radios are also subject to radiation interference caused by the nonlinearity of the shortwave radio's communication frequency band.

2. Research on intermodulation interference of transmitting and receiving systems

Among the many intermodulation parasitic products generated due to nonlinearity, the most typical and influential one is the third-order intermodulation (IM3) interference signal. The third-order intermodulation distortion signal is a powerful and difficult-to-filter-out parasitic product. When the receiver's input signal contains two closely spaced frequencies ω_1 and ω_2, its third-order intermodulation distortion signal frequency $\Delta\omega$ ($\Delta\omega = 2\omega_1 - \omega_2$) will fall into the receiving passband. Since $\Delta\omega$ is very close to ω_1 and ω_2, the receiving of the desired signal will be interfered. This section analyzes the third-order intermodulation distortion parameters of each module in the receiving system EMC modeling and derives the method to calculate the amplitude of the third-order intermodulation output signal at all levels based on the distortion signal voltage in the worst case of the cascade circuit. Then, we also propose an allocation method of third-order intermodulation distortion parameters for each module of the receiving equipment using the constrained relationship of third-order intermodulation signal strength. Finally, we discuss an analog predistortion circuit suppressing the third-order intermodulation signal for later improvement of the product.

(1) The TOI point cascade equation in the model of the transmitting and receiving system

Transmitters and receivers generally consist of antennas, RF systems, and modulation or demodulation systems. The most popular superheterodyne RF systems include sub-circuits with nonlinear characteristics including amplifiers and mixers. Section 5.4.2 has discussed the characteristics of the amplifiers and the mixers operating in the nonlinear region and the parameters that describe the nonlinear character-

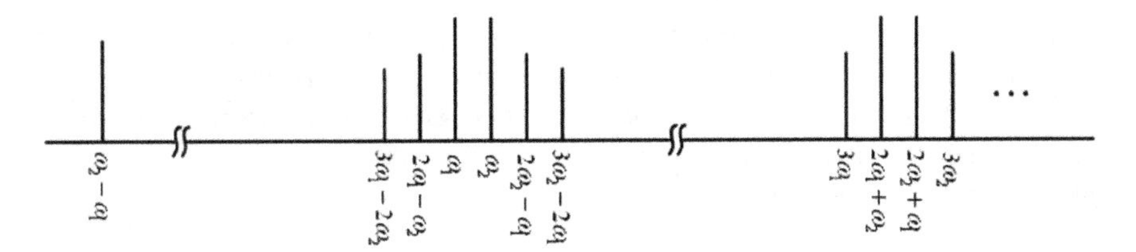

Fig. 5.73 Spectrum component after a dual-tone signal enters the nonlinear equipment

istics. The following discussion focuses on the third-order intermodulation distortion parameters (TOI point). The TOI point fully describes the third-order intermodulation distortion degree of the nonlinear amplifier and the mixer.

Taking two continuous wave signals as input, the frequency of the first signal is ω_1 and the frequency of the second signal is ω_2. The input voltage is v_i, and then there is

$$v_i = V_0(\cos\omega_1 t + \cos\omega_2 t) \tag{5.194}$$

where V_0 is the amplitude of the voltage of the continuous wave signal and t is time. Here, the Taylor series expansion is used to describe the nonlinear device and the strength of the third-order intermodulation signal is derived, i.e.,

$$v_o = a_0 + a_1 v_i + a_2 v_i^2 + a_3 v_i^3 + \cdots \tag{5.195}$$

where a_0, a_1, a_2, and a_3 are weighting factors, respectively.

Substituting (5.194) into (5.195), we have

$$\begin{aligned} v_o = {} & a_0 + a_1 V_0 \cos\omega_1 t + a_1 V_0 \cos\omega_2 t + \frac{1}{2} a_2 V_0^2 (1 + \cos 2\omega_1 t) + \frac{1}{2} a_2 V_0^2 (1 + \cos 2\omega_2 t) \\ & + a_2 V_0^2 \cos(\omega_1 - \omega_2)t + a_2 V_0^2 \cos(\omega_1 + \omega_2)t \\ & + a_3 V_0^3 \left(\frac{3}{4} \cos\omega_1 t + \frac{1}{4} \cos 3\omega_1 t \right) + a_3 V_0^3 \left(\frac{3}{4} \cos\omega_2 t + \frac{1}{4} \cos 3\omega_2 t \right) \\ & + a_3 V_0^3 \left[\frac{3}{2} \cos\omega_2 t + \frac{3}{4} \cos(2\omega_1 - \omega_2)t + \frac{3}{4} \cos(2\omega_1 + \omega_2)t \right] \\ & + a_3 V_0^3 \left[\frac{3}{2} \cos\omega_1 t + \frac{3}{4} \cos(2\omega_2 - \omega_1)t + \frac{3}{4} \cos(2\omega_2 + \omega_1)t \right] + \cdots \end{aligned} \tag{5.196}$$

It can be seen that after the dual-tone signal enters the nonlinear equipment, many parasitic products with different frequencies are generated. The frequency spectrum is shown in Fig. 5.73 [59].

If ω_1 and ω_2 are close to each other, all the even-order products will be far away from ω_1 and ω_2, such that they are easy to be filtered out from the output spectrum. The odd-order products, on the other hand, are likely to fall into the passband or near the passband. The third-order output signal is one of them, and it has the greatest effect on normal signals since the order is low and the power is strong.

Let P_{CW} be the output power of the signal at ω_1, and we can perform the approximate calculation ignoring the terms with small amplitude and with order higher than three. From Eq. (5.196), we can get

$$P_{CW} = \frac{1}{2} a_1^2 V_0^2 / Z_0 \tag{5.197}$$

where Z_0 is the impedance of the receiver RF system and can be set to 50Ω.

Similarly, let P_{OIM3} be the signal output power at the third-order intermodulation frequency $\Delta\omega$ ($\Delta\omega = 2\omega_1 - \omega_2$). According to (5.196), we have

$$P_{OIM3} = \frac{1}{2}\left(\frac{3}{4} a_3 V_0^3\right)^2 / Z_0 \tag{5.198}$$

When P_{CW} and P_{OIM3} are equal at the TOI point, the output signal voltage is

$$V_{OIM3} = \sqrt{\frac{4a_1}{3a_3}} \tag{5.199}$$

When the output power P_{OIP3} at the intercept point is equal to the linear response of P_{CW}, we have the following equation according to (5.197) and (5.198):

$$P_{OIP3} = P_{CW}\big|_{V_0 = V_{OIM3}} = \frac{2a_1^3}{3a_3} / Z_0 \tag{5.200}$$

Combining (5.197), (5.198), and (5.200), we can get the constraint among P_{OIM3}, P_{CW}, and P_{OIP3} as

$$P_{OIM3} = \frac{9a_3^2 V_0^6}{32} / Z_0 = \frac{a_1^6 V_0^6}{8Z_0} / \frac{4a_1^6}{9a_3^2 Z_0} = \frac{(P_{CW})^3}{(P_{OIP3})^2} \tag{5.201}$$

Then, we can obtain the constraint among V_{OIM3}, P_{CW}, and P_{OIP3} as

$$V_{OIM3} = \sqrt{P_{OIM3} Z_0} = \frac{\sqrt{(P_{CW})^3 Z_0}}{P_{OIP3}} \tag{5.202}$$

Assume G_1 and $P_{OIP3}^{(1)}$ are the power gain of the first stage and the output power of the first-stage TOI point, respectively; and assume G_2 and $P_{OIP3}^{(2)}$ are the power gain of the second stage and the output power of the second-stage TOI point, as shown in Fig. 5.74. From (5.201), we can get the third-order distortion power at the output of the first stages

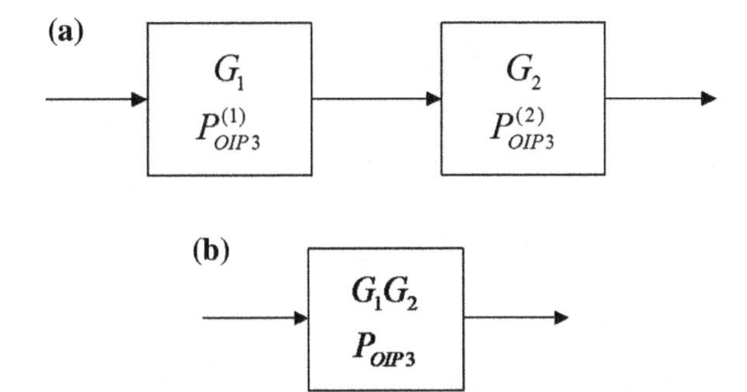

Fig. 5.74 Example of a TOI point equivalent to the cascade system. **a** Two cascade networks. **b** Equivalent network

$$P_{OIM3}^{(1)} = \frac{\left(P_{CW}^{(1)}\right)^3}{\left(P_{OIP3}^{(1)}\right)^2} \tag{5.203}$$

where $P_{CW}^{(1)}$ is the power of a single interference continuous wave output for the first stage.

The voltage associated with the third-order distortion power at the output of the first stage is

$$V_{OIM3}^{(1)} = \sqrt{P_{OIM3}^{(1)} Z_0} = \frac{\sqrt{\left(P_{CW}^{(1)}\right)^3 Z_0}}{P_{OIP3}^{(1)}} \tag{5.204}$$

In a cascaded system, the intermodulation products are deterministic signals (coherent signals), which cannot simply be a sum of power, but needs to be handled in voltage. These voltages are phase dependent, and there are phase delays at all levels, which may cause local cancelation. Here, in order to consider the worst-case interference signal strength, it is assumed that all signals do not have in-phase cancelation. The third-order intermodulation distortion voltage output $V_{OIM3}^{(2)}$ of the second stage is the third-order intermodulation distortion voltage output $V_{IM3}^{(1)}$ of the first stage multiplied by the voltage gain $\sqrt{G_2}$ of the second stage and the distortion voltage $\frac{\sqrt{\left(P_{CW}^{(2)}\right)^3 Z_0}}{P_{OIP3}^{(2)}}$ generated by the second stage, that is

$$V_{OIM3}^{(2)} = V_{OIM3}^{(1)}\sqrt{G_2} + \frac{\sqrt{\left(P_{CW}^{(2)}\right)^3 Z_0}}{P_{OIP3}^{(2)}} = \frac{\sqrt{G_2\left(P_{CW}^{(1)}\right)^3 Z_0}}{P_{OIP3}^{(1)}} + \frac{\sqrt{\left(P_{CW}^{(2)}\right)^3 Z_0}}{P_{OIP3}^{(2)}} \tag{5.205}$$

where $P_{CW}^{(2)}$ is the power of a single interference continuous wave output for the second stage; since $P_{CW}^{(2)} = G_2 P_{CW}^{(1)}$, then

$$V_{OIM3}^{(2)} = \left(\frac{1}{G_2 P_{OIP3}^{(1)}} + \frac{1}{P_{OIP3}^{(2)}}\right)\sqrt{\left(P_{CW}^{(2)}\right)^3 Z_0} \tag{5.206}$$

The output distortion power is

$$P_{OIM3}^{(2)} = \frac{\left(V_{OIM3}^{(2)}\right)^2}{Z_0} = \left(\frac{1}{G_2 P_{OIP3}^{(1)}} + \frac{1}{P_{OIP3}^{(2)}}\right)^2 \left(P_{CW}^{(2)}\right)^3 = \frac{\left(P_{CW}^{(2)}\right)^3}{\left(P_{OIP3}\right)^2} \tag{5.207}$$

Then, the TOI point of the cascade system is

$$P_{OIP3} = \left(\frac{1}{G_2 P_{OIP3}^{(1)}} + \frac{1}{P_{OIP3}^{(2)}}\right)^{-1} \tag{5.208}$$

By analogy, the constraint of the total distortion voltage of the receiver RF system and the TOI points at each level can be written as

$$V_{OIM3} = \left[\left(V_{OIM3}^{(1)}\sqrt{G_2} + \frac{\sqrt{\left(P_{CW}^{(2)}\right)^3 Z_0}}{P_{OIP3}^{(2)}}\right)\sqrt{G_3} + \cdots + \frac{\sqrt{\left(P_{CW}^{(N-1)}\right)^3 Z_0}}{P_{OIP3}^{(N-1)}}\right]\sqrt{G_N} + \frac{\sqrt{\left(P_{CW}^{(N)}\right)^3 Z_0}}{P_{OIP3}^{(N)}} \tag{5.209}$$

The corresponding output power of the cascade TOI point is

$$P_{OIP3} = \left(\frac{1}{G_N \cdots G_2 P_{OIP3}^{(1)}} + \frac{1}{G_N \cdots G_3 P_{OIP3}^{(2)}} + \cdots + \frac{1}{P_{OIP3}^{(N)}}\right)^{-1} \tag{5.210}$$

where $P_{OIP3}^{(i)}$ is the output TOI point power of the i-th sub-circuit; $P_{CW}^{(i)}$ is the output power of a single interfering continuous wave through the i-th sub-circuit, and it can be expressed as

$$P_{CW}^{(i)} = G_i P_{CW}^{(i-1)} \; (i = 2, 3, \ldots, N) \tag{5.211}$$

Usually, the passband of the filter in the prestage circuit of the superheterodyne receiver RF system is relatively wide. Assuming that both interference continuous waves fall within the filter passband, then there is no attenuation of the desired signal and the third-order intermodulation signal. The passband of the IF filter is narrow; thus, there will be attenuation $L_i (L_i < 1)$ to the interference continuous wave. The attenuation is introduced to correct the cascade equation of the TOI point.

Figure 5.75 is a cascade block diagram considering the selectivity of the filter in the RF system. Due to the selective suppression of the continuous wave interference signal by the filter module, Eq. (5.211) can be modified to

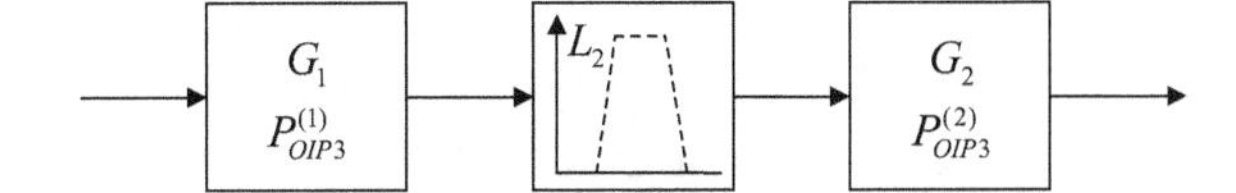

Fig. 5.75 Cascade block diagram considering filter selectivity in RF systems

$$P_{CW}^{(i)} = G_i L_i P_{CW}^{(i-1)} \; (i = 2, 3, \ldots, N)$$

where L_i is the attenuation of the continuous wave interference signals (the frequencies are ω_1 and ω_2) from the i-th filter module.

Here, when the module in the middle of Fig. 5.75 contains multiple filters, it will be unified to an attenuation value L_2.

Taking into account the module that contains the frequency selection characteristics, we can derive the following equation from (5.205) as

$$V_{OIM3}^{(2)} = \frac{\sqrt{G_2\left(P_{CW}^{(2)} / G_2 L_2\right)^3 Z_0}}{P_{OIP3}^{(1)}} + \frac{\sqrt{\left(P_{CW}^{(2)}\right)^3 Z_0}}{P_{OIP3}^{(2)}} = \left(\frac{1}{G_2 P_{OIP3}^{(1)} (L_2)^{3/2}} + \frac{1}{P_{OIP3}^{(2)}}\right)\sqrt{\left(P_{CW}^{(2)}\right)^3 Z_0} \tag{5.212}$$

The TOI point of the cascade system is

$$P_{OIP3} = \left(\frac{1}{G_2 P_{OIP3}^{(1)} (L_2)^{3/2}} + \frac{1}{P_{OIP3}^{(2)}}\right)^{-1} \tag{5.213}$$

Similarly, the corrected output power of the TOI point of the receiving equipment is

$$P_{OIP3} = \left(\frac{1}{G_N \cdots G_2 P_{OIP3}^{(1)} (L_N \cdots L_2)^{3/2}} + \frac{1}{G_N \cdots G_3 P_{OIP3}^{(2)} (L_N \cdots L_3)^{3/2}} + \cdots + \frac{1}{P_{OIP3}^{(N)}}\right)^{-1} \tag{5.214}$$

It can be seen from (5.214) that the filtering module will reduce the output power of the third-order intermodulation signal; when a filter with better selectivity is used in the latter stage ($L_i << 1$), the total output power of the TOI point of the receiving equipment is basically determined by the preceding stage.

(2) Method for allocating third-order intermodulation distortion parameters in system modeling

Whether to design a RF system of the transmitting/receiving equipment or to simulate and evaluate the system communication quality, the overall parameters provided by the RF system are signal-to-noise ratio (SNR), total noise figure (NF), bandwidth, dynamic range, gain, etc. The third-order intermodulation distortion parameters of the sub-circuit module are not included. Third-order intermodulation interference signals, which have a severe impact on normal signals, can often be generated and amplified in transmitting equipment and cause interference to other communication

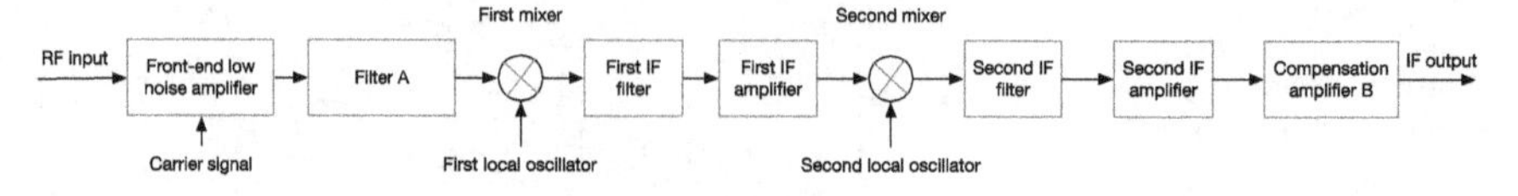

Fig. 5.76 Diagram of the ultrashort wave radio receiver from the case

platforms. If the design specifications of the receiving equipment do not include clear determinations of the third-order intermodulation distortion parameters, parasitic responses will be easily caused and the dynamic range will be greatly reduced. The accurate design of the third-order intermodulation distortion parameters of the sub-circuits in the RF system of the receiver can ensure that the receiver does not cause the degradation of the sensitivity and the deterioration of the communication quality due to the third-order intermodulation distortion signal being too large. Therefore, the third-order intermodulation distortion parameters of each sub-circuit module need to be reasonably allocated according to the overall performance indicator of the RF system.

We use an ultrashort wave radio as an example to describe the stepwise allocation of the third-order intermodulation distortion parameters of the receiving equipment from the top-level requirements to the sub-layers. The principle diagram is shown in Fig. 5.76. The RF system includes a front-end low-noise amplifier (LNA), the filter A, the first mixer, the first IF filter, the first IF amplifier, the second mixer, the second IF filter, the second IF amplifier, and the compensation amplifier B. The receiving equipment parameters are set as: The receiving frequency is 86 MHz, the total noise figure is 4.6 dB, the IF bandwidth is 68 kHz, and the receiving sensitivity is −93 dBm.

When multiple communication systems operate at the same time, the power of the signal entering the receiving system emitted from the adjacent channels is −30 dBm. During the analysis and design of the receiver, it is required that the maximum output power of the third-order intermodulation signal of the −30 dBm dual-tone signal through the receiving equipment is no larger than the maximum output noise power generated by the circuit within the system.

The two frequencies selected for the adjacent channel transmission signal are $\omega_1 = 85.475$ MHz and $\omega_2 = 84.925$ MHz, and the frequency of the third-order intermodulation distortion signal falling within the reception passband is $\Delta\omega = 2\omega_1 - \omega_2 = 86.025$ MHz. The attenuation of the two frequencies by filter A and the first IF filter is neglected. The sub-circuit modules that affect the power of the third-order intermodulation signal include the front-end LNA, the first mixer, the first IF amplifier, the second mixer, the second IF amplifier, the second IF filtering, and the compensation amplifier B.

The sensitivity of the receiver is −93 dBm. After the IF output, a high signal level is achieved. The total gain of this receiver RF system is 87 dB. The gain and noise figure of each module are shown in Table 5.15.

The total noise figure of the receiver is [19]

Table 5.15 Gain and noise figure of the modules in the receiving system

Sub-module	Gain G (dB)	Noise figure F (dB)
Front-end LNA module	8	2.8
First mixing module	6	7
First IF amplifier module	10	9.5
Second mixing module	6	7
Second IF amplifier module	25	9.5
Compensation amplifier module	32	8

$$F = F_1 + \frac{F_2 - 1}{G_1} + \frac{F_3 - 1}{G_1 G_2} + \cdots + \frac{F_6 - 1}{G_1 G_2 G_3 G_4 G_5} \tag{5.215}$$

where F is the ground truth of the total noise figure of the receiver RF system. The relationship between F and the figure from Table 5.15 is $\mathrm{F(dB)} = 10\log_{10}\mathrm{F}$ (ground truth).

According to Table 5.15, the total noise figure is 4.6 dB, the total gain is 87 dB, and the noise power output from the receiver RF system is

$$N_o = GkB\,[T_A + (F - 1)T_o] \tag{5.216}$$

where G is the total gain of the receiving RF system; k is the Boltzmann constant ($1.38 \times 10^{-23} J/K$); B is the IF bandwidth of the receiver RF system; T_A is the equivalent noise temperature fed to the receiver antenna; F is the total noise figure of the receiver RF system; T_0 is the temperature ($T_0 = 290$ K). Substituting the above value into (5.216),

$$N_o = 10^{8.7} \times 1.38 \times 10^{-23} \times 68 \times 10^3 \times [290 + (10^{0.46} - 1) \times 290] = 3.93 \times 10^{-7}\mathrm{W}$$

i.e., −34.05 dBm.

In order to suppress the influence of the transmitting signal from the adjacent channel to the receiver, two continuous waves at 84.925 and 85.475 MHz are used as the interference signal, to calculate the third-order intermodulation distortion that falls into the receiving passband after N_0 passes through the receiver RF system. It is required that the power intensity at the IF output end is no greater than the maximum output noise power generated by the circuit within the system; i.e., there is no spurious response. Therefore, the maximum output power of the total third-order intermodulation distortion signal allowed by the receiving RF system is −34.05 dBm. The converted voltage is $V_{OIM3} = 4.43 \times 10^{-3}$ V. Figure 5.77 is a behavioral simulation circuit model for an RF front-end system of an ultrashort wave receiver.

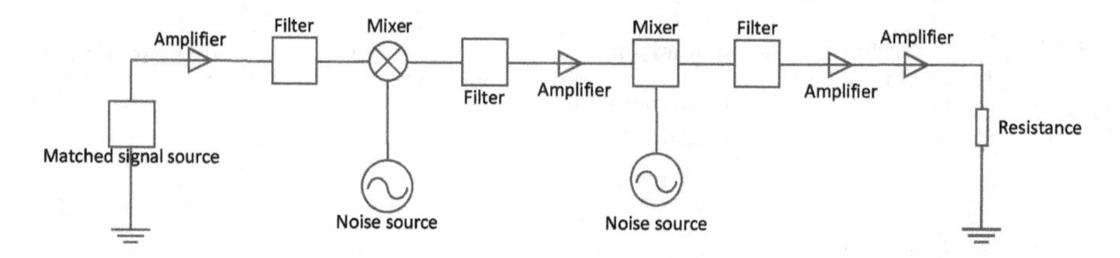

Fig. 5.77 Behavioral simulation model of the ultrashort wave receiving radio RF front end

Table 5.16 Output power of the TOI point and output voltage/power of the third-order intermodulation signal at each level of the receiving system

Sub-module	Output TOI point power (dBm)	Third-order intermodulation signal voltage (V)	Third-order intermodulation signal power (dBm)
Front-end LNA module	30	1.121×10^{-7}	−126.00
First mixing module	35	2.815×10^{-7}	−112.92
First IF amplifier module	50	1.879×10^{-6}	−101.51
Second mixing module	55	4.456×10^{-6}	−94.01
Second IF amplifier module	40	7.924×10^{-5}	−69.01
Compensation amplifier module	40	3.155×10^{-3}	−37.00

Considering the worst-case total third-order intermodulation distortion voltage, the third-order intermodulation distortion parameters are assigned to the modules with nonlinear characteristics inside the RF system circuit of the receiver according to (5.212). The power of the third-order intermodulation signal generated by each module is being strictly controlled as shown in Table 5.16.

It can be seen from Table 5.16 that the second IF filter suppresses the power of the two continuous wave signals by 75 dB, which is used to improve the TOI point of the subsequent stage. The output power of the TOI point from the second IF amplification can be adjusted to a smaller value. According to the assigned value, the calculated total third-order intermodulation distortion voltage is $V_{OIM3} = 3.155 \times 10^{-3}$ V (less than $V'_{OIM3} = 4.43 \times 10^{-3}$ V), which satisfies the requirement. It means that the generated third-order intermodulation distortion signal will not affect the normal communication of the receiver.

It is worth mentioning that although there is more than one method of allocating the third-order intermodulation distortion parameters, it is necessary to strictly follow the modified TOI point cascade equations to control the voltage of control signals step by step in the simulation model, so that the overall indicator will meet the requirements. Using the overall performance indicator, we can analyze the reasonable third-order intermodulation distortion parameters of all internal sub-circuits, such that with limited parameters available, we can guarantee the parasitic-response-free dynamic range of the receiver.

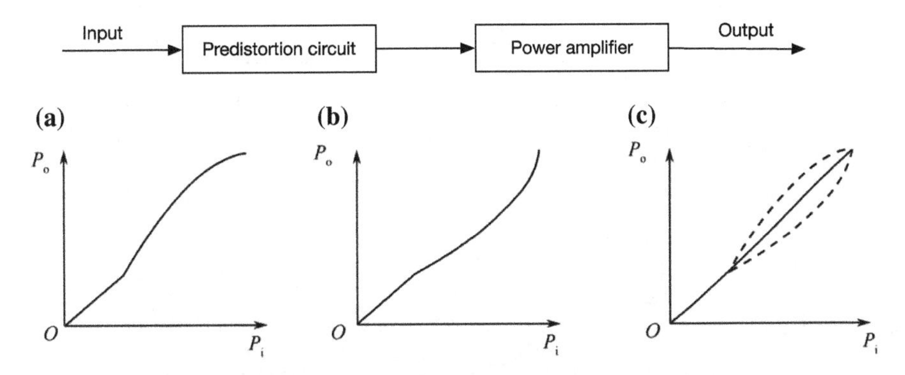

Fig. 5.78 Block diagram of the predistortion techniques and linearization. **a** Amplifier input/output transfer characteristics; **b** input/output transfer characteristics of the predistortion circuit; and **c** overall input/output transfer characteristics

(3) Analog predistortion circuits to suppress the low-order intermodulation

Last section focuses on the allocation of the third-order intermodulation distortion parameters of the RF system of the communication equipment using the overall requirements during the EMC design of the receiving system. The strength of the interfering signal gets controlled after entering the system and meets the system requirements. From the transmitting system point of view, this section provides a feasible solution to improve EMC, and a circuit that suppresses the third-order and fifth-order intermodulation distortions. The effectiveness of the circuit is then verified.

High-power signal radiation in the transmitting system is accompanied by a strong spurious signal group. The major source of these spurious signals is power amplifiers which usually operate in the nonlinear region to improve the efficiency [60]. There are many linearization techniques, such as feedforward, negative feedback, and predistortion to deal with the nonlinear problems in power amplifiers. Negative feedback technique has a simple circuit structure and is low cost, but it is instable. Feedforward technique can provide good linearity, but the complex and expensive analog equipment it uses are difficult to control. Thus, the predistortion technique is a low-cost technique which is also easy to model in simulation software [61].

The basic idea of the predistortion technique is to insert a piece of nonlinear equipment in front of the RF power amplifier so that the power amplifier's transmission characteristic is close to linear and the out-of-band diffusion is reduced. Its block diagram and transmission characteristics are shown in Fig. 5.78. The transform characteristics of the predistortion circuit are used to compensate for the nonlinear characteristics of the main power amplifier, such that the intermodulation distortion can be canceled and the nonlinearity of the power amplifier gets improved. The advantage of predistortion technology is that there is no stability problem and the price is low. Several carefully selected components are packaged into a single module and connected between the signal source and the power amplifier to form a predistortion linear power amplifier [62].

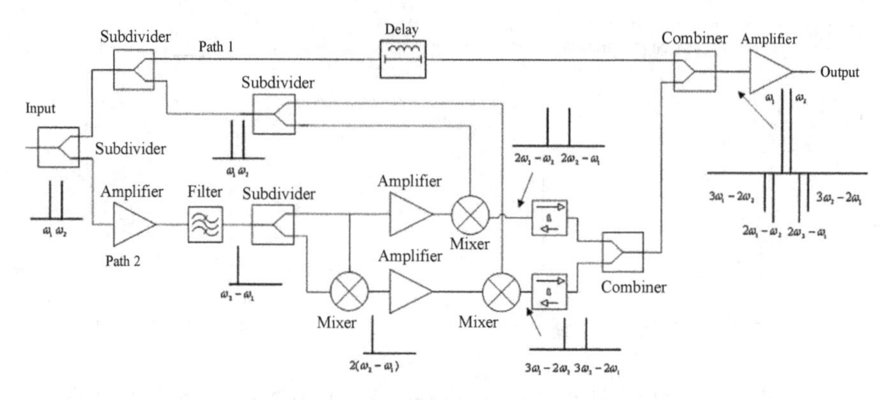

Fig. 5.79 Implementation block diagram of the predistortion circuit

The actual implementation of the analog predistortion circuit is shown in Fig. 5.79. The input dual-tone signal (ω_1, ω_2) is split by the power splitter into the fundamental and intermodulation distortion signals to generate two paths: linear path 1 and nonlinear path 2. The input signal in path 1 is delayed, and a nonlinear intermodulation distortion signal is generated in path 2, which enters the power amplifier together with the original dual-tone signal. Let the two signals be $A_1 \cos(\omega_1 t)$ and $A_2 \cos(\omega_2 t)$, and the signal of the input end of the predistortion circuit is:

$$x(t) = A_1 \cos(\omega_1 t) + A_2 \cos(\omega_2 t) \tag{5.217}$$

After passing through the first amplifier and filter in path 2, a second-order component $\omega_2 - \omega_1$ is obtained, and the signal strength is weak. After that, $\omega_2 - \omega_1$ is frequency doubled to generate component $2(\omega_2 - \omega_1)$. The components $2(\omega_2 - \omega_1)$ and $\omega_2 - \omega_1$ are then mixed with the fundamental waves ω_1 and ω_2, respectively, to generate $2\omega_2 - \omega_1$, $2\omega_1 - \omega_2$, $3\omega_2 - 2\omega_1$, and $3\omega_1 - 2\omega_2$. Finally, after amplitude and phase shift adjustment, the weak even-order signals are ignored. The output signal of path 2 is

$$\begin{aligned} b(t) =& A_{11} \cos(2\omega_1 t - \omega_2 t + \varphi_{11}) + A_{22} \cos(2\omega_2 t - \omega_1 t + \varphi_{22}) + A_{33} \cos(3\omega_1 t - 2\omega_2 t + \varphi_{33}) \\ &+ A_{44} \cos(3\omega_2 t - 2\omega_1 t + \varphi_{44}) \end{aligned} \tag{5.218}$$

where A_{11}, A_{22}, A_{33}, and A_{44} are the voltages for each component; φ_{11}, φ_{22}, φ_{33}, and φ_{44} are the corresponding phases.

The generated phase-inverted signal is used to cancel the low-order intermodulation distortion signal of the power amplifier. The actual input signal of the power amplifier is $x(t) + b(t)$.

Here, we provide an example in which a power amplifier with the predistortion circuits is designed, then tests are performed using Agilent ADS circuit analysis software, and finally the behavioral simulation method is used to implement the predistortion circuit and verify its effectiveness.

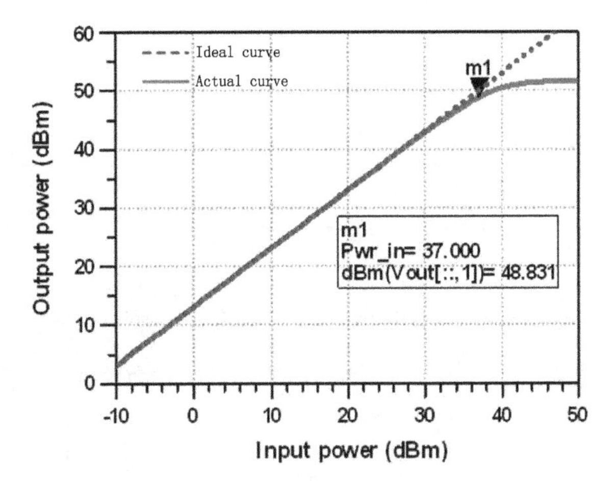

Fig. 5.80 Input and output characteristics of the power amplifier harmonic balance simulation

The power amplifier has a gain of 13 dB, and a harmonic balance simulator is used to calculate the behavioral model of the power amplifier. The input and output transfer characteristics are shown in Fig. 5.80. In the figure, m_1 is close to the 1 dB compression point. The signal power of m_1 is 37 dBm. The amplifier operates in the nonlinear region, and spurious components appear in the output signal. When there are multiple carrier signals in the input, the odd-order intermodulation interference signal is close to the normal signal, and it will be difficult to filter out.

In odd-order intermodulation signals, the low-order signal is strong. For the generated third-order and fifth-order intermodulation interference, we can add a simple predistortion circuit before the power amplifier. Figure 5.81 is a behavioral simulation model built according to the process in Fig. 5.79. Suppose the frequencies ω_1 and ω_2 of the dual-tone signal entering the power amplifier are 836 MHz and 842 MHz, respectively, and the second-order component with a frequency of 6 MHz ($\omega_2 - \omega_1$) is obtained in path 2 through a low-pass filter. Actually, the closer ω_1 and ω_2 are, the smaller $\omega_2 - \omega_1$ is, therefore, the easier it is to implement the low-frequency filter circuit design; the second-order component is then multiplied to obtain the 6 and 12 MHz signals simultaneously, and then the mixing, AM, and PM procedures are performed to the signal to obtain the third-order intermodulation component signals, of which the frequencies are 830 MHz ($2\omega_1 - \omega_2$) and 848 MHz ($2\omega_1 - \omega_2$). In addition, the fifth-order intermodulation component signals, of which the frequencies are 824 MHz ($3\omega_1 - 2\omega_2$) and 854 MHz ($3\omega_1 - 2\omega_2$), are generated. The signals generated in path 2 will enter the main power amplifier together with the original input signal to suppress the signal strength of the total output third-order and fifth-order intermodulation.

After the dual-tone signal in path 1 enters the power amplifier, the third-order and fifth-order intermodulation products with strong power are generated in addition to the main frequency. The intermodulation products generated in path 2 will output a phase-inverted signal waveform through the power amplifier; in other words, the waveform has a phase difference of approximately 180 degree from the intermodu-

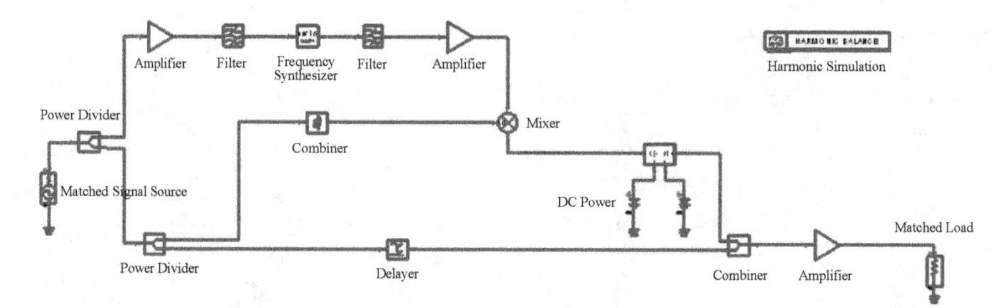

Fig. 5.81 Behavioral simulation model of the predistortion circuit

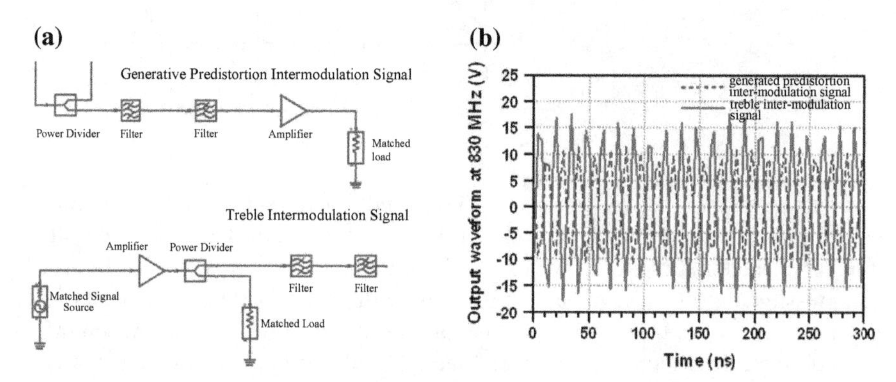

Fig. 5.82 Two-part third-order intermodulation signal simulation model and signal waveforms. **a** Third-order intermodulation signal extraction from path 1 and path 2 and **b** two third-order intermodulation signal waveforms of final output

lation signal generated in path 1. Here, we choose to observe the waveform of the third-order component $2\omega_1 - \omega_2$, which includes two parts: One is the third-order intermodulation signal which is generated by the predistortion and then amplified; the other is the third-order intermodulation signal generated by the main power amplifier. The simulation circuit generating the two parts of the signal is shown in Fig. 5.82a: Using the ideal filter, the intermodulation signal with a frequency of 830 MHz from the mixed signals before entering the power amplifier is obtained as shown in Fig. 5.81, and the output waveform is given after the power amplifier; the dual-tone signal is directly used to enter the power amplifier, and an ideal filter is used to obtain an intermodulation signal with a frequency only of 830 MHz. The waveforms of the two third-order intermodulation signals are shown in Fig. 5.82b. It can be seen that the phase difference between the two is about 180°.

The output spectrum of the power amplifier before and after the predistortion improvement is shown in Fig. 5.83. It can be seen that the third-order intermodulation can be improved at least 15 dB and the fifth-order intermodulation can be improved 9 dB.

The power output capability and linear performance of the RF power amplifier directly affect the quality of the wireless communication. In the design process, the

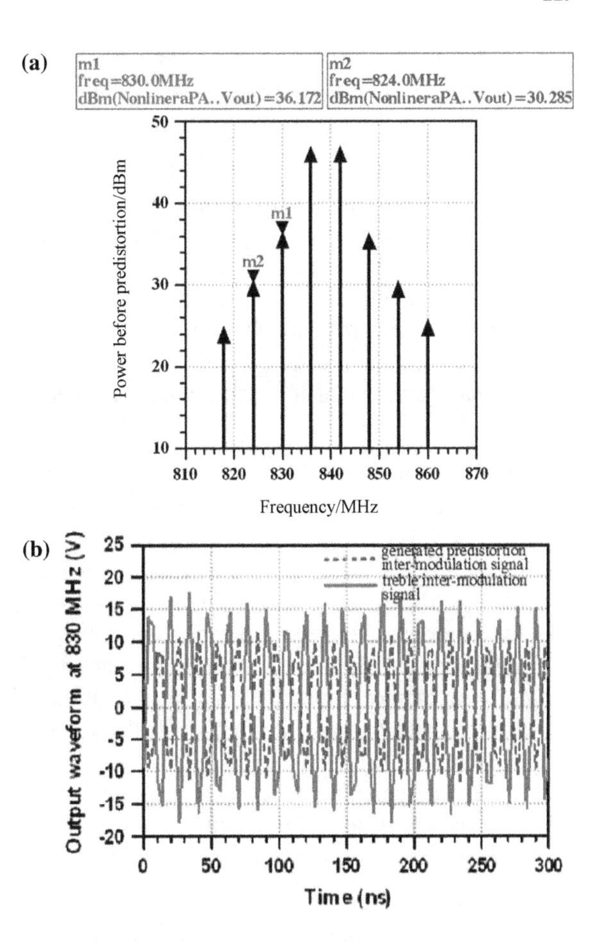

Fig. 5.83 Output spectrum of the power amplifier before and after using predistortion techniques. **a** Output spectrum before predistortion and **b** output spectrum after predistortion

predistortion circuit can be introduced if needed. Several components are selected to form a single module, connecting the low-order intermodulation output before being connected to the main power amplifier. In addition, during the analysis of the actual system EMC, the emission system that does not meet the regulation requirements after testing needs to suppress the intermodulation interference signals using necessary approaches. It is rare for existing equipment to be required to undergo massive troubleshooting. In order to meet the actual requirements of the system, we only need to build a predistortion circuit in front of the nonlinear amplifier in the predistortion model to reduce the strong third-order and fifth-order intermodulation signal interference caused by the nonlinear characteristics. The implementation is simple and effective.

5.4.3 Identification of Electromagnetic Emission Elements Based on Gray Model (Case 3)

The ultimate goal of the EMC study of electronic equipment is to make the equipment or system work properly in its electromagnetic environment without producing intolerable EMI to other equipment in the environment. In this section, we propose a method to analyze the EMC test spectrum [63], introduce the important electromagnetic emission elements, and identify potential causes of interference. The methods can be used to guide the equipment troubleshooting and quantitative design in early stage.

The characteristics of the electromagnetic spectrum obtained from test can be summarized as follows:

(1) Inherit attributes. As a causal system, the characteristics of the interference source can be extracted from the electromagnetic emission spectrum of the EUT.
(2) Discreteness. Although the electromagnetic emission of the equipment is continuous, the frequency spectrum measured by the EMI receiver is the emission field strength obtained at specific frequencies with a certain sampling interval.
(3) Complexity. Due to equipment nonlinearity, cable cross talk, case shielding, test environment, and other factors, the measured spectrum is extremely complicated, and there are interferences in the frequency range from several kilohertz to even megahertz.
(4) Observed spectrum. Electromagnetic emission spectrum is often as wide as hundreds of octaves, but a considerable part of the spectrum is the "observed spectrum" generated by the basic interference elements under nonlinear effects.

Since the spectrum obtained from equipment testing can always reflect the characteristics of the equipment's internal sources, studying the characteristics of the spectrum is very important for understanding the equipment's EMC. However, due to the complexity of the spectrum, the research process needs to be simplified and starts from the most typical spectrum. Considering the discrete nature of the frequency points, we first need to decide a method to analyze the discrete data.

From the analysis of the circuit mechanism, we know that the harmonics, as an interference element, is very important in electromagnetic spectrum. To better understand the test data, it is highly necessary to accurately extract the characteristics of the harmonics.

The goal of electromagnetic spectrum processing in view of harmonics is to extract the electromagnetic characteristics of the harmonic interference elements from the measured interference spectrum. Harmonic interference extraction process is shown in Fig. 5.84. First of all, the spectrum obtained from the test is preprocessed; that is, the overall trend component in the spectrum is extracted; then, the total

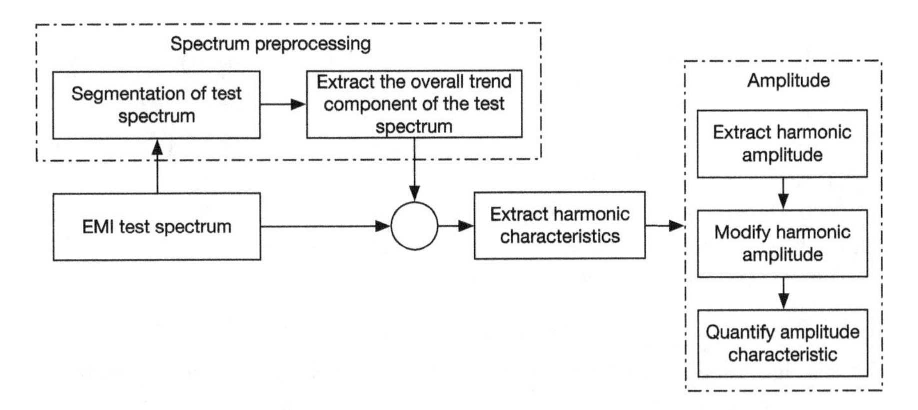

Fig. 5.84 Extraction process of the harmonic interference

trend component is subtracted from the total interference spectrum to extract the frequency of the harmonic component in the remaining spectrum; next, in accordance with the harmonic frequency, amplitude characteristics of the harmonic frequency are extracted; finally, we can get an overall understanding of the electromagnetic properties of the harmonics.

1. Spectral data preprocessing based on the gray model

First of all, the test data is preprocessed, and the gradual change components in the emission spectrum are extracted and eliminated. Then, we analyze the characteristics of the components with great impact on the spectrum.

We have tried to apply the gray system model to the spectrum sequence modeling in EMC based on our long-term study of the model. The GM (1, 1) model in the gray system theory is used to model the overall trend component.

The meanings of the symbol GM (1, 1) are as follows:

$$\begin{array}{cccc} G & M & (1, & 1) \\ \uparrow & \uparrow & \uparrow & \uparrow \\ \text{Gray} & \text{model} & \text{1st order function} & \text{1 variable} \end{array}$$

In order to simplify the analysis process, the spectrum obtained from the test is compared with the limit value of the EMI emission before the spectrum analysis, and the part beyond the regulation is studied.

(1) Spectrum segmentation

Assuming that the spectrum data of the EMI test is $X = \{x(1), x(2), \cdots, x(n)\}$, the mean sequence of each small segment is calculated using (5.219), i.e.,

$$S(1+(K-1)l_1 : kl_1) = \sum_{i=1}^{kl_1} X(i)/l_1 \quad (k = 1, 2, \ldots, n/l_1) \tag{5.219}$$

The length of the selected segment, denoted as l_1, can be adjusted according to the length of the test sequence.

Then, we can find the segment points in the generated ladder-shaped curves and extract the segment points in the analysis spectrum as $x(s_1), x(s_2), \cdots, x(s_m)$. The spectrum is thus divided into $(s_m + 1)$ segments.

(2) Overall trend component modeling

The test data usually has a certain overall trend, which reflects the superposition of the equipment's broadband interference and the gradual change components in the environment level. The contribution of this component to the spectrum is not significant. The overall trend component of the broadband needs to be eliminated to extract other forms of interference.

Before the establishment of the overall trend component, the sequence to be modeled must be cumulatively generated first.

The original modeling sequence is $X^{(0)} = (x^{(0)}(1), x^{(0)}(2), \ldots, x^{(0)}(n))$. After accumulation, the generated sequence is $X^{(1)} = (x^{(1)}(1), x^{(1)}(2), \ldots, x^{(1)}(n))$. According to the GM (1, 1) model building method, the parameter values in the model are estimated after the cumulatively generated sequence is generated in consecutive neighbors. The definition of the consecutive neighbor generation sequence is $Z^{(1)} = (z^{(1)}(2), z^{(1)}(3), \ldots, z^{(1)}(n))$, where $z^{(1)}(k)(k = 2, 3, 4, \ldots, n)$ is the background value.

The relationship between the consecutive neighbor-generated value in the interval [k, k+1] and the accumulated generated value of the endpoint is shown in Fig. 5.85. When the sequence data change is mild (low-growth indicator), the estimated value of the parameters constructed with the consecutive neighbor-generated mean sequence is more accurate and the model error is small [64]; when the sequence data changes drastically (large growth indicator), the modeling error increases and it is necessary to process the sequence so that the one-accumulation generation curve of the modeling sequence meets the requirement of the low exponential growth.

Now, we take the data sequence shown in Fig. 5.86 as an example to model the overall trend component. An accumulation of the sequence in Fig. 5.86 is shown by the solid line in Fig. 5.87. The dashed line in Fig. 5.87a is the background value curve of the construction parameter column used when the GM (1, 1) model is directly established for the original data. The dashed line in Fig. 5.87b is the background curve generated using the segmentation method of this book. There are two segmentation points on the curve, which are denoted with circles. From the comparison of the two figures, it can be seen that the segmented background value is closer to the original sequence cumulatively generated value curve. The establishment of a GM (1, 1) model for each segment will better represent the overall trend of the original sequence.

Fig. 5.85 Accumulation generation and consecutive neighbor generation

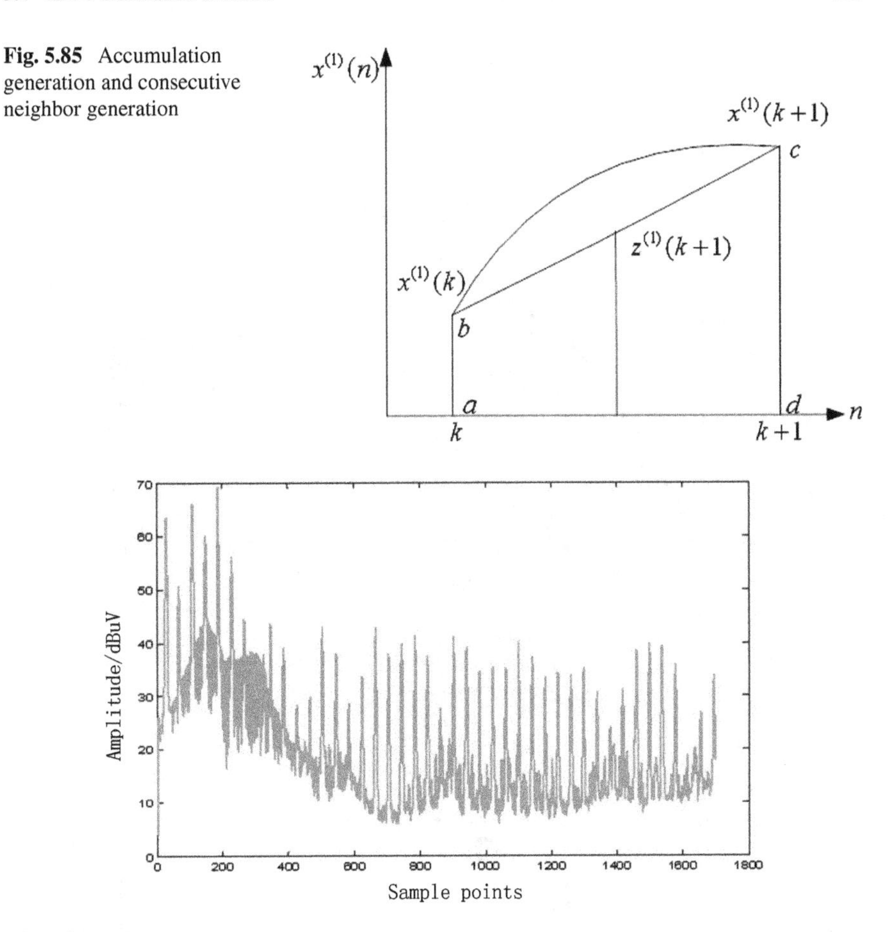

Fig. 5.86 Original data sequence curve

We perform direct modeling and subsection modeling, respectively, of the original sequence and obtain the model curve as shown in Fig. 5.88. The errors in the direct modeling and the segmented modeling are calculated. The modeling error in Fig. 5.88a is 35.93%. In Fig. 5.88b, the modeling error becomes 28.75%. Therefore, when the consistency of the slope of one-accumulation generation curve is poor, as shown in Fig. 5.87a, we should use subsection modeling to achieve a better overall trend component fitting.

Based on the theoretical analysis and experimental validation, it can be seen that: Before modeling the overall trend component of the spectrum, we need to test whether the spectrum sequence needs to be segmented; if necessary, then we shall model the overall trend component in the segments. It is because that the segmented sequence shows a certain trend, and there is consistency among segments; we can use the GM (1, 1) model to build models for each of the $(s_m + 1)$ segments separately and obtain the estimated value of each segment $\{\hat{x}_{s_1}^{(0)}, \hat{x}_{s_2}^{(0)}, \cdots, \hat{x}_{s_m}^{(0)}, \hat{x}_{s_{m+1}}^{(0)}\}$. Then, we sum up the

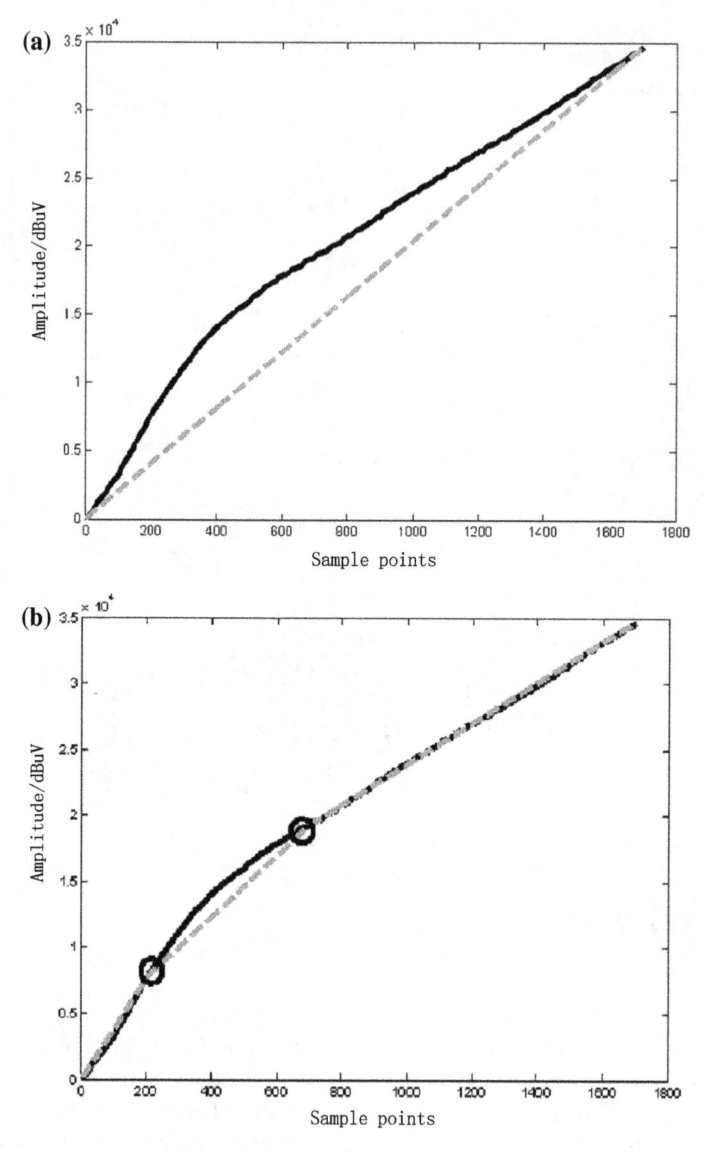

Fig. 5.87 Accumulation generation sequence

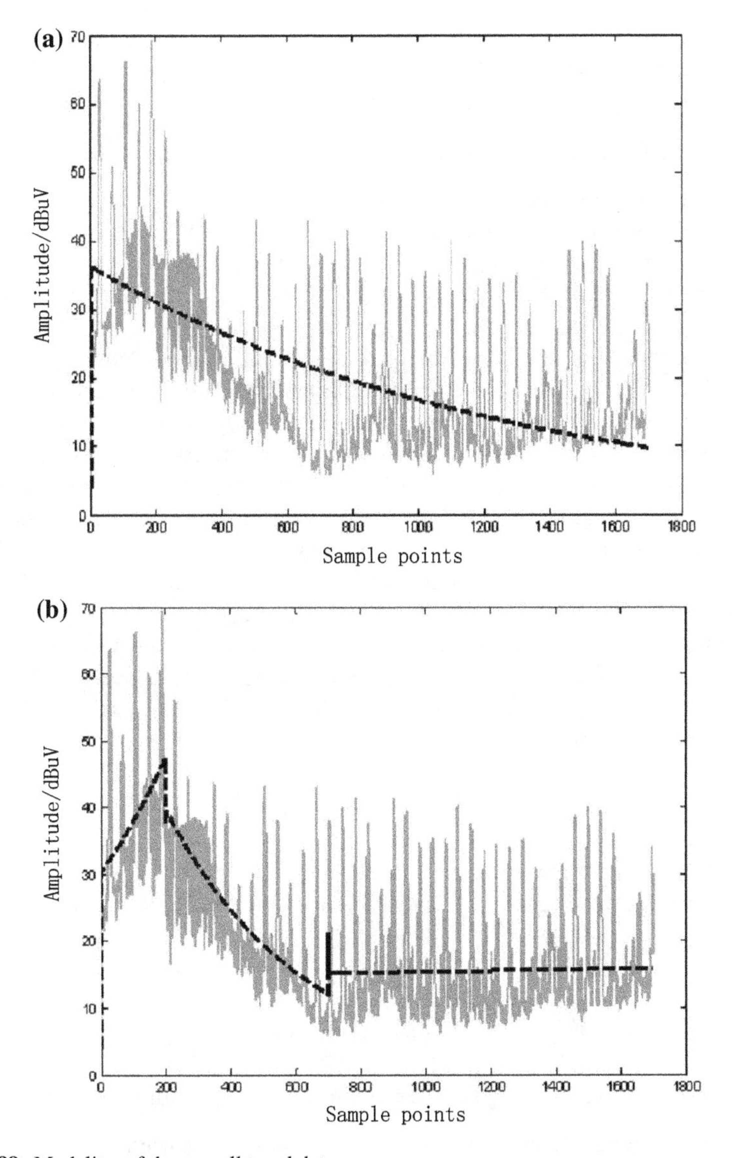

Fig. 5.88 Modeling of the overall trend data sequence

estimated value from the start to the end and obtain the estimation of the overall trend, i.e., $X^{(0)} = (x^{(0)}(1), x^{(0)}(2), \ldots, x^{(0)}(n))$, of the entire spectrum sequence that needs to be analyzed, i.e., $\hat{X} = (\hat{x}(1), \hat{x}(2), \ldots \hat{x}(n))$.

2. Harmonic frequency characteristics extraction from the spectrum

After extracting the overall trend component, the residual component can be considered as the emission component that has the main contribution in the spectrum. The analysis of these components is performed from both the frequency and the amplitude perspectives. Here, we will focus on the extraction of the frequency characteristics of harmonic interference components. On the one hand, the frequency of harmonic components can be extracted to determine the fundamental frequency to initially locate the interference source. On the other hand, the amplitude at the frequency point can be extracted based on the frequency information, which serves as the basis of the amplitude characteristics of the interference source.

The spectral data containing harmonics includes not only a fundamental wave component of a certain frequency but also a harmonic component of the fundamental frequency multiplication; therefore, the harmonic wave component is periodic. For such a data sequence, a gray system model and a periodic extension model need to be used at the same time. That is, the overall trend component in the spectrum is first modeled, and the modeling residual (the difference between the original test sequence and the GM (1, 1) model sequence) is modeled periodically.

The spectral residual sequence is obtained by removing the overall trend components in the test spectrum. It can be written as

$$x1'(k) = x^{(0)}(k) - \hat{x}(k) \tag{5.220}$$

Based on the analysis of the physical meaning of the spectrum, before the residual series is modeled, the negative items in the series are set to zero to obtain a modified residual sequence $x'(k)$, which satisfies

$$x'(k) = \begin{cases} x1'(k), & x1'(k) \geq 0 \\ 0, & x1'(k) < 0 \end{cases} \tag{5.221}$$

Then, we can find the periodic extension model of the residual series $x'(k)$.

3. Harmonic amplitude characteristics extraction from spectrum

In the study of EMC, we can often derive the characteristics of the waveform that produce interference from the envelope characteristics of electromagnetic emission test data [65]. Therefore, this book adopts the Fourier envelope theory to analyze various types of radiation waveforms to obtain some simple rules.

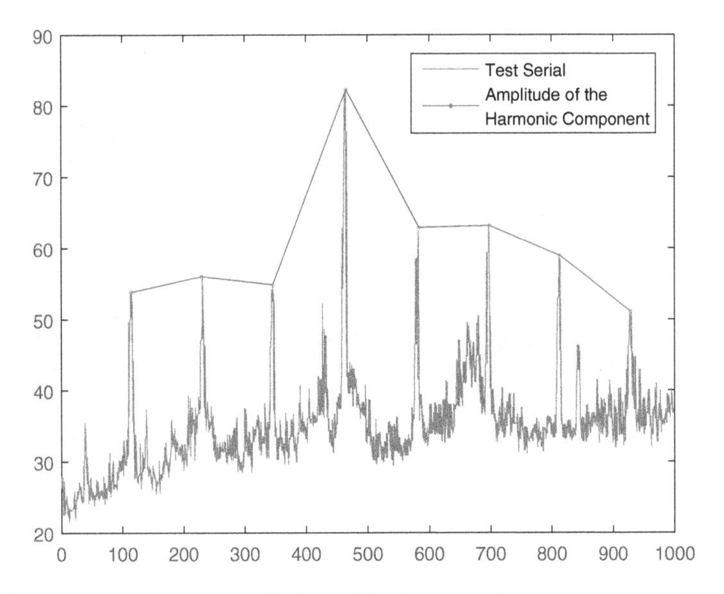

Fig. 5.89 Harmonic component amplitude modification extraction

(1) Extraction of harmonic amplitude points

According to the previous analysis, the harmonic frequencies are obtained, and the corresponding frequencies are selected for amplitude extraction of the harmonic points. In this situation, the frequency during the measurement is obtained by discrete sampling, so the measured signal peak and the actual peak value may not be exactly the same in frequency, which can also cause deviations at the fundamental frequency. If this deviation is not taken into consideration, the deviation will increase with the increase of the multiple when calculating the fundamental frequency multiplication, which may result in incorrect results. For this reason, it is necessary to calculate the possible range of frequency instead of a simple frequency point. In practice, we can take the left and right ten frequency points of the selected signal frequency as the possible range of the signal frequency and select the signal peak in this range as the envelope of the harmonic components, as shown in Fig. 5.89.

(2) Modification of the radiation emission amplitude

It is almost impossible to restore the exact radiation field intensity generated by the original signal directly from the emission spectrum. Therefore, some simple formulas are used to describe the characteristics of the radiation field, and a certain modification is made to the test data to obtain the approximated characteristics of the emission source [66]. These simplifications include: keeping only the value of the field in the most suitable direction; aligning the receiving antenna with the direction of maximum polarization; assuming a uniform current distribution in the cable, it is acceptable to use an average equivalent current rather than the maximum; and ignoring the dielectric and impedance loss of the wire. In view of the above simplification, only

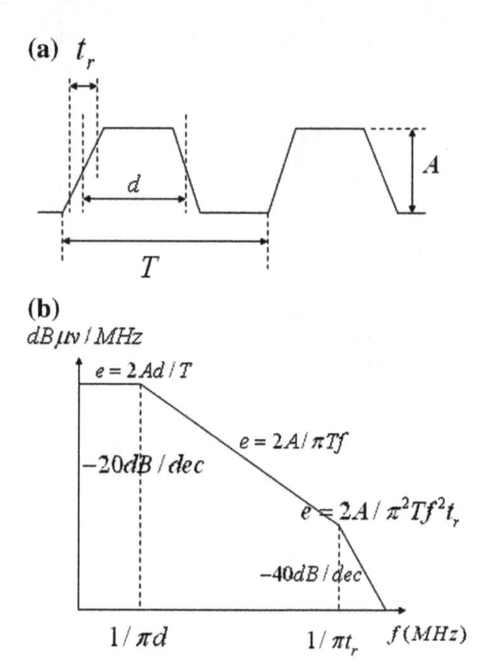

Fig. 5.90 Trapezoidal wave and its spectral envelope. **a** Trapezoidal waves. **b** Spectrum envelope

the change in the slope of the envelope is used when analyzing the amplitude envelope of the harmonics. That is, only certain quantized characteristics of the waveform can be obtained, but the accurate amplitude values of the radiation source cannot be obtained. Practical applications show that the simplified formula can solve most practical problems.

(3) Quantitative characteristics of the harmonic spectrum amplitude

The corresponding amplitude for the extracted harmonic frequency points is obtained. The x-axis is then converted to a logarithmic coordinate, and the harmonic envelope is plotted. Then, we shall find the turning points on the envelope, from which the waveform characteristics of the corresponding source of harmonic components can be calculated.

Now, we consider the waveform shown in Fig. 5.90a. Its spectral envelope is shown in Fig. 5.90b. From the beginning to $F_1 = 1/\pi d$, the envelope of the spectrum is constant, then decreases with a slope of -20 dB/dec until $F_2 = 1/\pi \tau_r$, and then falls again with a slope of -40 dB/dec.

The amplitude peak A (W, A, etc.), the pulse width d, the signal period T, and the rising time t_r (the time corresponding to a signal amplitude change from 10 to 90%, and if the falling and rising times are different, the shorter one is selected as t_r) can be read from Fig. 5.90b. Therefore, once the envelope of the spectrum is obtained, some characteristics of the original waveform can be determined based on the slope turning points. The width of the pulse waveform can be obtained from the first slope turning point, and the rising time of the waveform can be calculated from the second turning point. If we can get all the exact values of the spectrum envelope, which is very difficult, we could also calculate the amplitude of the waveform.

4. Correspondence between the interference elements and the actual equipment

To connect the extracted interference elements with a certain device/circuit in the equipment to understand the constituent elements of the spectrum, it is necessary to correlate the characteristics of the interference elements. The corresponding of the characteristics is started with the frequency characteristics, and frequency match is directly compared. If there are no matching devices, the amplitude characteristics are then considered, as shown in Fig. 5.91. This way, the spectral characteristics obtained from the test can be completely matched with the characteristics of the device/circuit in the equipment.

5. Examples and verification of electromagnetic emission element identification

(1) Examples of conduction emission

This section analyzes three examples from different data analysis perspectives: The importance of modifying the residuals is explained through the light source system conduction analysis; the frequency of the harmonic components in the frequency spectrum is analyzed through the switching mode power supply (SMPS) conduction emission; through testing equipment with unknown internal composition, its frequency and amplitude are analyzed using the methods presented in this section. The measured data in this section was all obtained from testing.

(1) Conduction analysis of the light source system. In Fig. 5.92, the solid line shows the conduction emission (CE) measurement result of the positive line of a light source system's power module. The dashed line is the overall trend component extracted using the method proposed in this book. The dashed line in Fig. 5.93 is a gray-periodic model obtained by directly modeling the residual sequence with an extracted periodic interval of 59.

The residual sequence is modified as shown in Eq. (5.221). Only harmonic components that contribute positively to the frequency spectrum are considered. Then, the sequence of the modified residual sequence is modeled as shown in the dashed line in Fig. 5.94. The extracted period of the model is 118. Since the frequency sampling interval set in the test is 5 kHz, the direct modeling harmonic frequency is 295 kHz and the error is 12.46%. Furthermore, the modified modeling harmonic frequency is 590 kHz and the error is reduced to 10.85%. If we want to further reduce the error of the model sequence, we can model the residual with negative values, but the results of the analysis do not contribute much to the understanding of the characteristics of the key components in the spectrum. Therefore, considering the result and efficiency, we do not perform more accurate spectrum modeling.

The switching frequency of the power module of the actual light source system is 590 kHz. It is obvious that the result after the above modification of the residual

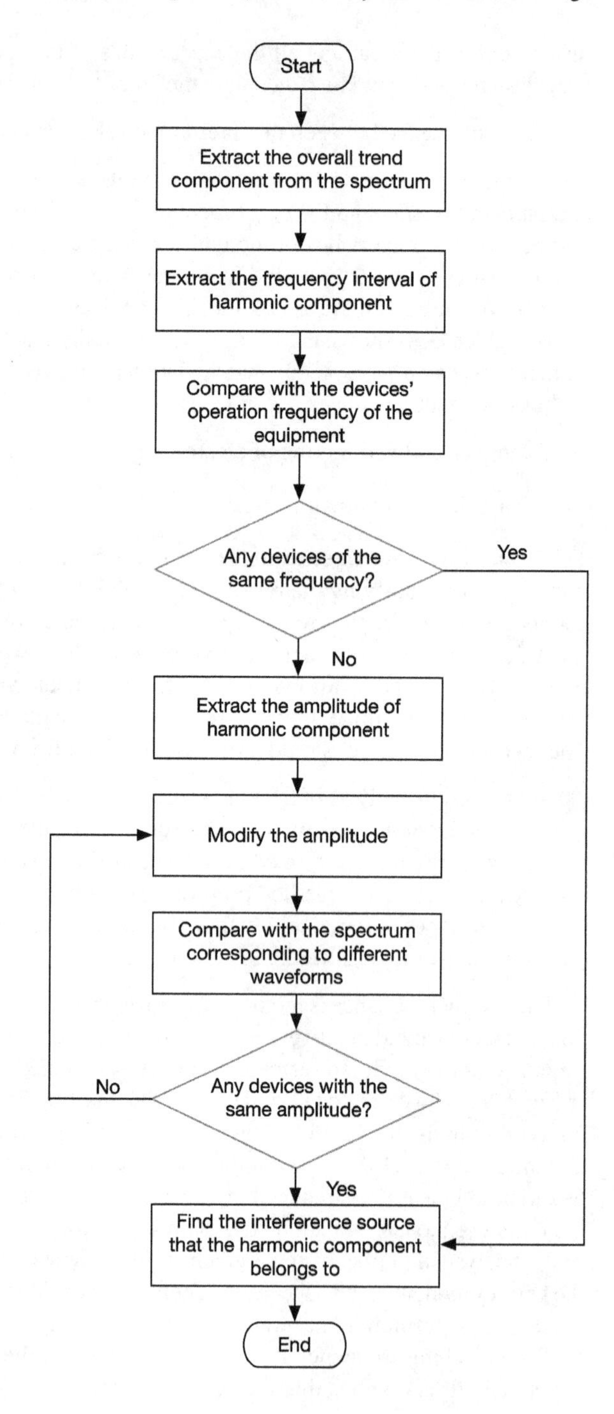

Fig. 5.91 Flowchart of the correspondence between the interference components and equipment

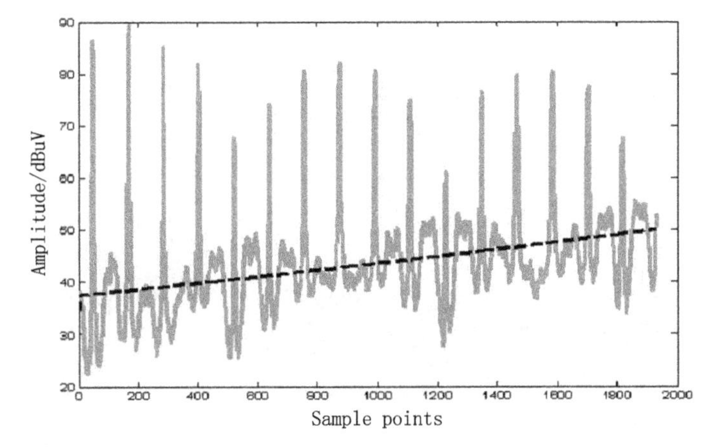

Fig. 5.92 Light source system test spectrum sequence and overall trend modeling sequence

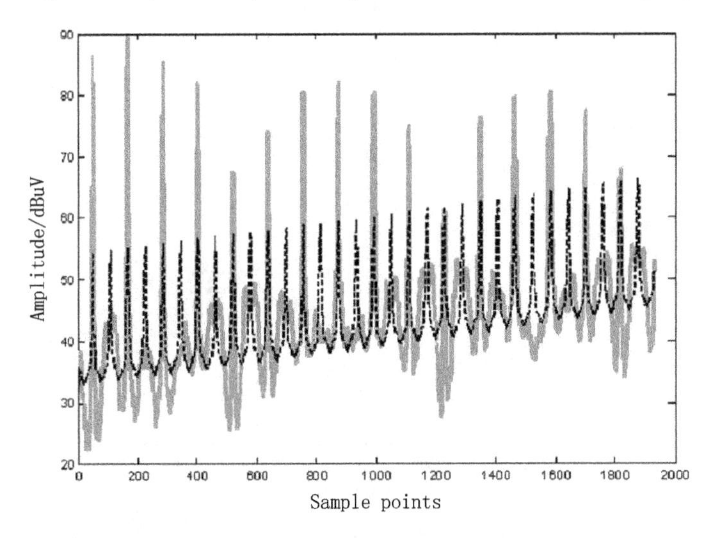

Fig. 5.93 Direct modeling of the spectral residuals of light source systems

better satisfies the accuracy requirement. It is also found that the cause of harmonics emission of the switch section is the insufficiency of the power supply filter. (2) Conduction emission analysis of SMPS. Figure 5.95 shows the frequency-domain conduction emission (CE102) measurement result of the SMPS' positive line. The 250 kHz to 8.5 MHz part of the spectrum is modeled and analyzed, the x-axis frequency display is labeled with serial numbers, and then the spectrum analysis is performed.

First, we use the trapezoidal envelope of the spectrum to perform spectrum segmentation. The trapezoidal envelope shown in Fig. 5.96 in the dashed line is cal-

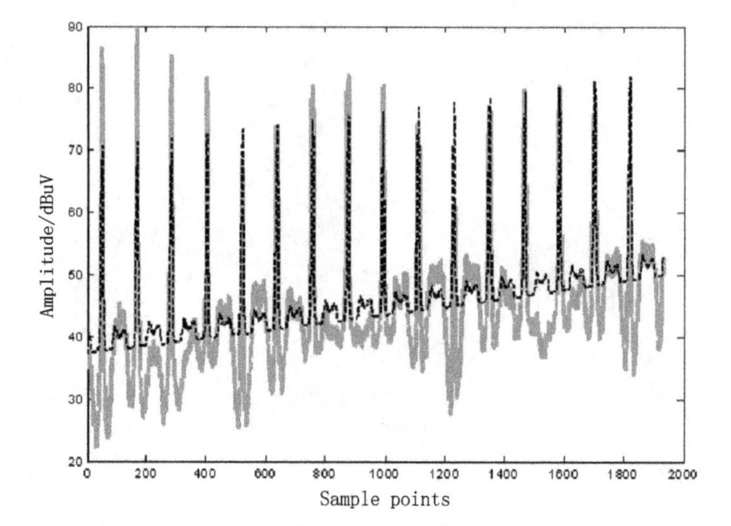

Fig. 5.94 Periodic modeling of residuals after modification

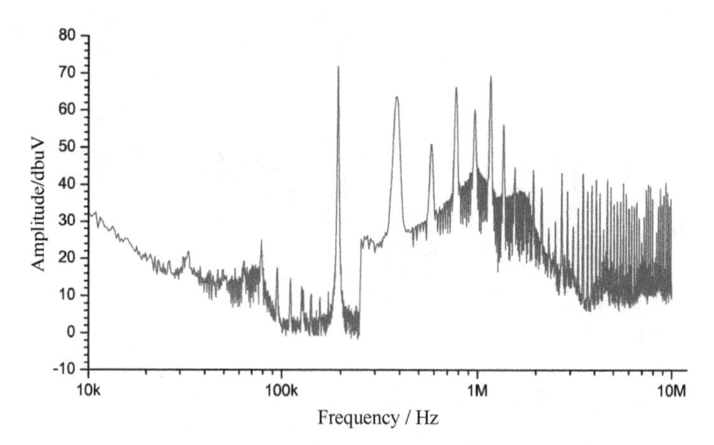

Fig. 5.95 SMPS power conduction emission (CE102) test result

culated using Eq. (5.219). The spectrum segments calculated from the difference sequence are listed in Table 5.17.

Table 5.17 shows that the two calculated frequency spectrum segmentation points are No. 200 and No. 700. Therefore, the spectrum shown in Fig. 5.96 will be modeled in three segments divided by these two segment points.

The gray GM (1, 1) model is used to model the overall trend component of the spectrum as shown by the dashed line in Fig. 5.97a. Then, we shall use the gray-periodic modeling method to extract the frequency of the harmonics and the obtained sequence interval is 40. The model is shown in Fig. 5.97b. Multiplying the sequence interval with the test interval of 4875 Hz, we can get a harmonic interval of 195 kHz.

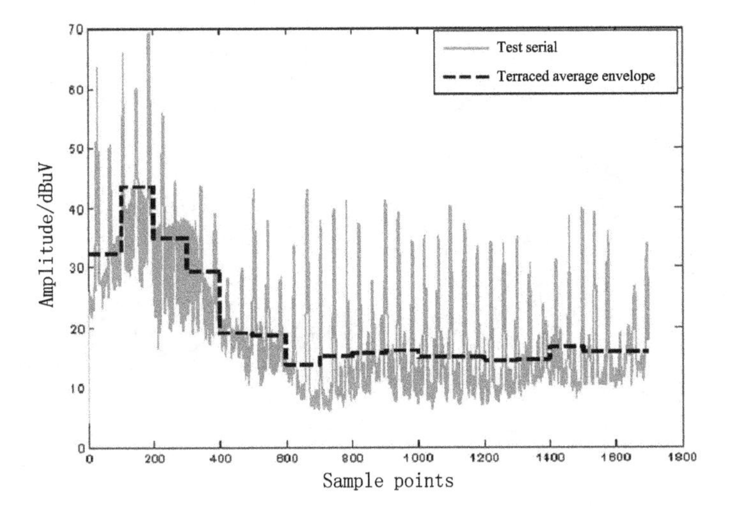

Fig. 5.96 Test data sequence and the trapezoidal envelope

Table 5.17 Segmentations of the conduction spectrum of the power supply

Serial number of spectrum (*100)	1	2	3	4	5	6	7	8
Trapezoidal envelope difference sequence	1	0	0	0	0	0	1	1
Whether a segmentation point	No	Yes	No	No	No	No	Yes	No
Serial number of spectrum (*100)	9	10	11	12	13	14	15	16
Trapezoidal envelope difference sequence	1	0	1	0	1	1	0	1
Whether a segmentation point	No	No	No	No	No	No	No	No

Since the on–off frequency of the transistor of the SMPS under test is 195 kHz, it can be inferred that the excessive portion in the frequency spectrum is mainly caused by the frequent on–off of the transistor. Therefore, we can add a filtering device to the power supply end in the circuit including the transistor or add the power factor correction circuit to change the emission of harmonics.

(3) Conduction emission analysis of a rotary motor. The comparison of the conduction emission spectrum of a rotary motor and the emission limit is shown in Fig. 5.98. Spectral analysis is performed by extracting the spectrum from 10 k to 250 kHz that are severely over standard. The extracted frequency spectrum is shown in Fig. 5.99, where the dashed line is the trapezoidal curve obtained by dividing it into segments and averaging the segments.

The same calculation method as in the previous example is used to obtain the difference sequence represented by 0/1, as listed in Table 5.18.

According to the calculation procedure, only one segmentation point is selected which is also the first point of the sequence. Therefore, this spectrum does not need to be segmented. We can build its gray GM (1, 1) model as shown in Fig. 5.100.

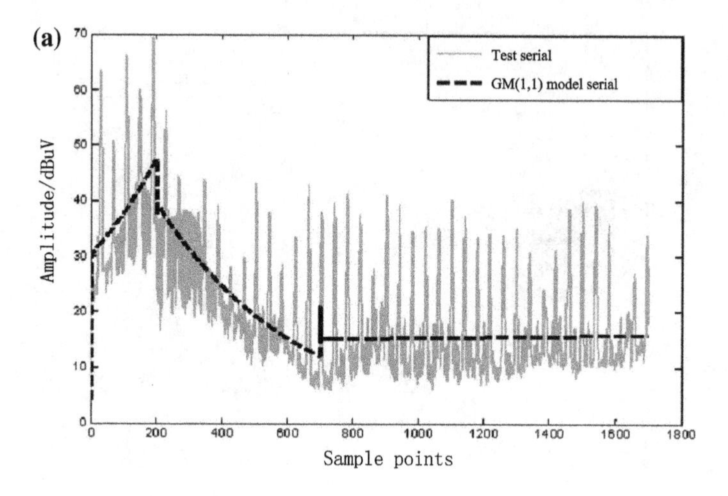

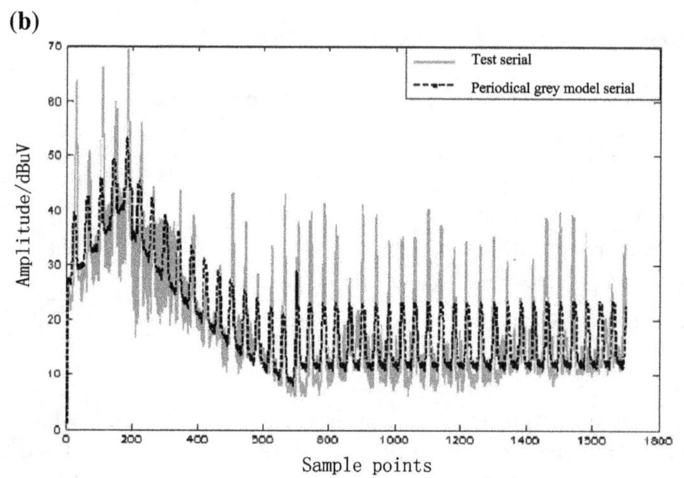

Fig. 5.97 Test data series and the overall trend components and the gray-periodic model sequence. **a** Test data series and the overall trend components. **b** Test data sequence and the gray-periodic model sequence

Table 5.18 Spectrum segmentations of the rotary motor

Serial number of spectrum (>100)	1	2	3	4	5	6	7	8	9
Trapezoidal envelope difference sequence	1	1	1	1	1	0	1	0	0
Whether a segmentation point	Yes	No	No	No	No	No	No	No	No

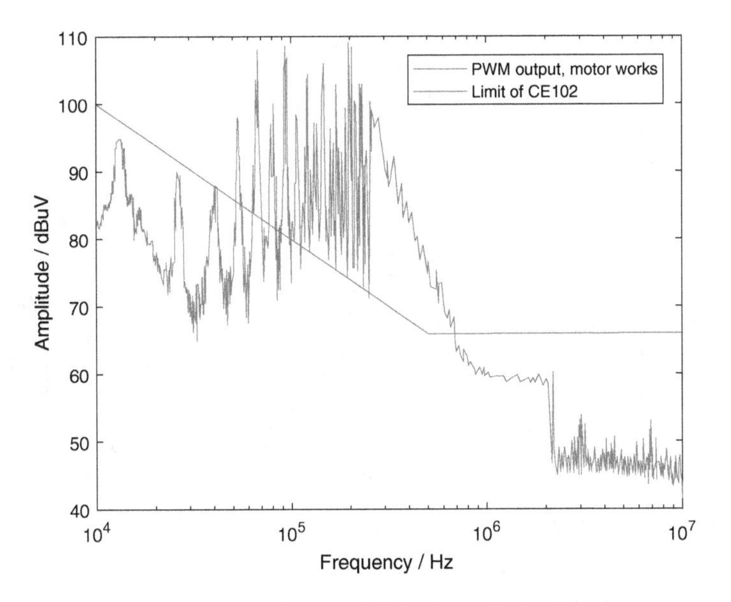

Fig. 5.98 Comparison of output waveforms and the emission limit when the rotary motors with PWM are switched on

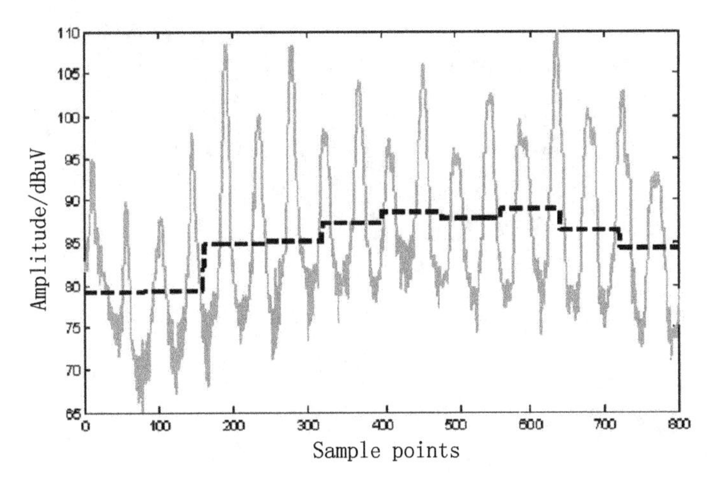

Fig. 5.99 Emission spectrum and its trapezoidal envelope of the rotary motor

Then, we build a periodic model of the residual sequence and add it to the gray GM (1, 1) sequence to get

$$\hat{x}^{(0)}(k) = 81.5775e^{0.00011(k-1)} + f_{44}(k) \tag{5.222}$$

It can be seen from (5.222) that the period parameter is 44. Then, we multiply it with the spectrum test interval 300 Hz and get the fundamental frequency of the

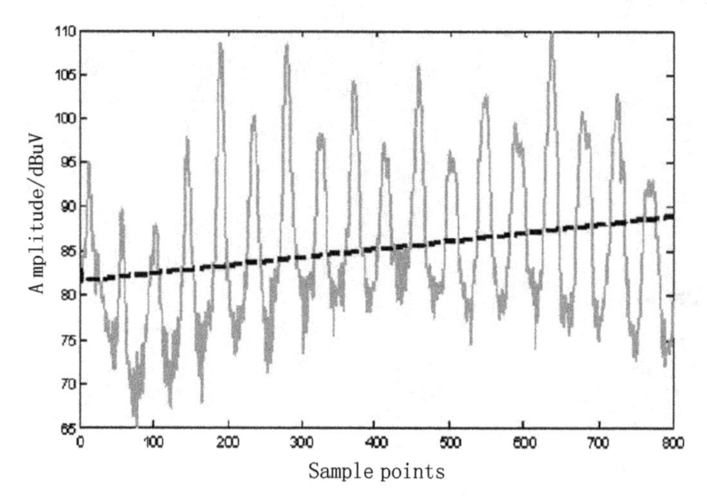

Fig. 5.100 Rotary motor test sequence and its overall trend component

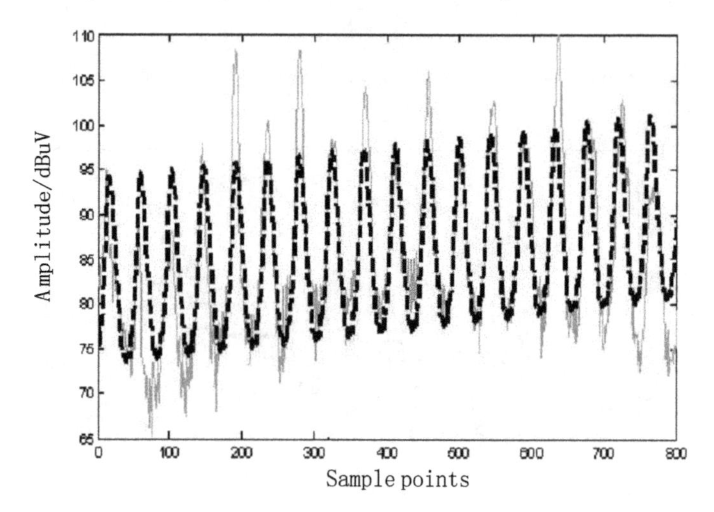

Fig. 5.101 Test sequence and gray-periodic model sequence

harmonics as 13.2 kHz. The established model and original sequence are shown in Fig. 5.101.

The amplitude of the harmonic component is extracted. Since the data is obtained from the conduction test, no amplitude modification is required. The envelope of its harmonic component is drawn directly with logarithmic coordinates as shown in Fig. 5.102. Then, we can remove the outliers in the harmonics, trace the envelope of the harmonics using the thick line in Fig. 5.102, and detect the slope turning point of the tenfold octave in the envelope using Fourier envelope analysis.

The harmonic frequency is 13.2 kHz, and the conduction emission test starts at 10 kHz. Since the complete information of the waveform cannot be obtained,

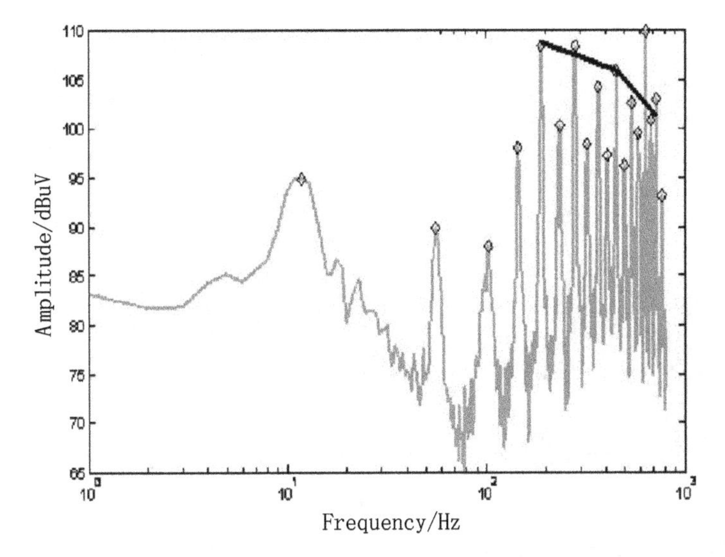

Fig. 5.102 Harmonic envelope of the test sequence

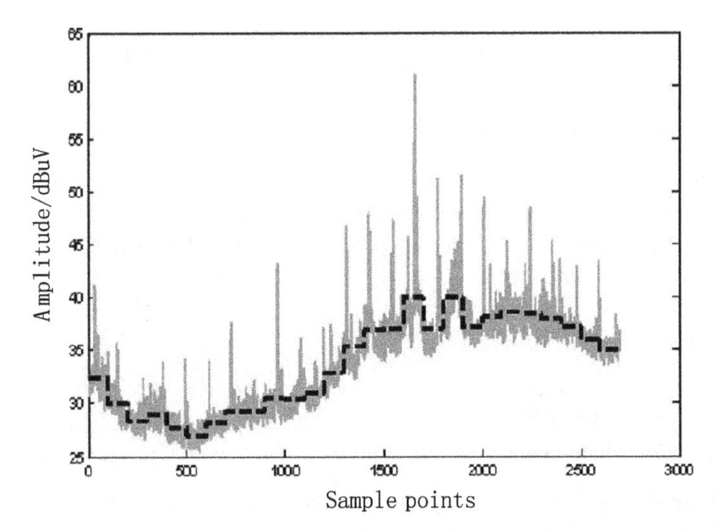

Fig. 5.103 Test data and its trapezoidal envelope

only the second turning point frequency of 146.5 kHz can be detected, of which the corresponding waveform indicates that the pulse rising time is 2.2 μs. The rotating frequency of the rotary motor is 13.2 kHz, and the triggering time of the PWM wave that controls its rotation is extremely short, which is the same as the analysis result in this section. Therefore, we can conclude that the interference emissions in the spectrum are generated by the PWM signal. The overshoot of the conduction emissions can be improved by setting the PWM control chip. If we cannot change

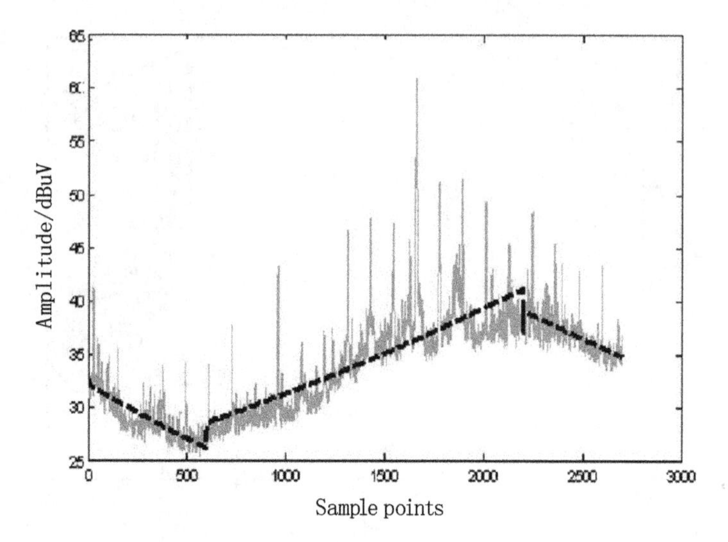

Fig. 5.104 Test data and the overall trend components

the time to control its rotation, we can try better shielding the chip or filtering the output signal.

(2) Radiation emission examples

The electromagnetic interference generated inside the electronic equipment can radiate to the space through PCB, the internal interconnection wires within the equipment, and various interface cables connecting the equipment to the outside.

The examples analyze the emission spectrum of an electronic equipment containing one harmonic component and multiple harmonic components. The measured data was all obtained from the one-meter radiation emission test.

(1) Analysis of radiation emission of a piece of navigation equipment. The radiation emission spectrum of a piece of navigation equipment is shown in Fig. 5.103. The dashed line is the trapezoidal curve obtained by averaging the segments.

The envelope difference sequence is calculated as listed in Table 5.19.

Equation (5.219) is used to calculate the three turning points with 0/1. However, because the identifier of the second point is the same as the first point, it can be omitted. Thus, there are only two segmentation points remaining, of which one is the 600th point of the sequence, and the other is the 2200th point. The sequence is divided into three parts by the two points.

The GM (1, 1) model is established separately for these three parts, and then the segments are connected. Finally, the overall trend component of the spectrum sequence is obtained as shown in Fig. 5.104.

Then, combining the periodic model with the gray system model, we obtain a gray-periodic combination model. The parameter in the model is calculated to be

Table 5.19 Segmentations of the conduction spectrum of the navigation equipment

Serial number of spectrum (>100)	1	2	3	4	5	6	7	8	9	10	11	12	13
Trapezoidal envelope difference sequence	0	0	1	0	0	1	1	1	1	0	1	1	1
Whether a segmentation point	No	No	No	No	No	No	No	No	No	No	Yes	No	No
Serial number of spectrum (>100)	14	15	16	17	18	19	20	21	22	23	24	25	26
Trapezoidal envelope difference sequence	1	1	1	0	1	0	1	1	0	0	0	0	0
Whether a segmentation point	No	No	No	No	No	No	No	No	Yes	No	No	No	No

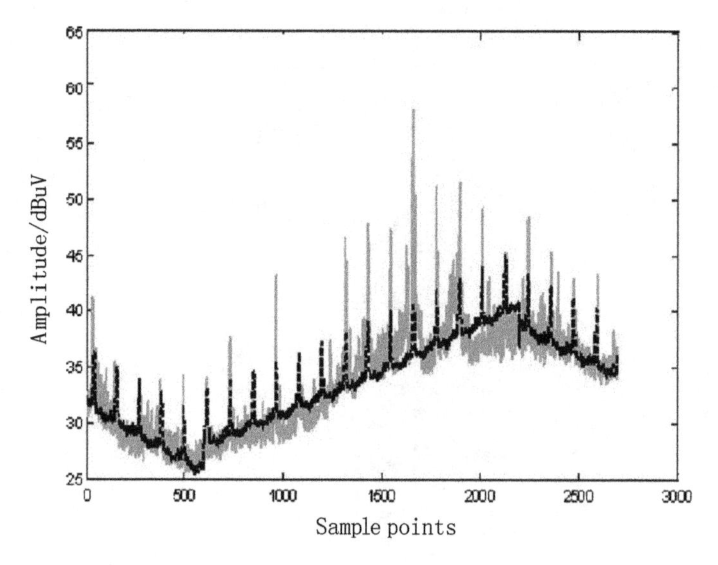

Fig. 5.105 Test sequence and the gray-periodic model sequence

$$\hat{x}^{(0)}(k) = 32.1701e^{-0.00034(k_1-1)} + 28.5493e^{0.00023(k_2-1)} + 39.1213e^{-0.00023(k_3-1)} + f_{117}(k) \tag{5.223}$$

The curve of the model sequence is shown in Fig. 5.105. It can be seen from the figure that the frequency of the harmonics obtained from the modeling agrees well with the actual measured curve. From (5.223), we can see that the periodic sequence interval is 117. Multiplying it with the test interval 47 kHz, we can get the fundamental frequency of the harmonic components as 5.5 MHz.

According to the obtained fundamental frequency information, we can get the amplitudes of the various harmonics. Then, we can use the "modification method of the radiation emission amplitude" in the section "harmonic amplitude characteristics extraction from spectrum" (case 3) to modify the amplitude. The spectrum and the values of the harmonics after modification are shown in Fig. 5.106a. Then, we draw the envelope and display it in logarithmic coordinates, as shown in Fig. 5.106b.

The fundamental frequency obtained from the modeling is 5.5 MHz. The radiation emission test starts from 3 MHz. Since the complete information of the waveform cannot be obtained, the information of the first turning point is lost. Only the second turning point of the envelope can be detected, and its corresponding frequency is 115.5666667 MHz. Based on the frequency value, we can calculate the rising time of the pulse waveform of the source and get a result of 2.75 ns. From the fundamental frequency information, the period of the waveform can be calculated as 181 ns. Assuming that the duty cycle of the waveform is 50%, it can be seen that the rising time of the pulse is too short, which leads to harmonic interference of the funda-

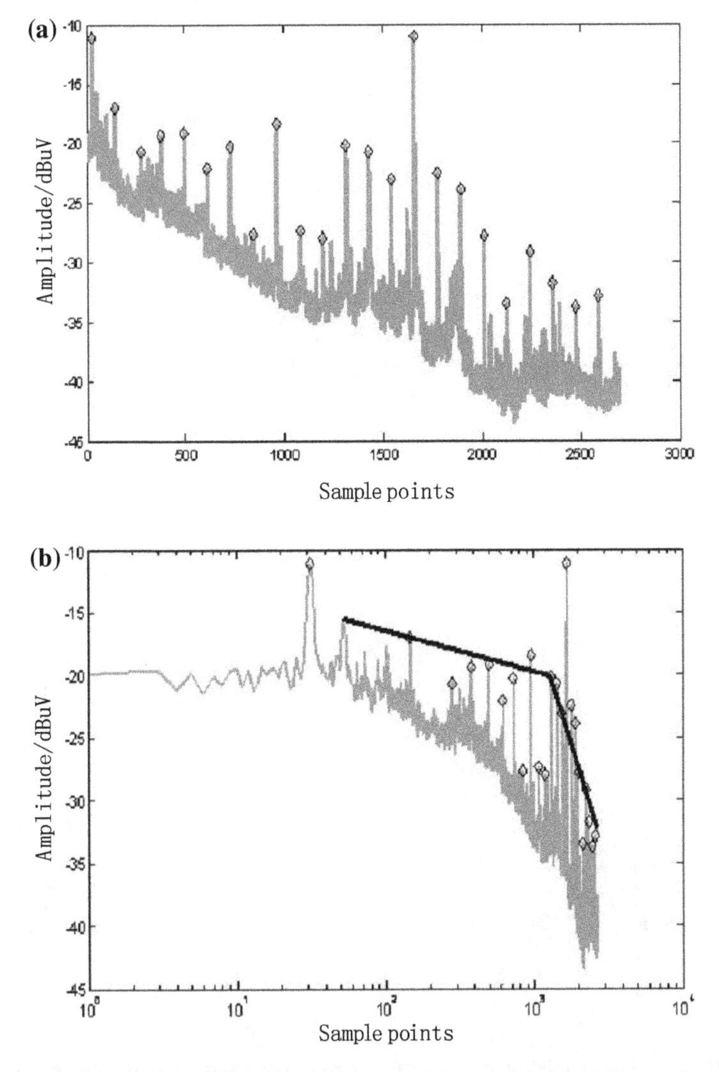

Fig. 5.106 Harmonics and harmonic envelope curves of the test sequence. **a** Harmonics of the test sequence after modification and **b** harmonic envelope curves after modification

mental frequency in an extremely wide frequency band. During the troubleshooting procedure of EMC, we can suppress the harmonics of the output waveform of the crystal clock or increase the rising time. For the later production of this product, it is necessary to select a crystal with better performance during the design phase and perform better chip-level filtering.

(2) Spectrum analysis of multi-harmonic equipment. In some tests, the spectrum data may contain harmonics generated by a plurality of different fundamental

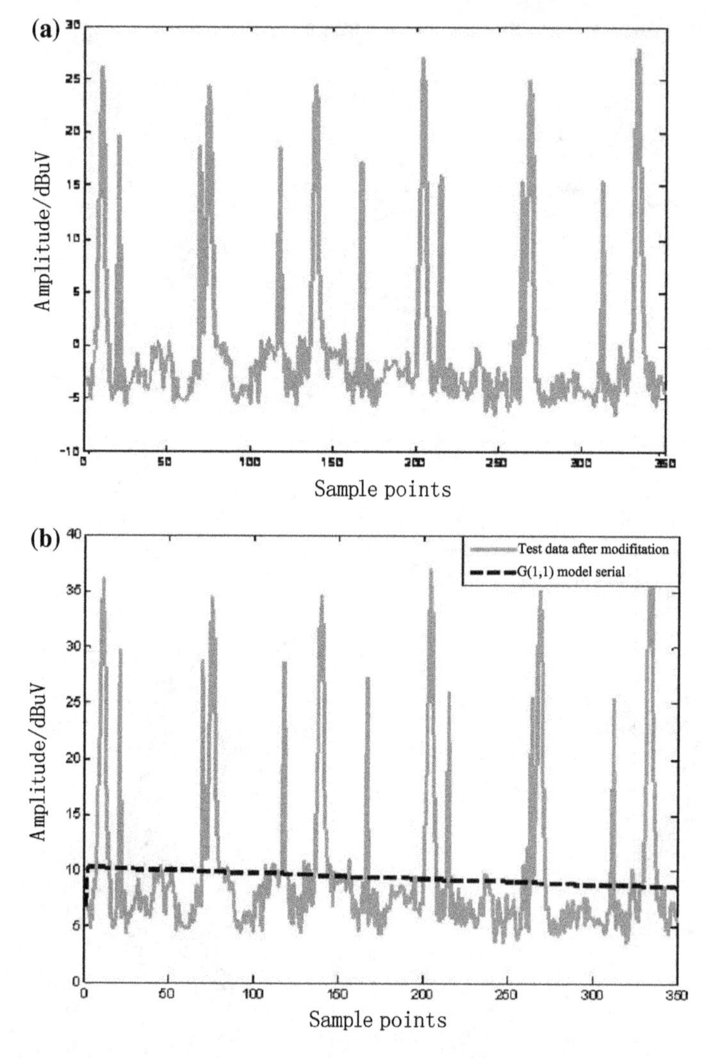

Fig. 5.107 GM (1, 1) model of the original test spectrum sequence (modified test sequence). **a** Original test spectrum sequence and **b** test sequence and the GM (1, 1) model

frequency components. For the analysis of this spectrum, a "multiple harmonic frequency extraction" method is used to extract the harmonic components of the spectrum.

First, we observe the test spectrum sequence which is shown in Fig. 5.107a. From the figure, we can see that the sequence is positive and negative staggered. Since the condition of gray modeling is that the sequence is nonnegative, and the effect of amplitude can be temporarily ignored when extracting the frequency characteristics, the minimum value in the spectrum sequence is detected as −6.3836. Therefore, the

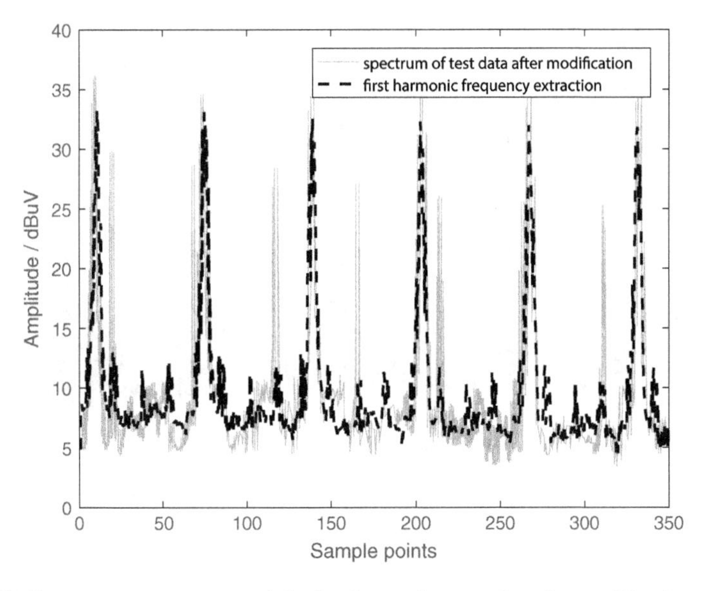

Fig. 5.108 Test spectrum sequence and the first harmonic extraction after modification

values of the entire sequence are added by 10 to ensure that all values are nonnegative, as shown by the solid line in Fig. 5.107b. Then, the overall trend component of the modified sequence is modeled, and the fitting function of the GM (1, 1) model is

$$\begin{aligned}\hat{x}^{(0)}(k) &= (1 - e^{a})(x^{(0)}(1) - \frac{b}{a})e^{-a(k-1)} \\ &= 10.2879e^{-0.0005(k-1)}\end{aligned} \quad (5.224)$$

The dominant period is calculated using the method of variance analysis in combination with the maximum deviation square sum between each group and the mean value, and the simulation sequence shown in Fig. 5.108 is obtained. This sequence contains a harmonic with an interval of 64. The amplitude of the harmonic component in the sequence containing a single harmonic of the model is set to be a random value within the range of the total trend component, and the range is calculated using Eq. (5.224) to be (6.8175, 10.2826). Therefore, we set the harmonic of cycle 64 to follow the uniform distribution of U (5, 11).

The resulting new model sequence is shown in Fig. 5.109.

The second harmonic of the new spectrum sequence is extracted to obtain a period that satisfies the conditions of variance analysis. The serial number is 48. The result of the second harmonic extraction is shown as the dashed line in Fig. 5.109.

The superposition of the two harmonics is compared with the original test spectrum sequence, as shown in Fig. 5.110. No other dominant period is found after detection of the residual sequence. At this point, the harmonic extraction modeling process ends.

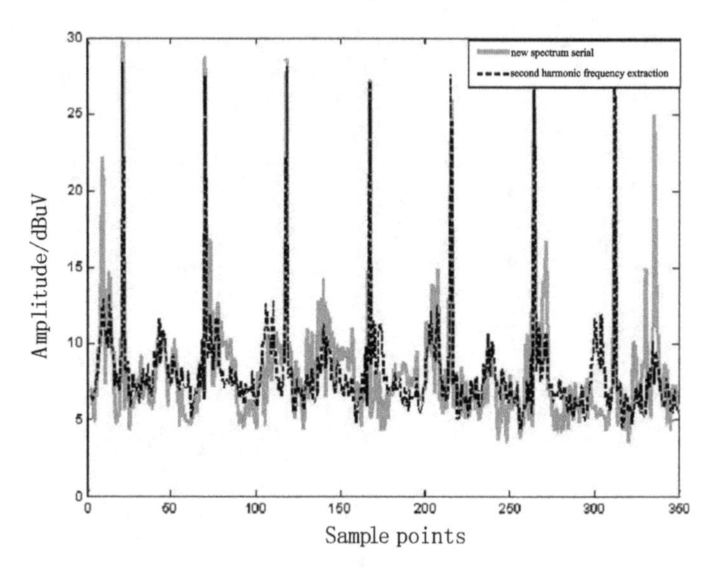

Fig. 5.109 New spectrum sequence and the second harmonic extraction

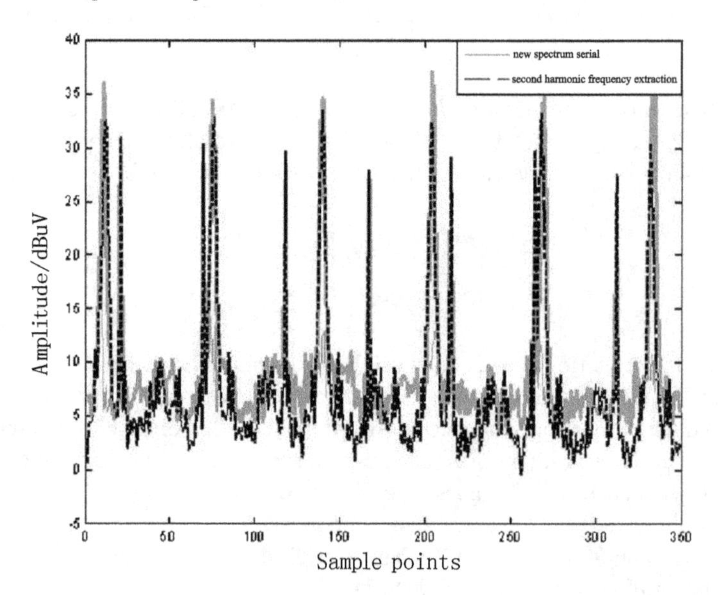

Fig. 5.110 Comparison of the original test spectrum sequences with multiple harmonic extractions

Chapter 6
Application Cases of System-Level EMC Quantitative Design and Analysis

Taking the system-level EMC quantitative design and evaluation methodology and software developed by the EMC Technology Institute of Beihang University as an example (Fig. 6.1), this chapter explains in detail about how to design and evaluate EMC quantitatively on system level, including: geometric modeling methods for EMC of the aircraft platform, quantitative evaluation of system-level EMC behavioral model, and the quantitative design of the system-level EMC behavioral model.

6.1 EMC Geometric Modeling Method for Aircraft Platform

To ensure the accuracy and efficiency of the quantitative system-level EMC design, the geometric model must be extracted and preprocessed to satisfy the requirements for simulation.

Aircraft platforms generally use CAD models including Catia, Pro-E, and UG as the input format. The required structure information can be extracted from the model files for reconstruction and fixing. Three models—wireframe, surface, and solid—are widely used in the CAD software currently. However, only solid model can represent 3D objects completely and unambiguously, while the other two cannot maintain the complete shape information.

1. Analysis of the Model Defects

The defects of the model need to be modified using different methods based on their manifestations. Most defects of the model are caused by the system precision. Because the topology and geometry of the model are separated in the CAD model, the defects caused by precision not only appear in the geometrical expression, but also in the topological expression. To modify the defects in topological expression, it is necessary to eliminate the nonregular defects at first; thus, the precision defects can be modified by using the adjacent boundary discretization method or surface intersection method to determine whether it is a geometric defect or a topological

© National Defense Industry Press and Springer Nature Singapore Pte Ltd. 2019

D. Su et al., *Theory and Methods of Quantification Design on System-Level Electromagnetic Compatibility*, https://doi.org/10.1007/978-981-13-3690-4_6

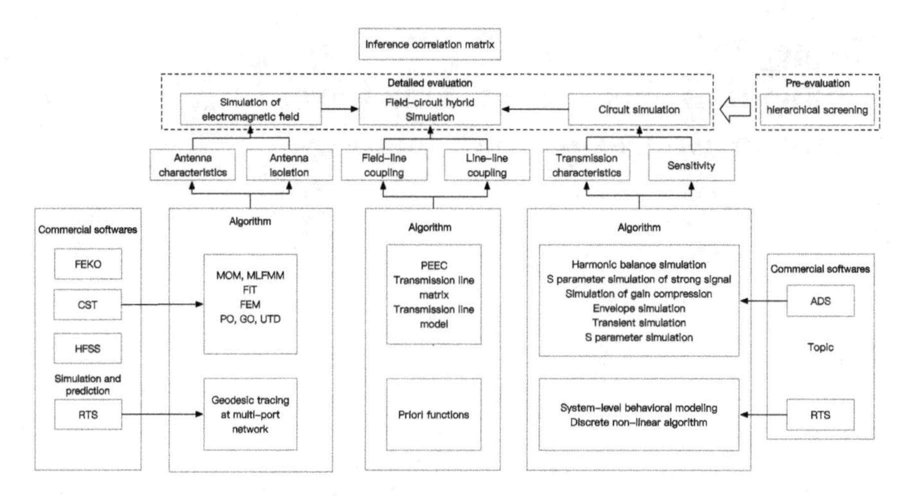

Fig. 6.1 Function framework of the system-level EMC quantitative design and evaluation software

defect. If the precision defect is topological, then the model can be fixed by re-intersection and topological reconstructing. If the precision defect is geometric, the surface must be fixed by modifying the mathematical definition of the geometric object, as shown in Fig. 6.2.

2. Structure Processing and Optimization of Electromagnetic Model used by Moment Method

The electromagnetic model for simulation analysis using moment method (MoM) can be obtained by building the electromagnetic model of the antenna, chamfering the geometric structure, and equivalenting typical radiation source and material. Then, structure processing and optimization can be performed toward the electromagnetic model.

When the MoM is used for simulation, structure processing and optimization of the electromagnetic model can be done as follows:

(1) Dipole and other linear antennas are all suitable to be equivalent to line model.
(2) Minimize geometry chamfering; if inevitable, simplify the chamfering properly during the electromagnetic modeling process.
(3) The triangulation patch model of the surface should follow the minimum criterion based on the square sum of the difference between any two sides as much as possible.
(4) When using the multi-level fast multi-pole algorithm (MLFMA), the iterative residuals of 0.01 is enough to meet the general engineering requirements.
(5) When choosing the integral equation, the combined field integral equation is preferred. If the target does not satisfy the closed condition, we should fill the target properly to make it applicable to the combined field integral equation. The combined factor could be other than 0.02.

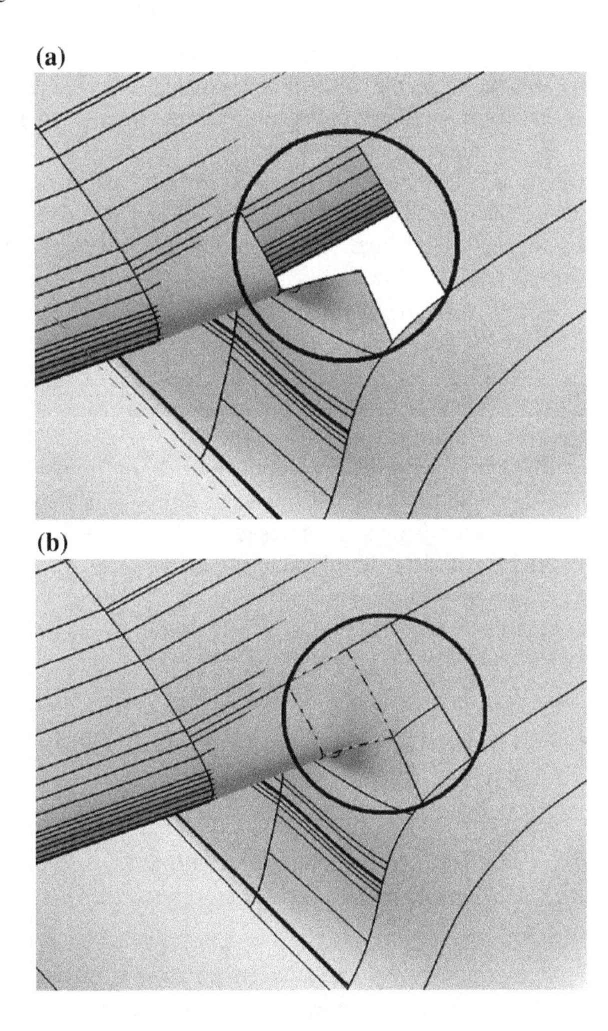

Fig. 6.2 Model comparison before and after fixing, **a** before fixing; **b** after fixing

(6) Typical antenna excitations, such as dipoles, can be simulated using discrete ports.

3. Structure Processing and Optimization Method of Electromagnetic Modeling for the Physical Optics Simulation

When building the electromagnetic model using high-frequency approximation method, if the electrical size of the local structure in the model does not conform to the precondition of the high-frequency approximation method, a large error would occur. Therefore, small structures in the model (such as small electronic equipment on the aircraft) can be simplified or ignored, because they do not satisfy the precondition of the high-frequency method and will bring in unnecessary densified grids for description in calculation.

When using the high-frequency method, small structures can be simplified or ignored based on the following assumption of physical optics: The curvature radius of the object on the surface is much larger than the incident wavelength.

When the wavelength of the electromagnetic wave is very short and the spatial variation of the mediums is slow enough, it can be assumed that the properties of the electromagnetic wave field in the local region are the same as those in the homogeneous medium. This way, the local wave surface of the electromagnetic wave is equivalent to the wave surface of a plane wave.

Specifically, the sphere can be used as a reference target for discrimination. For a sphere, its radius of curvature equals to its own radius. Therefore, the maximum size should satisfy $D \geq 10\lambda_0$. That is to say, when solving the scattering problem for electromagnetic wave of 10 GHz, small structures with a maximum size of 0.3 m could all be ignored from the target.

The structure processing and optimization method of electromagnetic modeling for simulation and analysis when using high-frequency approximation method (which is mainly physical optics method) could be summarized as follows: (1) Small structures could be simplified or ignored;

(2) The identification of small structure is based on the inference of the first basic hypothesis in physical optics $\left(\sqrt{\rho} >> \sqrt{\lambda_0}\right)$.

6.2 Quantitative Evaluation Methods for System-Level EMC

6.2.1 Interference Pair Determination and Interference Calculation

First, a coarse evaluation between antennas is carried out [57]. The purpose of the coarse evaluation is to eliminate the transceiver pairs that are clearly free of interference, thereby reducing the work required for the detailed evaluation in the next stage. The basic method is to perform simulation evaluation using a conservative simulation model and a loose electromagnetic emission limit.

Then, the method of screening by levels is adopted to predict EMC. It consists of two parts: (1) one-to-one transmission/receiving equipment pair prediction; that is, by calculating the interference margin between each transmitting and receiving pair, we can determine whether the interference margin meets the requirements and predict whether there is mutual interference between the two pieces of devices; (2) multi-to-one intermodulation interference prediction; that is, we analyze the working frequency bands of each piece of equipment in the platform, focusing on the interference of two signals and three signals intermodulation to decide whether the intermodulated signal will interfere with the receiver, as shown in Fig. 6.3.

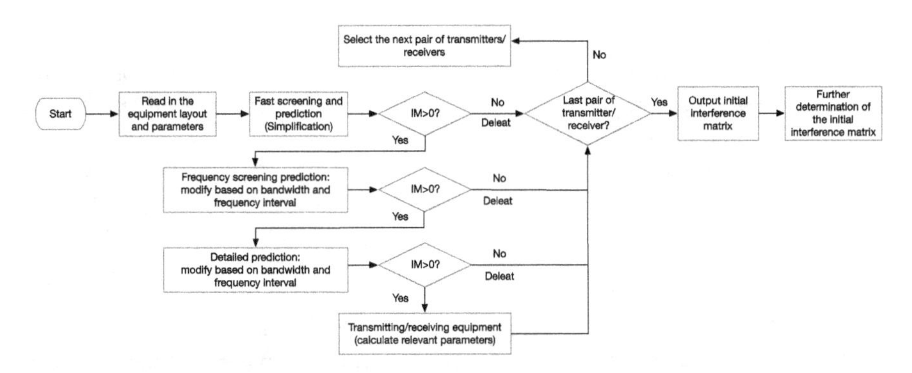

Fig. 6.3 Interference matrix generation process based on hierarchical screening prediction technology

1. Model and method of fast screening

The main purpose of the fast screening is to make a preliminary judgment from the frequency point of view, delete the transmitting/receiving pairs that do not interfere with each other, and predict the type of interference that may exist in the transmitting/receiving pairs, such that we can provide evidence for model selection for the interference margin calculation in the amplitude screening. According to our experience of EMC engineering design, when performing fast screening, the following frequency limit assumptions for transmitter and receiver are made:

(1) Assume the minimum spurious frequency of the transmitter is $(f_T)_{\min}$, and it can be denoted as $0.1 \times f_T$.
(2) Assume the maximum spurious frequency of the transmitter is $(f_T)_{\max}$, and it can be denoted as $10 \times f_T$.
(3) Assume the minimum spurious frequency of the receiver is $(f_R)_{\min}$, and it can be denoted as $0.1 \times f_R$.
(4) Assume the maximum spurious frequency of the receiver is $(f_R)_{\max}$, and it can be denoted as $10 \times f_R$.

Based on the above assumptions, the following judgments can be made:

(1) If $|f_T - f_R| < 0.2 \times f_R$, there is a fundamental interference margin;
(2) If $0.1 \times f_R < f_T < 10 \times f_R$, there is a transmitter interference margin;
(3) If $0.1 \times f_T < f_T < 10 \times f_T$, there is a receiver interference margin;
(4) If $0.1 \times f_T < 10 \times f_R$ or $0.1 \times f_R < 10 \times f_T$, there is a clutter interference margin;

where f_T is the transmitter center frequency and f_R is the receiver center frequency.

2. Model and method of amplitude screening

The purpose of the amplitude screening is: Based on the amplitude of the interference signal, calculate and analyze the interference margin between the transmitting source and the susceptive device, remove as many obvious noninterference conditions as possible, perform frequency screening on equipment with large interference margins, and correct the interference margins. Interference margins appear in four forms: fundamental interference margin (FIM), transmitter interference margin (TIM), receiver interference margin (RIM), and spurious interference margin (SIM).

When it is determined that there is interference according to the fast screening, the corresponding interference margin calculation is performed according to the type of interference. The fundamental interference margin level is first calculated. If the interference margin is less than the screening level, there is no need to calculate the other three cases because their margin is even lower. Conversely, if the interference margin of the fundamental interference exceeds −6 dB, the TIM and RIM need to be calculated. If either of the interference margins generated by either of them exceeds −6 dB, the SIM needs to be calculated. When expressed in dB, the model for calculating the interference margin is

$$IM = P_T + G_T - L_{TR} + G_R - P_R + CF \tag{6.1}$$

where P_T is the transmitting power of the transmitting equipment T_m at the transmitting frequency f_T, and this indicator needs to be confirmed according to the transmitter design indicator; G_T is the gain of the transmitting antenna at f_T; L_{TR} is the isolation between the transmitting antenna and the receiving antenna, which can be calculated using an electromagnetic field calculation software. It can also be simplified to free space attenuation and get solved using the geodesic method; P_R is the susceptibility threshold level of the receiver R_n when the receiving frequency is f_T; CF is the correction coefficient for the interference margin, which is not considered in the amplitude analysis. CF is set to zero in the amplitude screening and will be calculated in the frequency screening.

In order to calculate the interference margin, it is necessary to guild a transmitter model, an antenna model, a transmission model, and a receiver model. The models are built based on the baseband characteristics such that we can obtain the amplitude–frequency characteristics including harmonics and out-band characteristics. Then, we can calculate the value of each item in formula (6.1), and substitute it into the calculation to get the interference margin.

The isolation L_{TR} is calculated using the geodesic method based on the minimum angle.

3. Model and method of frequency screening

Correction factor CF is confirmed in the following two cases:

(1) Tuning. $\Delta f \leq (B_T + B_R)/2$, where $\Delta f = |f_e - f_r|$, B_T is the 3 dB bandwidth of the transmitter, B_R is the 3 dB bandwidth of the receiver. The tuning case is shown in Fig. 6.4a.

In this case, if the receiver bandwidth B_R is greater than or equal to the transmitter bandwidth B_T, then the correction factor CF is zero. If the receiver bandwidth B_R is less than the transmitter bandwidth B_T, then the calculation model for the correction factor CF is calculated using $CF = K \log(B_R/B_T)$ (dB), where K is the constant

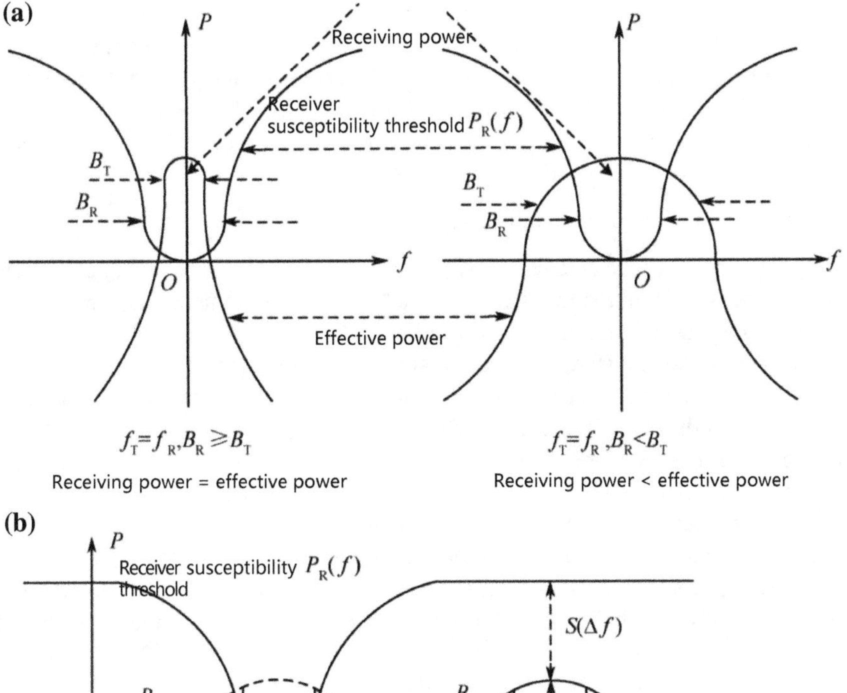

Fig. 6.4 Tuning and detuning, **a** tuning; **b** detuning

for a particular transmitting/receiving pair (when $B_R \geq B_T$, or aligned with the same channel, $K = 0$; when the noise-like signal of the RMS level is used, and $B_R < B_T$, there is $K = 10$; when using the peak level of the pulse signal, and $B_R < B_T$, then $K = 20$). (2) Detuning. When $\Delta f > (B_R + B_T)/2$, i.e., the transmitter and receiver center frequency deviates, the diagram of the detuning is shown in Fig. 6.4b.

In this situation, the transmitter power can enter the receiver in two ways. ① The transmitter transmits a modulation sideband into the receiver at the primary response frequency. The correction factor is $CF_t = K \log(B_R/B_T) + M(\Delta f)$ (dB), where $M(\Delta f)$ is the modulation sideband level (dB) above the transmitter power at the frequency interval Δf; K is the same as the tuning case. ② The power of the transmitter's main output frequency enters the receiver's detuned response, and then $CF_r(\Delta f) = -S(\Delta f)$ (dB), where $S(\Delta f)$ is higher than the receiver's selectivity (dB) of the fundamental sensitivity when the frequency interval is Δf. The correction factor added to the interference margin model is $CF = \max(CF_t, CF_r)$.

4. Models and methods of detailed screening

The main purpose of the detailed analysis is to analyze the electromagnetic interference effects (such as intermodulation) of each piece of equipment and consider the corresponding receiver desensitization and other factors.

Intermodulation interference means that two or more signals are mixed at a nonlinear component to produce new signal frequency components that can be amplified and detected if they are close enough to the receiver's tuning frequency. The mechanism is the same as functional signals and thus may cause performance degradation. They are equivalent to a linear combination of integer multiples of each signal frequency, and the interference frequency is predictable.

Desensitization means that when the receiver receives one or more strong nonfunctional signals in a channel adjacent to the tuning frequency of the receiver, the nonlinearity of the front end of the receiver causes the gain of the functional signal to decrease.

The signal-to-noise ratio without interference is

$$S/N = P_D - P_{REF} + (S/N)_{REF}(\text{dB})$$

where P_D is the function signal level (dBm); P_{REF} is the reference signal level (dBm); $(S/N)_{REF}$ is the signal-to-noise ratio of the reference signal level.

When the interfering signal power is greater than the saturation threshold of the receiver front end, the signal-to-noise ratio begins to decrease. The desensitization signal-to-noise ratio is

$$(S/N)' = S/N - (P_A - P_{SAT})/R(\text{dB})$$

where S/N is the signal-to-noise ratio (dB) generated by the level of the functional signal without interference; P_A is the power of the interference signal at the input of the receiver (dBm); P_{SAT} is the saturation level of the receiver front end (dBm); R is the rate of desensitization.

The saturation level of the receiver front end is

$$P_{SAT} = P_B + 10lg(\Delta f/f_{OR})$$

where P_B is the saturation reference power (dBm) of the receiver; $\Delta f/f_{OR}$ is the interference transmission frequency interval relative to the receiver's tuning frequency.

The amount of desensitization is

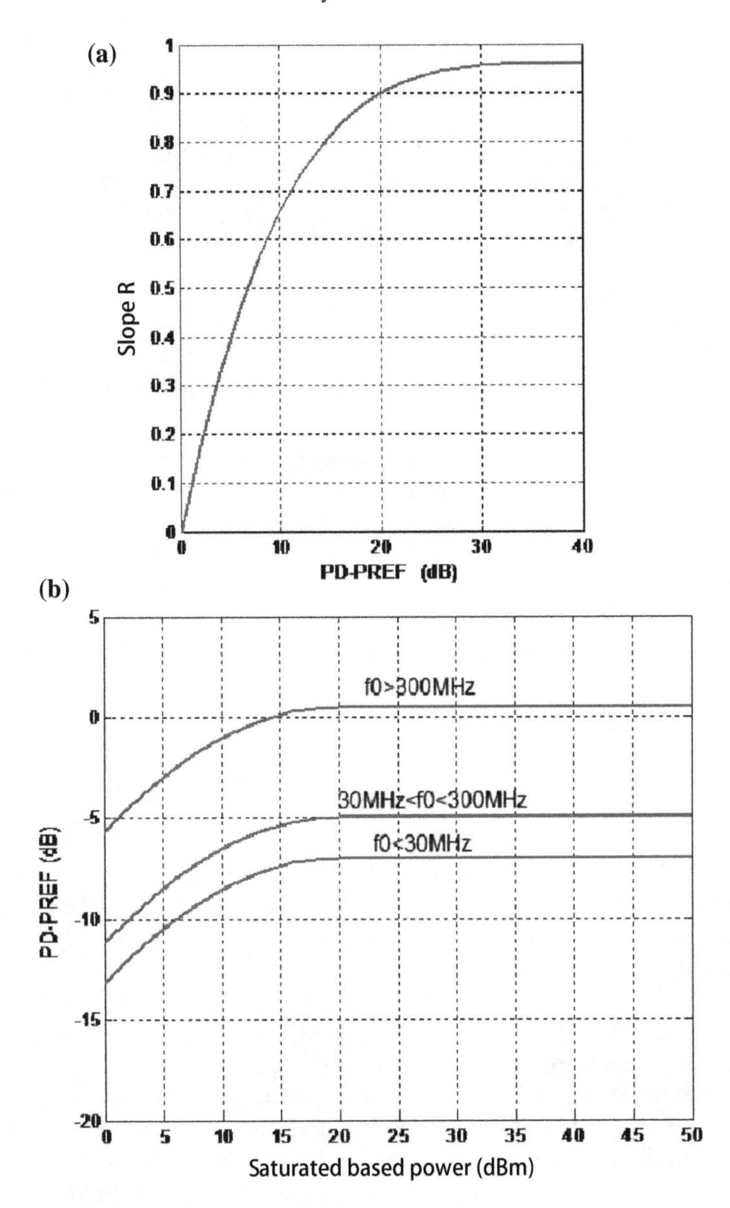

Fig. 6.5 Desensitization rate and saturation reference power curve, **a** desensitization rate; **b** saturation reference power curve

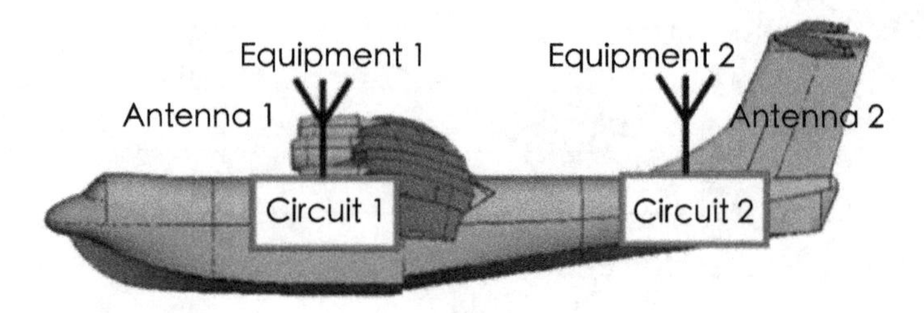

Fig. 6.6 Diagram of the equipment installation

$$P_{SAT} = P_B + 10\lg(\Delta f / f_{OR})$$

where P_A, P_B, $\Delta f / f_{OR}$ and (R) are the same as previously defined.

Since the level of the functional signal P_D is usually unknown, it is considered that $P_D - P_{REF}$ is in a dynamic range, and the amount of desensitization calculated in this situation will be smaller. According to the desensitization curve shown in Fig. 6.5a, the slope R can be obtained using $P_D - P_{REF}$; according to the saturated reference power curve shown in Fig. 6.5b, the saturation reference power P_B of the receiver can be calculated using $P_D - P_{REF}$; then, we can obtain the desensitization value.

6.2.2 Field–Circuit Collaborative Evaluation Technique

The electromagnetic field generated by external electromagnetic field, antenna radiation field, case and cable can be equivalent to sources at circuit level. Then, we can do comprehensive EMC simulation at circuit level. This technology is called field–circuit hybrid technology.

Aircraft platforms, airborne equipment, installed cables, and antennas form complex conduction or radiation coupling cross-linking relationships, resulting in complex field–circuit coupling relationships among aircraft equipment/subsystems. Therefore, it is necessary to carry out comprehensive simulation analysis of the system from the aspect of electromagnetic field and circuit; i.e., we need to perform corresponding electromagnetic simulation, and fully consider the influence of electromagnetic environment in equipment-level circuit simulation analysis.

Figure 6.6 shows the pair of transmitting/receiving equipment that are likely to interfere with each other in the interference pair decision. Both equipment 1 and equipment 2 are composed of an antenna and a circuit, respectively. When analyzing the mutual interference situation of this transmitting–receiving equipment pair, it is necessary to separately model and simulate the antenna and circuit of the two equipment, and equivalent the field simulation result into a circuit module, which is later imported into the circuit for synthesis simulation.

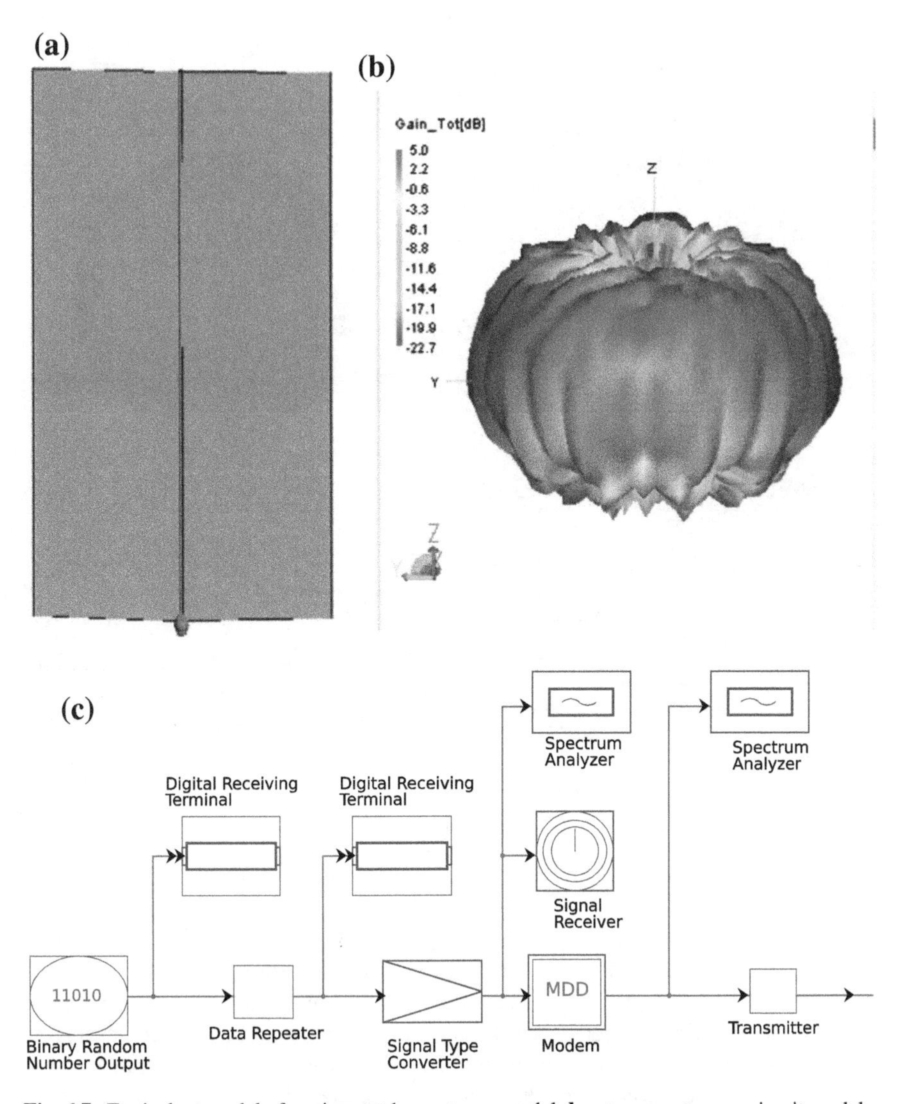

Fig. 6.7 Equivalent model of equipment 1, **a** antenna model; **b** antenna pattern; **c** circuit model

The antenna of equipment 1 is modeled as shown in Fig. 6.7a. The three-dimensional far-field characteristics of antenna 1 are shown in Fig. 6.7b. The circuit of equipment 1 is modeled as shown in Fig. 6.7c.

The antenna of the equipment 2 is modeled as shown in Fig. 6.8a, and the three-dimensional far-field characteristics of antenna 2 are obtained by simulation as shown in Fig. 6.8b; the circuit of the equipment 2 is modeled as shown in Fig. 6.8c.

In the comprehensive simulation model in Fig. 6.9, the antenna simulation results, transmission channel model, and equipment behavioral circuit model of the two

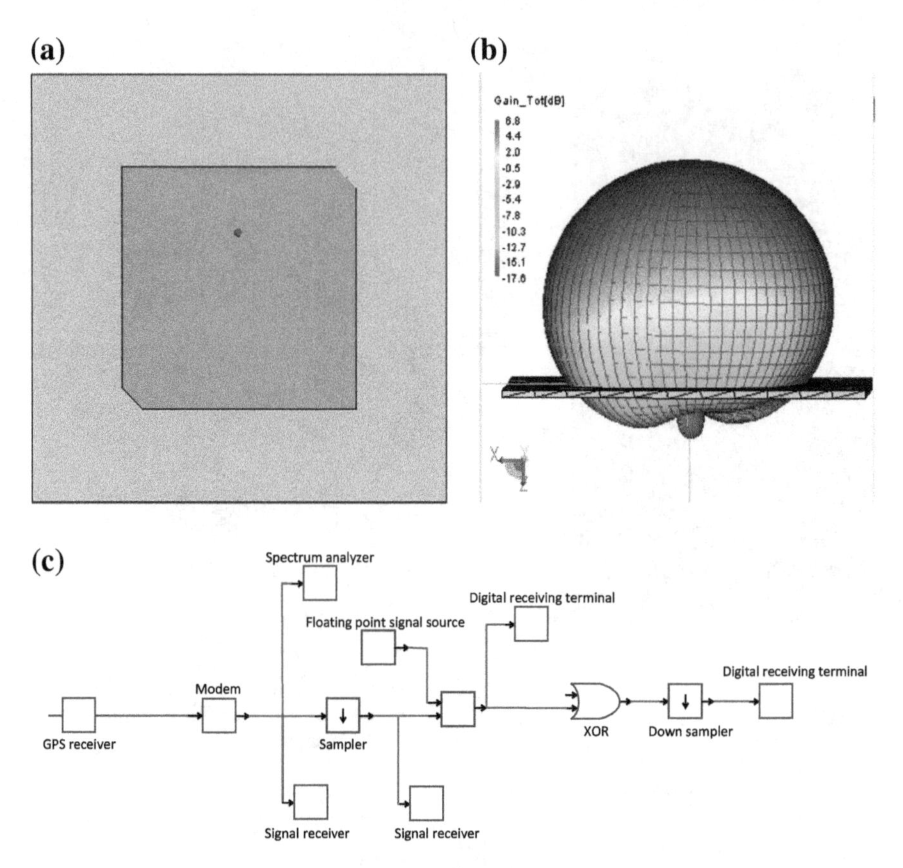

Fig. 6.8 Equivalent model of equipment 2, **a** antenna model; **b** antenna pattern; **c** circuit model

equipment are comprehensively analyzed. Two aspects are required for the analysis, which are the "field" aspect and the "equivalent circuit" aspect. Using the "field" method, we can directly solve the system electromagnetic radiation problems according to Maxwell's equations using the finite difference method, transmission line matrix method, moment method, and time-domain integral equation method. This kind of methods is theoretically strict; thus, the calculation results are accurate. However, the calculation time and memory requirements are relatively high in practical applications. The "equivalent circuit" method mainly refers to the circuit equivalent to the transmission model. With certain approximation, the mutual coupling relationship between equipment is simplified into an equivalent transmission module in circuit analysis by simplifying the coupling channel model and the radiation emission characteristics of the antenna. The transmission module is then converted into a circuit simulation module that can be imported into the circuit simulation to perform the complete analysis of the field of the system.

In the transmission module channel of Fig. 6.9, the isolation between equipment is calculated. It mainly includes two aspects: first, the calculation of the antenna gains

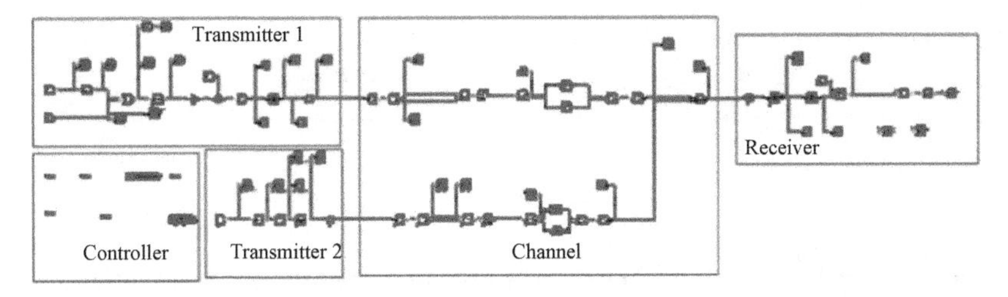

Fig. 6.9 Field–circuit collaborative simulation model of equipment 1 and equipment 2

in the corresponding directions of the two antennas; and second, the calculation of the path attenuation. Then, the isolation can be expressed as $L_{total} = G_1 - L + G_2$. The gain patterns simulated by the antenna 1 and the antenna 2 are, respectively, imported into the program, and according to the installed position (X_1, Y_1, Z_1) of the antenna 1 and the installed position (X, Y_2, Z_2) of the antenna 2, the antenna gain of the antenna 1 in the direction of the antenna 2 which is denoted as G_1 and the gain of antenna 2 in the direction of antenna 1 which is denoted as G_2 are calculated, respectively. Then, the spatial isolation between the antennas is calculated according to the type of the selected channel; thus, the isolation between the equipment is obtained. When the antenna needs local optimization, because the range for position adjustment is small, the change of the antenna gain pattern can be neglected. Therefore, the isolation between the two equipment after the position adjustment can be quickly obtained, and the scheme can be quickly adjusted accordingly. When the antenna needs to be adjusted in a large range of positions, it is necessary to reperform the field analysis, import the new gain pattern into the program, and then calculate the corresponding isolation.

The technology of directly incorporating the equivalent circuit model of the antenna into the system circuit model and participating in the system-level EMC simulation is still in research [16].

6.2.3 *The Method of EMC Coordination Evaluation*

After the interference prediction, if there is mutual interference between some RF equipment in the aircraft platform, the design scheme needs to be adjusted. In the adjustment, it is necessary to analyze the degree of mutual influence of these equipment, so as to identify equipment that requires more protection or suppression [67]. The DEMATEL method is a suitable system analysis method, which has been well applied in many fields; it is based on the matrix operation to obtain the mutual influence relationship, which matches the idea of interference correlation matrix in interference prediction. Thus, the DEMATEL method is used to analyze the influence degree and degree of being influenced by the equipment in the scheme. Through analysis, the main interference sources and susceptive equipment in the onboard RF

Table 6.1 Mapping of interference correlation matrix and quantitative correlation matrix

C_{ij}	No interference	<−30 dB	<−16.5 dB	<−6 dB	≤0 dB	>0 dB
M_{ij}	0	1	2	3	4	5

equipment can be identified. The steps to analyze the influence the airborne RF equipment gets and the influence it has toward other equipment are as follows:

1. The quantization correlation matrix **M** is obtained by mapping the interference correlation matrix **C**.

Each element in the interference correlation matrix represents the interference margin of the corresponding equipment pair and reflects the one-to-one interference correlation relationship between the two equipment. After the interference correlation matrix **C** is obtained, we need to perform numerical mapping using the mapping relationship listed in Table 6.1. The mapping relationship is based on the margin requirements of the EMC Engineering Design Manual and GJB 1389. When the margin is larger than 0 dB, there is interference; when the margin is −6 to 0 dB, the interference is at a critical state. The interference margin requirement of common equipment is less than −6 dB; for pyrotechnics, the requirement is less than −16.5 dB; for safety equipment, the requirement is less than −30 dB. This book uses six levels {0, 1, 2, 3, 4, 5} to map the six corresponding interference margin ranges. The higher the level, the greater the possibility of mutual interference between the two equipment.

According to the quantitative correlation relationship showed in Table 6.1, the interference correlation matrix can be transformed into a quantitative correlation matrix, as follows:

$$\mathbf{C} = \begin{bmatrix} C_{11} & C_{12} & \cdots & C_{1q} \\ C_{21} & C_{22} & \cdots & C_{2q} \\ \vdots & \vdots & \ddots & \vdots \\ C_{p1} & C_{p2} & \cdots & C_{pq} \end{bmatrix} \tag{6.2}$$

$$\mathbf{M} = \begin{bmatrix} M_{11} & M_{12} & \cdots & M_{1q} \\ M_{21} & M_{22} & \cdots & M_{2q} \\ \vdots & \vdots & \ddots & \vdots \\ M_{p1} & M_{p2} & \cdots & M_{pq} \end{bmatrix} \tag{6.3}$$

2. Expand the quantitative correlation matrix **M** to obtain the direct influence matrix **D**.

Since the number of transmitting equipment and receiving equipment in an aircraft is usually not equal, the size of matrix **M** is p × q, and it is usually not a square matrix. However, the operation of the DEMATEL method requires a square matrix, so the quantitative correlation matrix **M** needs to be expanded to the direct influence matrix **D**. The principle of expansion is as follows: The equipment set TR contains p transmitting equipment and q receiving equipment. The total number of equipment pieces in TR is n (equipment that is used both for transmitting and receiving is counted as 1 in TR). The matrix element of the existing equipment pair is the interference margin. When there is no equipment pair, the corresponding element is set to 0.

The direct influence matrix can be obtained by expansion as

$$\mathbf{D} = \begin{bmatrix} D_{11} & D_{12} & \cdots & D_{1n} \\ D_{21} & D_{22} & \cdots & D_{2n} \\ \vdots & \vdots & \ddots & \vdots \\ D_{n1} & D_{n2} & \cdots & D_{nn} \end{bmatrix} \tag{6.3}$$

Then, the comprehensive influence matrix T can be derived from the direct influence matrix **D**. The direct influence matrix **D** can be normalized to be

$$\mathbf{G} = \frac{1}{\max\limits_{1 \le i \le n} \sum_{j=1}^{n} D_{ij}} \mathbf{D} \tag{6.1}$$

Then, the comprehensive influence matrix can be calculated as

$$\mathbf{T} = \mathbf{G}(\mathbf{I} - \mathbf{G})^{-1}.$$

where **I** is the identity matrix.

(3) $\mathbf{T_f}$ is the vector describing the degree of influence the equipment has on others. $\mathbf{T_e}$ is the vector describing the degree of influence the equipment gets. $\mathbf{T_f}$ and $\mathbf{T_e}$ can be obtained from the comprehensive influence matrix **T**.

The influence degree of the equipment S_i can be calculated using $f_i = \sum_{j=1}^{n} T_{ij}$. The influence degree of all equipment constitutes a vector $\mathbf{T_f}$ ($\mathbf{T_f}$ is the influence degree vector of the equipment, which represents the order of influence of the equipment on other equipment. It is used to determine the major interference source). Then, we can calculate the degree of influence the equipment S_j received using $e_i = \sum_{j=1}^{n} T_{ji}$. The degree of influence that all equipment received constitutes a vector $\mathbf{T_e}$ ($\mathbf{T_e}$ is a vector describing the degree of influence the equipment receives, which represents the order of the influence strength the equipment receives. It is used to determine the major susceptive equipment.).

4. Examples

There are 12 transmitting equipment and 17 receiving equipment in an aircraft. 11 of them are used both for transmitting and receiving and are denoted as $\{TR_1\ TR_2 \cdots TR_{11}\}$; there is one equipment that is only used for transmitting, which is denoted as $\{T_1\}$; there are six equipment that are only used for receiving, which are denoted as $\{R_1\ R_2 \cdots R_6\}$. Tables 6.2 and 6.3 provided partial parameters for the transmitting and receiving equipment, respectively. In the parameter list, the working frequency input is the central operating frequency of these equipment; the working bandwidth input is the maximum operating frequency range of the equipment; the transmitting power refers to the rated power, and the installed position refers to the location of the equipment on the same Cartesian coordinate system of the aircraft. The azimuth refers to the horizontal center direction of the main lobe of the antenna. The beam width of the azimuth refers to the lobe width of the antenna in the horizontal direction. The elevation angle refers to the vertical center direction of the main lobe of the antenna. The beam width of the elevation angle refers to the lobe width of the antenna in the vertical direction.

The transmitting equipment parameters and receiving equipment parameters in Tables 6.2 and 6.3 are imported into the software. Then, the interference residual is calculated for each equipment pair based on the hierarchical screening method, and the interference correlation matrix as shown in Table 6.4 is obtained.

In the interference correlation matrix, "Y" indicates no interference. There are three scenarios: (1) The equipment both transmits and receives, but does not interfere with itself; (2) there is not any possibility that the transmitting and receiving equipment overlap in spectrum based on the initial analysis of frequency, so they will not interfere with each other; (3) the two pieces of equipment do not overlap in the working time based on the working time analysis in the detailed screening, so the two pieces of equipment do not have the possibility of mutual interference.

The rest of the interference correlation matrix is the interference margin of the two pieces of equipment. The margin is calculated by amplitude filtering and corrected by frequency filtering. It can be seen from the interference correlation matrix that 204 pairs of transmitting and receiving equipment are formed by the RF equipment. Through calculation, we find that there are 24 equipment pairs that exhibited the possibility of interference, 3 pairs of equipment have critical interference, and 177 pairs do not interfere with each other. It can be seen that there is a lot of equipment with interference possibility, and the range of interference margin varied greatly. For this example, the results of the analysis are reasonable. It can be seen from the parameter list of the transmitting equipment and the receiving equipment that many pieces of equipment have a large working bandwidth, and the bandwidth here refers to in which the equipment can always work, rather than the bandwidth used for only one operation. For example, if the shortwave can work at 2–30 MHz, the working bandwidth in the list is 28 MHz, but the actual shortwave communication uses only a very narrow bandwidth. As a result, some pieces of equipment do not have fundamental interference in actual operation, but is determined to have fundamental interference as in the example. Therefore, the calculated interference margin is large, and there are many equipment pairs with possible interference. The interference margin for some equipment pairs is small because the interference level lower than the background noise has not been considered as the background noise level, but the theoretical value calculated from the model is used, which helps to determine the possibility of mutual interference of equipment.

Table 6.2 Parameters of the transmitting equipment

Equipment name	Operation frequency/MHz	Operation bandwidth/MHz	Transmission power/dBm	Modulation mode	Gain/dB	Polarization mode	Installation location/m (X, Y, Z)	(Azimuth, beam width)/(°)	(Elevation angle, beam width)/(°)
TR1	16	28	53	FM	3	Vertical	(116.973, −0.645, 2)	(0, 360)	(0, 180)
TR2	141	66	44	FM	1.5	Vertical	(99.272, −0.645, 1.602)	(0, 360)	(80, 100)
TR3	157	2	39	FM	2	Vertical	(106.075, −17.591, 1.7)	(0, 360)	(80, 100)
TR4	245.5	5	38	FM	2.4	Vertical	(106.074, −12.377, 1.7)	(0, 360)	(80, 100)
TR5	1087.5	125	57	PM	3	Vertical	(106.074, 10.309, 1.7)	(0, 360)	(100, 100)
TR6	1090	3	57	FM	1.5	Vertical	(123.929, −0.645, 9.2)	(0, 360)	(80, 100)
TR7	1090	10	60	FM	4	Vertical	(92.753, −0.645, 1.6)	(0, 360)	(80, 100)
TR8	1621	10	42	FM	12	Right handed circular	(110.740, −0.645, 1.6)	(0, 360)	(90, 30)
TR9	4300	200	18.5	FM	12	Horizontal	(123.625, −2.228, 1.7)	(180, 120)	(90, 30)
TR10	9310	500	60	FM	0	Horizontal	(117.611, −2.228, 1.7)	(180, 120)	(90, 30)
TR11	9375	100	70	FM	28	Horizontal	(89.964, −0.645, 0)	(180, 20)	(90, 20)
T1	121.5	5	17	FM	2	Vertical	(111.984, −0.645, 1.6)	(0, 360)	(80, 100)

Table 6.3 Parameters of the receiving equipment

Equipment name	Operation frequency/MHz	Operation bandwidth/MHz	Sensitivity/dBm	Modulation mode	Gain/dB	Polarization mode	Installation location (X, Y, Z) (m)	(Azimuth, beam width) (°)	(Elevation angle, beam width) (°)
TR1	16	28	−113	FM	3	Vertical	(116.937, −0.645, 2)	(0, 360)	(0, 180)
TR2	141	66	−101	FM	1.5	Vertical	(99.272, −0.645, 1.602)	(0, 360)	(80, 100)
TR3	157	2	−107	FM	2	Vertical	(106.075, −17.591, 1.7)	(0, 360)	(80, 100)
TR4	245.5	5	−103	FM	2.4	Vertical	(106.074, 12.377, 1.7)	(0, 360)	(80, 100)
TR5	1070	256	−90	PM	3	Vertical	(106.076, 10.309, 1.7)	(0, 360)	(80, 100)
TR6	1030	0.01	−121	FM	1.5	Vertical	(123.929, −0.645, 9.2)	(0, 360)	(80, 100)
TR7	1030	10	−77	FM	4	Vertical	(92.753, −0.645, 1.6)	(0, 360)	(80, 100)
TR8	1621	10	−107	FM	12	right handed circular	(110.74, −0.645, 1.6)	(0, 360)	(0, 30)
TR9	4300	200	−121	FM	12	Horizontal	(123.625, 4.062, 8)	(0, 360)	(180, 10
TR10	9375	500	−61	FM	0	Horizontal	(117.611, −2.228, 1.7)	(180, 120)	(90, 30)

(continued)

Table 6.3 (continued)

Equipment name	Operation frequency/MHz	Operation bandwidth/MHz	Sensitivity/dBm	Modulation mode	Gain/dB	Polarization mode	Installation location (X, Y, Z) (m)	(Azimuth, beam width) (°)	(Elevation angle, beam width) (°)
TR11	9375	50	−105	FM	28	Horizontal	(89.964, −0.645, 0)	(180, 20)	(90, 20)
R1	0.95	1.6	−69	FM	0.5	Vertical	(97.501, −0.645, 1.602)	(0, 360)	(80, 100)
R2	75	10	−61	FM	8	Horizontal	(122.942, 2.332, 8)	(0,360)	(180, 60)
R3	110	4	−101	FM	2	Horizontal	(121.648, −0.645, 7.695)	(0, 360)	(180, 60)
R4	159	6	−107	FM	1.2	Vertical	(106.073, −14.155, 1.7)	(0, 360)	(80, 100)
R5	327.5	15	−87	FM	2	Horizontal	(121.648, −0.645, 7.695)	(0, 360)	(180, 60)
R6	1575	10	−133	FM	3	Right−handed circular	(99.274, −0.645, 1.602)	(90, 360)	(0, 60)

Table 6.4 Interference correlation matrix

Transmitting/receiving	TR1	TR2	TR3	TR4	TR5	TR6	TR7	TR8	TR9	TR10	TR11	T1
TR1	Y	−63.7	−77.3	−118.2	−87.1	−133.2	−137.9	Y	Y	Y	Y	−78.1
TR2	26	Y	101.9	3.10	−12	−77.4	−62.5	−145	−206.9	−182.8	−174.2	85.3
TR3	15.6	98.6	Y	−25.8	−40.4	−69.1	−64.5	−141	−190.3	−173.6	−172.8	−54.3
TR4	−22.9	−43.4	−50.9	Y	−0.4	−64	−59.1	−98.8	−189.3	−167.5	−163.8	24.6
TR5	−61.6	2.7	−7.6	−2.5	Y	91.4	99.1	−47.8	−150.4	−120	−118.4	−24.8
TR6	−59.8	−48.2	−75.2	−70.8	40.5	Y	23.2	−67.7	−111.1	−85	−94.7	−94.3
TR7	−80.5	−50.3	−117.7	−110.9	61.2	−23.8	Y	−100.6	−166.2	−136	−115.4	−140.9
TR8	Y	−133.8	−140.1	−113.7	−1.6	−56.6	−55.1	Y	−118.8	−85.2	−91.7	−136.5
TR9	Y	−67	−69.2	−63.7	−15.4	−66.6	−79.9	−91.1	Y	−57.9	−65.3	−94
TR10	Y	−161	−165.6	−149.2	−82.4	−90.3	−94.3	−122.7	−107.4	Y	40.3	−183.2
TR11	Y	−152.3	−182.5	−127.9	−54.5	−124	−97.7	−124.5	−139.2	64.3	Y	−147.8
R1	−36.6	Y	Y	Y	Y	Y	Y	Y	Y	Y	Y	Y
R2	4	−97.7	−107.2	−122.5	−140.9	−168.9	−182.7	−201.3	−243	Y	Y	−110.8
R3	26.8	40.6	−56.2	−71.3	−59.5	−74.8	−98.4	−154.5	−192.6	−166.9	−177	−20.3
R4	20.4	104.3	125.6	−18.9	−30.2	−68.9	−64	−139	−189.3	−172.1	−172.4	−54
R5	−31.2	−44.2	−18.3	−76.9	−38.7	−68.3	−89.5	−120.3	−171	−152.4	−162.6	−117.3
R6	−55.4	22	−12.7	−80.8	23.3	−37.2	−22.3	32.8	−101.8	−71	−62.4	−140.5

Table 6.4 is the calculated interference correlation matrix **C** of the airborne RF equipment. Then, a numerical quantitative mapping is performed to obtain a quantitative correlation matrix **M**.

$$
M = \begin{bmatrix}
0 & 1 & 1 & 1 & 1 & 1 & 1 & 0 & 0 & 0 & 0 & 1 \\
5 & 0 & 5 & 5 & 3 & 1 & 1 & 1 & 1 & 1 & 1 & 5 \\
5 & 5 & 0 & 2 & 1 & 1 & 1 & 1 & 1 & 1 & 1 & 1 \\
2 & 1 & 1 & 0 & 4 & 1 & 1 & 1 & 1 & 1 & 1 & 5 \\
1 & 5 & 3 & 4 & 0 & 5 & 5 & 1 & 1 & 1 & 1 & 2 \\
1 & 1 & 1 & 1 & 5 & 0 & 5 & 1 & 1 & 1 & 1 & 1 \\
1 & 1 & 1 & 1 & 5 & 2 & 0 & 1 & 1 & 1 & 1 & 1 \\
0 & 1 & 1 & 1 & 4 & 1 & 1 & 0 & 1 & 1 & 1 & 1 \\
0 & 1 & 1 & 1 & 3 & 1 & 1 & 1 & 0 & 1 & 1 & 1 \\
0 & 1 & 1 & 1 & 1 & 1 & 1 & 1 & 1 & 0 & 5 & 1 \\
0 & 1 & 1 & 1 & 1 & 1 & 1 & 1 & 1 & 5 & 0 & 1 \\
1 & 0 & 0 & 0 & 0 & 0 & 0 & 0 & 0 & 0 & 0 & 0 \\
5 & 1 & 1 & 1 & 1 & 1 & 1 & 1 & 1 & 0 & 0 & 1 \\
5 & 5 & 1 & 1 & 1 & 1 & 1 & 1 & 1 & 1 & 1 & 2 \\
5 & 5 & 5 & 2 & 1 & 1 & 1 & 1 & 1 & 1 & 1 & 1 \\
1 & 1 & 2 & 1 & 1 & 1 & 1 & 1 & 1 & 1 & 1 & 1 \\
1 & 5 & 2 & 1 & 5 & 1 & 2 & 5 & 1 & 1 & 1 & 1
\end{bmatrix}
$$

The size of the matrix **M** is 17×12, but the DEMATEL method is based on a square matrix, so the quantization correlation matrix **M** needs to be expanded to the direct influence matrix **D**. The principle of expansion is as follows: From the 12 pieces of transmitting equipment and 17 pieces of receiving equipment, the total equipment set TR is obtained. The total number of equipment pieces in TR is 18, which includes 11 transceivers, 1 transmitting only equipment, and 6 receiving only equipment. For the newly added transmitting–transmitting, receiving–receiving, and receiving–transmitting equipment pairs, we can get the direct influence matrix **D** by adding zeros, i.e.,

$$
\mathrm{D}=\begin{bmatrix}
0&1&1&1&1&1&1&0&0&0&0&1&0&0&0&0&0&0\\
5&0&5&5&3&1&1&1&1&1&1&5&0&0&0&0&0&0\\
5&5&0&2&1&1&1&1&1&1&1&1&0&0&0&0&0&0\\
2&1&1&0&4&1&1&1&1&1&1&5&0&0&0&0&0&0\\
1&5&3&4&0&5&5&1&1&1&1&2&0&0&0&0&0&0\\
1&1&1&1&5&0&5&1&1&1&1&1&0&0&0&0&0&0\\
1&1&1&1&5&2&0&1&1&1&1&1&0&0&0&0&0&0\\
0&1&1&1&4&1&1&0&1&1&1&1&0&0&0&0&0&0\\
0&1&1&1&3&1&1&1&0&1&1&1&0&0&0&0&0&0\\
0&1&1&1&1&1&1&1&1&0&5&1&0&0&0&0&0&0\\
0&1&1&1&1&1&1&1&1&5&0&1&0&0&0&0&0&0\\
0&0&0&0&0&0&0&0&0&0&0&0&0&0&0&0&0&0\\
1&0&0&0&0&0&0&0&0&0&0&0&0&0&0&0&0&0\\
5&1&1&1&1&1&1&1&1&0&0&1&0&0&0&0&0&0\\
5&5&1&1&1&1&1&1&1&1&1&2&0&0&0&0&0&0\\
5&5&5&2&1&1&1&1&1&1&1&1&0&0&0&0&0&0\\
1&1&2&1&1&1&1&1&1&1&1&1&0&0&0&0&0&0\\
1&5&2&1&5&1&2&5&1&1&1&1&0&0&0&0&0&0
\end{bmatrix}
$$

The direct influence matrix $\mathbf{D}$ is normalized into matrix $\mathbf{G} = \frac{1}{\max\limits_{1 \le i \le n} \sum_{j=1}^{n} D_{ij}} \mathbf{D}$, which can be further derived into the comprehensive influence matrix $\mathbf{T} = \mathbf{G}(\mathbf{I} - \mathbf{G})^{-1}$, where $\mathbf{I}$ is the identity matrix.

The influence degree of the equipment S_i can be calculated using $f_i = \sum_{j=1}^{n} T_{ij}$, and the influence degree of all equipment constitutes the influence vector $\mathbf{T_f}$. Then, we can calculate the degree of influence the equipment S_j received using $e_i = \sum_{j=1}^{n} T_{ji}$ and the degree of influence all equipment receives constitutes the influence vector $\mathbf{T_e}$. Then, there is

$$
\mathbf{T_f} = \begin{bmatrix} 1.48\ 1.61\ 1.35\ 1.33\ 1.86\ 1.10\ 1.29\ 0.78\ 0.68\ 0.84\ 0.84\ 1.43\ 0\ 0\ 0\ 0\ 0\ 0 \end{bmatrix}
$$

$$
\mathbf{T_e} = \begin{bmatrix} 0.36\ 1.33\ 0.98\ 0.88\ 1.47\ 1.00\ 0.85\ 0.70\ 0.64\ 0.68\ 0.68\ 0\ 0.04\ 0.64\ 1.00\ 1.25\ 0.64\ 1.44 \end{bmatrix}
$$

From the equipment influence degree vector T_f, it can be concluded that the equipment that has a serious influence on other equipment in the aircraft platform (i.e., the main equipment that causes the system to have EMC problems) are $TR_5, TR_2, TR_1, T_1, TR_3, TR_4$, and we need to take good control over their radiation. Similarly, it can be concluded from the influence degree vector T_e that the equipment pieces that are more seriously affected by others (i.e., the equipment pieces that is more susceptive to the interference in the system) are TR_5, R_6, TR_2, R_4, which need to be in good protection.

6.3 Method for System-Level EMC Quantitative Design

Using the degree of influence and the degree of being influenced, we can identify the equipment that needs to be specifically designed or protected in the system-level EMC design. Since the aircraft is a complex system, the design scheme to achieve good EMC of the aircraft is not unique. This book uses the Analytic Network Process-Technique for Order of Preference by Similarity to Ideal Solution (ANP-TOPSIS) to analyze the weight vector of each piece of onboard electronic equipment and its influence on the EMC of the whole aircraft. Furthermore, the different design schemes of each piece of equipment are sorted by its advantages and disadvantages. Finally, the equipment's weight and its advantage ranking of equipment in different schemes are being combined to obtain the optimal design scheme of the EMC of the whole aircraft.

This section introduces the system-level EMC design method from three aspects: the quantification of indicator weighting, the optimization design of individual indicator, and the collaborative optimization design of multiple indicators [68].

6.3.1 The Quantification Method of EMC Indicator Weight

The quantitative design of the whole system is based on the quantitative evaluation of the advantages and disadvantages of the EMC design scheme, which further relies on the evaluation indicators. However, the EMC assessment indicators are numerous and the relationship between them is inextricably linked. To solve this problem, based on the research results of general system-level EMC indicator system, and through the appropriate tailoring and refinement, we have applied the general EMC indicator system to the aviation platform.

We have proposed an indicator weight calculation method based on the indicator system research results. Combining the method with the actual characteristics of the scheme, we further quantify the indicators using the analytic network process (ANP). This way we obtain the final indicator weights, which serve as the input and basis for the subsequent indicator decomposition.

1. Aircraft EMC indicator system

Through the in-depth analysis of the factors affecting the EMC of the aircraft, we establish a system for indicator selection. The indicator affecting the EMC of the aircraft can be divided into the following categories: transmitter performance indicators, receiver performance indicators, antenna layout performance indicators, interconnection performance indicators, etc., as shown in Fig. 6.10.

2. Calculation of EMC indicator weights

The ANP can describe complex systems. It is a scientific decision-making method to quantify each factor with a scale of 1–9, compare them in pairs, and finally rank

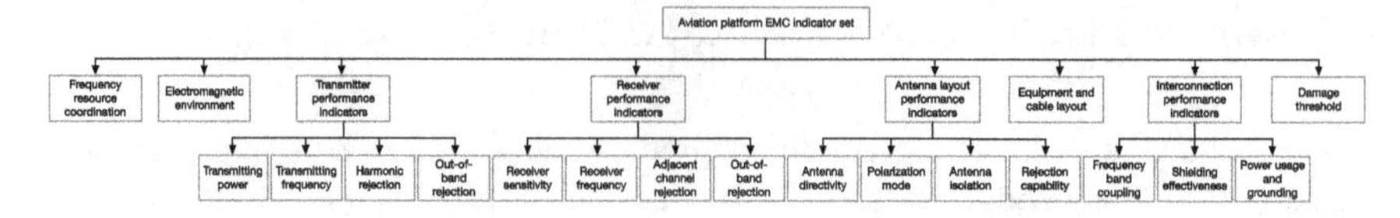

Fig. 6.10 Aircraft EMC performance indicator system

Fig. 6.11 Basic flowchart of ANP

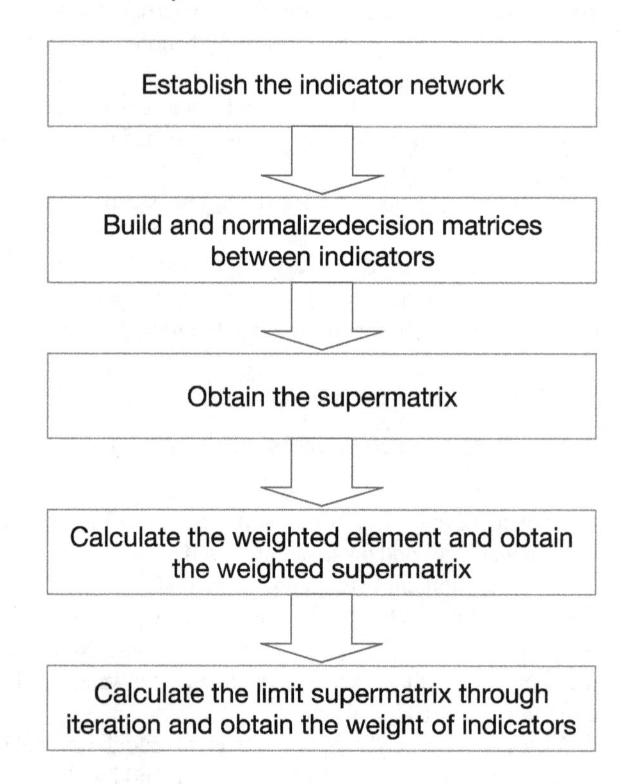

them. ANP expresses complex problems in the form of a network. The basic principle of ANP is the same as the analytic hierarchy process (AHP). The first difference between the two lies in the model structure, and the second difference is that ANP introduces the application and analysis of the supermatrix.

First of all, ANP classified the elements in two parts. The first part is called the control factor layer, which generally includes the target of the problem and the decision criteria. All the decision criteria are considered to be independent and only subject to the target element. The second part is the network layer which is composed of elements determined by the upper elements.

The basic flowchart of the ANP method is shown in Fig. 6.11.

Assuming the EMC design of an aircraft prioritizes the layout of the radio, we use this EMC design as an example to illustrate the effectiveness of the method. First, EMC of the communication station is measured by the following indicators: the receiver sensitivity (C_{11}), IF rejection ratio (C_{12}), transmitting power (C_{13}), antenna isolation (C_{21}), VSWR of antenna (C_{22}), shielding effectiveness (C_{31}), frequency band coupling (C_{32}), where C_{11}, C_{12}, and C_{13} belong to the equipment's own performance indicator set C_1; C_{21} and C_{22} belong to the antenna performance indicator set C_2; C_{31} and C_{32} belong to the interconnection performance indicator set C_3.

(1) Calculate the unweighted supermatrix. The EMC performance P of the communication station is taken as the criterion. The elements in the antenna layout performance in the network layer C_{21} are taken as the secondary criterion. The advantages of the influence degree of each element in the performance of the equipment C_1 toward C_{21} are compared. Then, we can obtain the judgment matrix and normalize the matrix.

Thus, the unweighted supermatrix under criterion C is

$$\mathbf{W} = \begin{pmatrix} W_{11} & W_{12} & W_{13} \\ W_{21} & W_{22} & W_{23} \\ W_{31} & W_{32} & W_{33} \end{pmatrix} = \begin{bmatrix} 0.0000 & 0.0000 & 0.0000 & 0.6369 & 0.6586 & 0.5525 & 0.5584 \\ 0.0000 & 0.0000 & 0.0000 & 0.1047 & 0.1562 & 0.1701 & 0.1210 \\ 0.0000 & 0.0000 & 0.0000 & 0.2583 & 0.1852 & 0.2803 & 0.3196 \\ 0.7500 & 0.5000 & 0.8000 & 0.0000 & 0.0000 & 0.8750 & 0.8333 \\ 0.2499 & 0.5000 & 0.2000 & 0.0000 & 0.0000 & 0.1249 & 0.1667 \\ 0.7500 & 0.5000 & 0.8571 & 0.9000 & 0.5000 & 0.5000 & 0.8000 \\ 0.2500 & 0.5000 & 0.1429 & 0.1000 & 0.5000 & 0.5000 & 0.2000 \end{bmatrix} \tag{6.6}$$

where the sub-block W_{ij} of the supermatrix $\mathbf{W}$ is normalized, while $\mathbf{W}$ is not. Therefore, we need to construct a weight matrix to normalize $\mathbf{W}$.

(2) Calculate the weighted supermatrix. When calculating the weighted supermatrix, the weight matrix under the target criterion needs to be calculated first. For this reason, the importance of each element is compared under the target criterion based on the target P (as shown in Table 6.5). By normalizing the feature vectors, a weight matrix can be obtained, where the order vector component corresponding to the element group unrelated to C_i is zero.

The weight matrix is

$$\mathbf{A} = \begin{pmatrix} 0.0000 & 0.5528 & 0.1971 \\ 0.6496 & 0.0000 & 0.1309 \\ 0.3504 & 0.4481 & 0.6719 \end{pmatrix} \tag{6.7}$$

Table 6.5 Element group C_i

C_i	C_1	C_2	C_3
C_1	0.0000	0.5528	0.1971
C_2	0.6496	0.0000	0.1309
C_3	0.3504	0.4481	0.6719

Table 6.6 Weight of each indicator

Receiver sensitivity/dBm	IF rejection ratio/dB	Transmitter power/W	Antenna isolation/W	SWR of antenna	Shielding effectivity	Frequency coupling
0.2204	0.1233	0.1774	0.1742	0.1342	0.112	0.0585

When the weight matrix **A** has been obtained, we perform $\overline{W_{ij}} = a_{ij} W_{ij}, i = 1, 2, 3, 4, j = 1, 2, 3, 4$ to get the weight matrix $\overline{\mathbf{W}}$. The column sum of matrix $\overline{\mathbf{W}}$ is 1. $\overline{\mathbf{W}}$ is also known as column random matrix, i.e.,

$$\overline{\mathbf{W}} = \begin{bmatrix} 0.0000 & 0.0000 & 0.0000 & 0.3521 & 0.3641 & 0.1089 & 0.1101 \\ 0.0000 & 0.0000 & 0.0000 & 0.0579 & 0.0863 & 0.0335 & 0.0238 \\ 0.0000 & 0.0000 & 0.0000 & 0.1428 & 0.1024 & 0.0552 & 0.0630 \\ 0.4872 & 0.3248 & 0.5197 & 0.0000 & 0.0000 & 0.1145 & 0.1091 \\ 0.1623 & 0.3248 & 0.1299 & 0.0000 & 0.0000 & 0.1249 & 0.1667 \\ 0.2628 & 0.1752 & 0.3003 & 0.4033 & 0.2241 & 0.3360 & 0.5375 \\ 0.0876 & 0.1752 & 0.0501 & 0.0448 & 0.2241 & 0.3360 & 0.1344 \end{bmatrix} \tag{6.8}$$

(3) Calculate the limit weighted supermatrix. After the weighted supermatrix $\overline{\mathbf{W}}$ is obtained, $\mathbf{W}_s^l = \lim_{k\to\infty} \overline{W}^k$ is performed ($k \to \infty$). The result is a long-term stable matrix. The value of each row of the matrix $\mathbf{W}_s^l$ is the same, which is the weight value of each element, as shown in Table 6.6.

6.3.2 The Optimization Method of Single EMC Indicator

The process of the single EMC indicator optimization method is as follows: (1) Access the evaluation indicators to obtain the EMC indicator weights for the specific project; (2) then classify the evaluation indicators according to the weights to obtain the design indicators priority and the design margin; (3) the indicators are statically designed; (4) after all the designs are completed, the individual design indicators are adjusted according to the test data, and the top-down adjustment design is conducted. The indicator allocation method is shown in Fig. 6.12.

The indicators are divided into frequency domain, margin, and time domain. Now, we introduce the method of the single-indicator allocation of frequency domain and margin.

1. Frequency-domain design method

There are many pieces of airborne spectrum-dependent equipment, so spectrum is a valuable resource on an aircraft. Spectrum must be allocated reasonably to save resources and to reduce interference caused by frequency overlap between equipment. Generally speaking, equipment should work in different frequency bands.

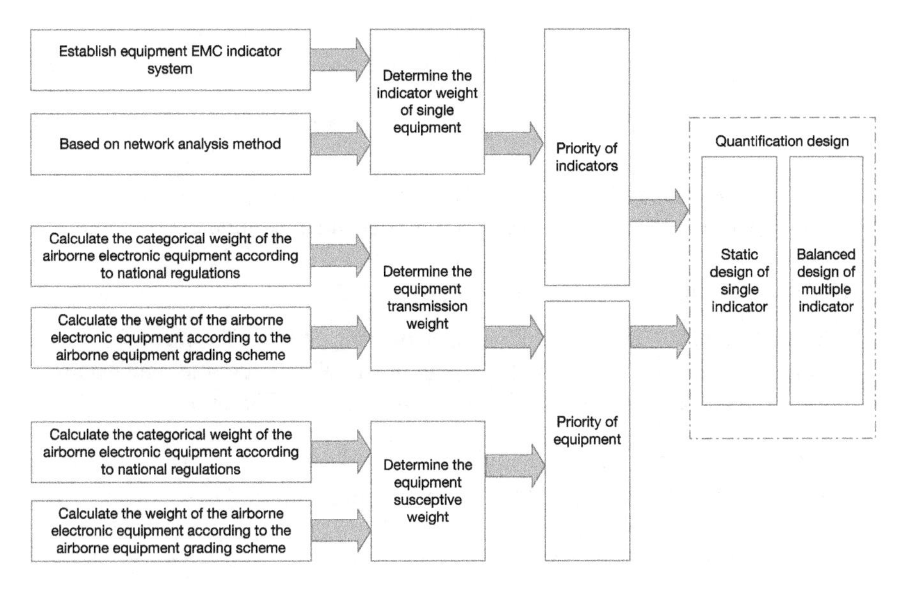

Fig. 6.12 Indicator allocation method

For equipment that must work in the same frequency band, the farther apart, the better. Therefore, it is necessary to allocate and optimize the frequency of all spectrum-dependent equipment. Radio equipment are usually subject to frequency adjustment. Since the radios usually use a wide bandwidth and some separated frequency points, it is necessary to select the schemes for the N pieces of radio equipment that the interference among them is the minimum. The mathematical model of the problem is as follows:

Equipment 1: The frequency range is $\left(f_L^1, f_H^1\right)$ and the frequency interval is Δf_1.
Equipment 2: The frequency range is $\left(f_L^2, f_H^2\right)$ and the frequency interval is Δf_2.
Equipment 3: The frequency range is $\left(f_L^3, f_H^3\right)$ and the frequency interval is Δf_3.

....

Let $f_i \in \left\{f_L^i, f_L^i + \Delta f_i, f_L^i + 2\Delta f_i, f_L^i + 3\Delta f_i, \ldots, f_H^i\right\}$ and $f_j \in \left\{f_L^j, f_L^j + \Delta f_j, f_L^j + 2\Delta f_j, f_L^j + 3\Delta f_j, \ldots, f_H^j\right\}$, and then we have

$$EMC\left(f_i, f_j\right) = \begin{cases} 1 \ (equipment\ i\ and\ equipment\ j\ are\ compatible) \\ 0 \ (equipment\ i\ and\ equipment\ j\ are\ not\ compatible) \end{cases}$$

In an EMC design scheme, there are M pieces of spectrum-dependent equipment and N ($M \geq N$) pieces of equipment that need to be allocated with frequencies. The expression of the total number of interferences is

$$I(f_1, f_2, \ldots, f_N, f_{N+1}, \ldots, f_M) = \sum_{i=2}^{N}\sum_{j=1}^{i-1} EMC(f_i, f_j) + \sum_{i=2}^{N}\sum_{j=N1}^{M} EMC(f_i, f_j) \tag{6.9}$$

Therefore, frequency allocation is to select the appropriate frequency points within the frequency range available for the spectrum-dependent equipment to ensure that the interference among all equipment is minimized.

The software implementation process is as follows:

(1) Assume that there are N_1 pieces of spectrum-dependent equipment.
(2) Among them, there are N_2 pieces of equipment with adjustable frequencies.
(3) The adjustable frequencies of each equipment is: Equipment 1: F_E1_1, F_E1_2; F_E1_3; F_E1_4; F_E1_5; F_E1_6 ... (from small to large, coded as 0, 1, 2, 3, 4, 5, 6, 7, 8 ..., respectively); Equipment 2: F_ E2_1, F_ E2_2; F_ E2_3; F_ E2_4; F_ E2_5; F_ E2_6 ... (from small to large, coded as 0, 1, 2,3,4,5,6,7,8..., respectively); Equipment N_2: F_EN2_1, F_EN2_2; F_EN2_3; F_EN2_4; F_EN2_5; F_EN2_6 ... (from small to large, coded as 0, 1, 2, 3, 4, 5, 6, 7, 8 ..., respectively);
(4) The available frequencies of each equipment are encoded from small to large;
(5) Input the population size (*popsize*) (It is determined based on the product of the number of available frequency points of all equipment, and the maximum number is the value of the product, ranging from 50 to 500. If 500 is greater than the maximum value, then change the number 500 to the maximum value; if 50 is greater than the maximum value, then the range is changed to 1—the maximum); the crossover probability (*pc*) (0.2–0.99); the mutation probability (*pm*) is (0.001–0.1); the maximum number of iterations (*maxgeneration*) is (100–300);
(6) The *popsize* number of initial schemes is randomly selected from X number of schemes (the product of the number of available frequency points of all equipment). The frequency allocation method is shown in Fig. 6.13.

2. Margin design method

Referring to the *EMC-Analyze user's manual*, the multi-port interference and coupling are analyzed to perform margin design. First, we shall define the interference coupling ports and susceptive coupling ports of the equipment/subsystem, including the antenna port, cable port, and case slot port. Then, we call the simulation prediction algorithm to calculate the interference margin of the susceptive object under each coupling channel, and then calculate the cumulative interference margin. Finally, according to a certain indicator allocation method, we can complete the design of the margin. The process is shown in Fig. 6.14.

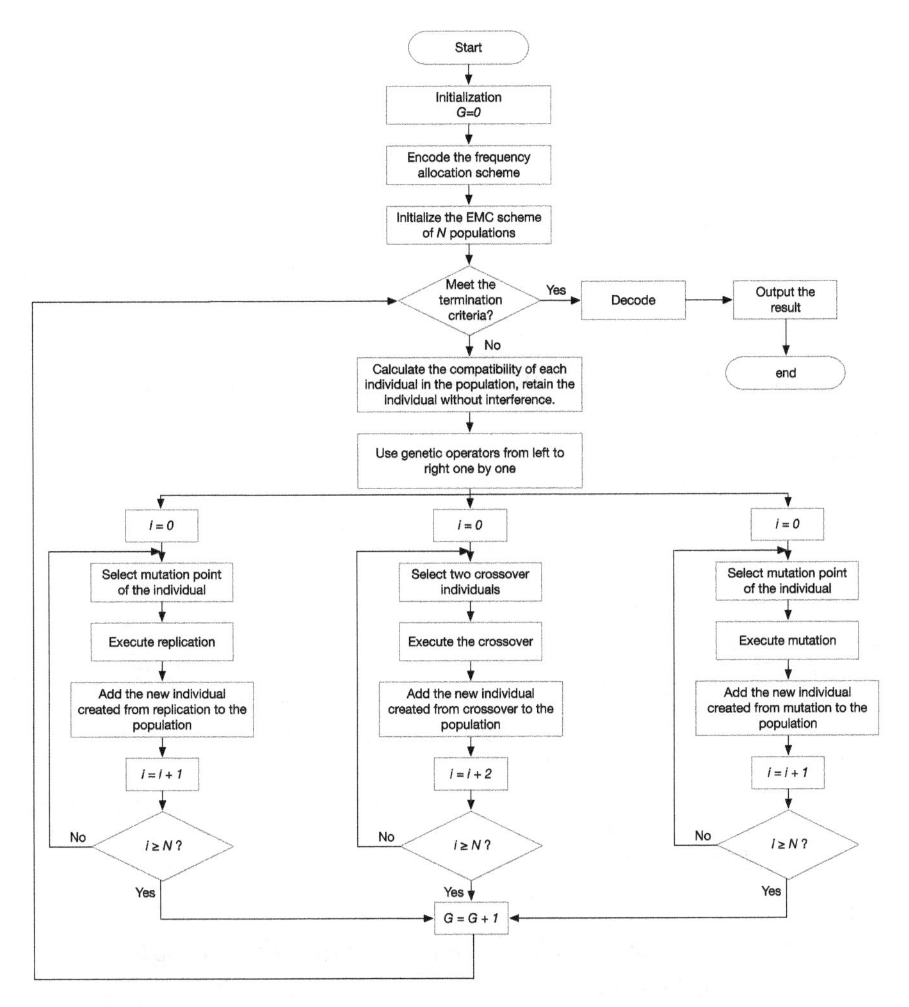

Fig. 6.13 Frequency allocation method

After calculating the emission spectrum that makes an equipment compatible, the spectrum of the transmitting equipment at the receiver shall be compared with the red standard limit. When there is interference, the allocation method in Sect. 4.3.6 can be followed to perform quantitative allocation of indicators.

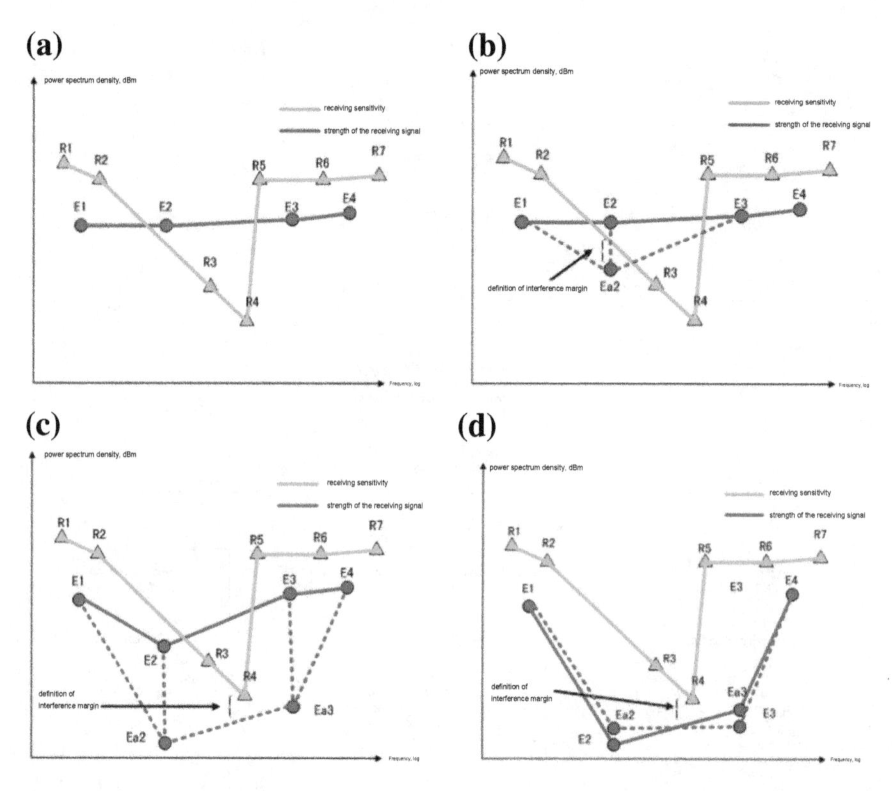

Fig. 6.14 Margin design method, **a** transmitting spectrum and receiving spectrum, **b** adjustment to make E2 compatible, **c** parallel adjustment of E2–E3 to make R4 compatible, **d** E2–E3 rotation adjustment around the frequency point where R4 is located for compatibility

6.3.3 The Collaborative Optimization Method for Multiple EMC Indicators

The multi-indicator collaborative dynamic design method refers to taking into account and collaboratively adjust multiple EMC design indicators according to equipment weights and indicator weights, and select an optimal solution to make the design compatible. Obviously, compared with the single-indicator design method, the multi-indicator collaborative design tends to be more in line with the actual situation, and the global optimization can be achieved with the minimum cost-effectiveness. The collaborative design method of indicators is shown in Fig. 6.15.

After generating multiple sets of EMC design schemes, the ANP-TOPSIS method is used for scheme evaluation.

The ANP method has been introduced in Sect. 6.3.1. The Technique for Order of Preference by Similarity to Ideal Solution (TOPSIS) is a commonly used evaluation method. Its main idea is to measure the Euclidean distance of a multi-dimensional

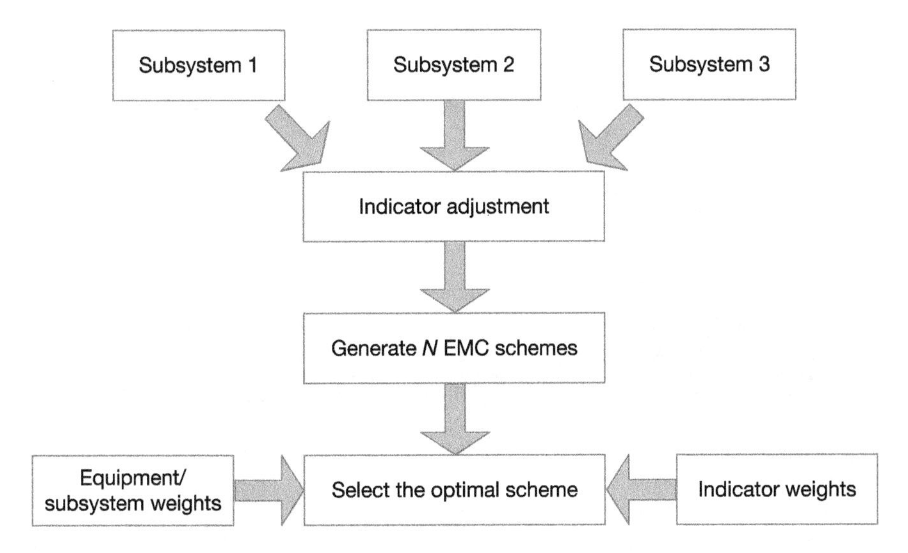

Fig. 6.15 Collaborative design method of indicator

indicator space to rank the schemes according to their advantages and disadvantages. The general steps for ranking the schemes using the TOPSIS method are as follows:

(1) Construct an initial evaluation matrix **C**.
(2) Normalize the decision matrix.
(3) Calculating a weight matrix for the normalized matrix in (1).
(4) Determine the positive ideal point $\mathbf{X}^+$ and the negative ideal point $\mathbf{X}^-$.
(5) Calculate the Euclidean distance of each scheme to the positive ideal point d_i^+ and the negative ideal point d_i^-.
(6) Calculate the closeness of each solution to the ideal solution d_i.
(7) Rank the schemes in descending order d_i.

ANP-TOPSIS scheme evaluation (Fig. 6.16): First, identify the indicators to be involved in the evaluation and design; then construct the network (ANP) of the evaluation system, including the calculation of the unweighted supermatrix, the weighted supermatrix and the limit weighted supermatrix; finally, rank the schemes according to TOPSIS, including constructing the normalized decision supermatrix, determining the positive ideal point and the negative ideal point of the weighted normalization matrix, calculating the Euclidean distance from each scheme to the positive ideal point and the negative ideal points, calculating the closeness of each scheme to the ideal point, and ranking the schemes according to the closeness. The parameters in the ranked scheme are the parameters of the indicator decomposition, as shown in Fig. 6.16.

Since the aircraft EMC indicators appear in many types such as precise number, interval number, and fuzzy number [69], the evaluation of aircraft EMC is a multi-attribute decision-making problem of indicators. For the convenience of our readers,

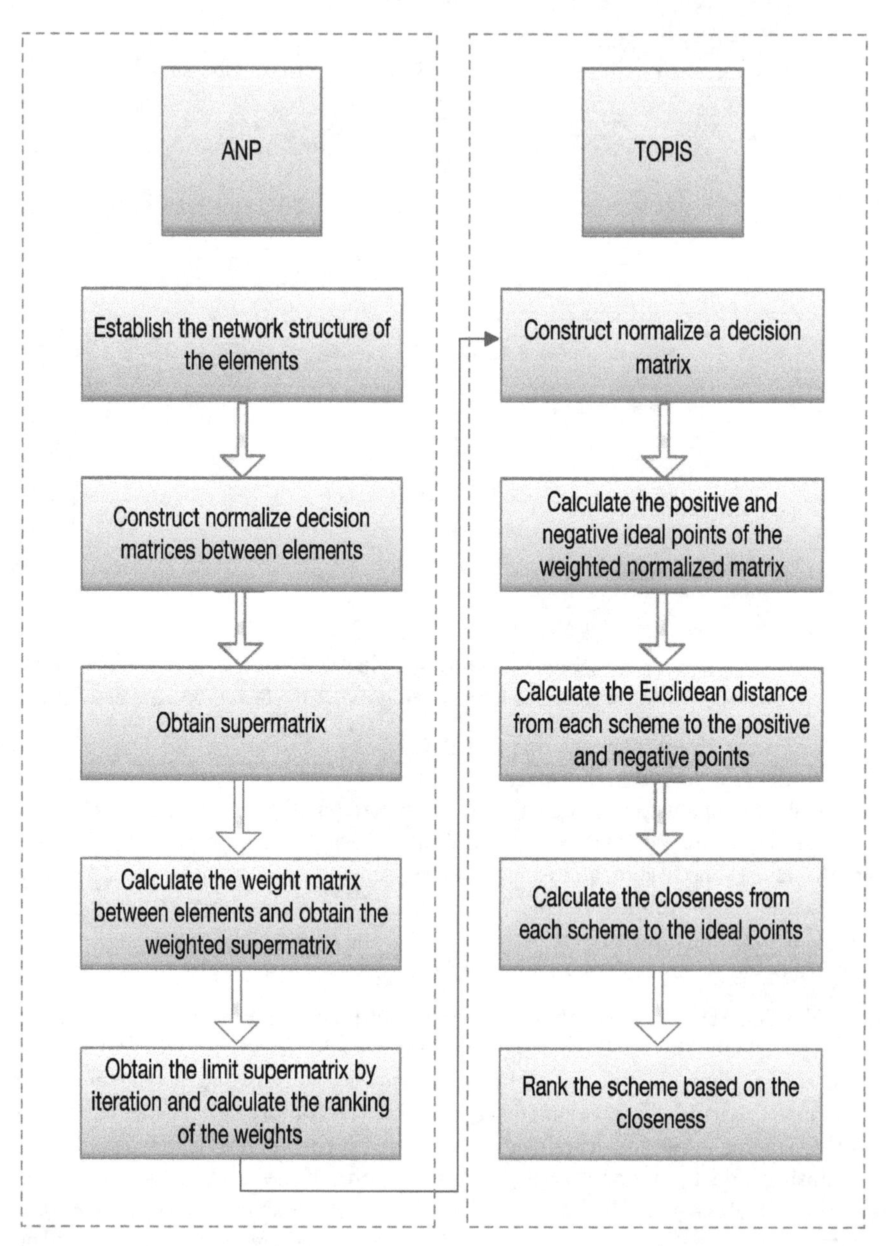

Fig. 6.16 Flowchart of scheme evaluation

here we explain the type of indicators and their metrics. Since the precise number type is self-explained, we focus on the interval number type and fuzzy number type.

1. The definition and operation rule of interval numbers

Definition 1 $a = [a^L, a^U]$ is called the number with closed intervals, where $a^L \in R, a^U \in R$, and $a^L \leq a^U$. The total number of closed intervals on R is denoted as $\overline{R}$.

The operation rule of the interval number: Let $a = [a^L, a^U]$ and $b = [b^L, b^U]$ be two arbitrary positive closed interval numbers, and the interval number can be calculated using

$$\begin{cases} a + b = [a^L + b^L, a^U + b^U] \\ a \times b = [a^L b^L, a^U b^U] \\ a \div b = \left[\frac{a^L}{b^L}, \frac{a^U}{b^U}\right] \\ ka = [ka^L, ka^U] \, (k > 0) \\ \frac{1}{a} = \left[\frac{1}{a^U}, \frac{1}{a^L}\right] \end{cases} \tag{6.10}$$

2. The definition and operation rule of fuzzy numbers

Definition 2 $a = (a^l, a^m, a^u)$ is called a triangular fuzzy number, if its membership function is $u_a(x) : R \to [0, 1]$, i.e.,

$$u_a(x) = \begin{cases} 0 & (x \leq a^l) \\ \frac{x - a^l}{a^m - a^l} & (a^l < x < a^m) \\ \frac{x - a^u}{a^m - a^u} & (a^m < x < a^u) \\ 0 & (x > a^u) \end{cases} \tag{6.11}$$

where $x \in R$, $0 < a^l \leq a^m \leq a^u \in \Psi$, $\Psi = [l, L]$ is any real interval; when $a^l = a^m = a^u$, a represents a nonfuzzy number.

The triangular fuzzy number is intuitive and easy to use. It can express multiple language variables. If the experts have higher evaluation of an object, we shall choose the larger number Ψ for a^L, a^M, a^U; if the evaluation is lower, we shall choose the smaller number. For the evaluation of indicator attributes, seven variables, "Extra Low" (EL), "Very Low" (VL), "Low" (L), "Medium" (M), "High" (H), "Very High" (VH), "Extra High" (EH), can be used and they can be recorded as

$$EL = (0, 0, 0.1), VL = (0.1, 0.2, 0.3), L = (0.2, 0.3, 0.4), M = (0.4, 0.5, 0.6), H = (0., 0.7, 0.8), \\ VH = (0.8, 0.9, 1.0), EH = (0.9, 1.0, 1.0)$$

Considering any two triangular fuzzy numbers $a = (a^l, a^m, a^u)$ and $b = (b^l, b^m, b^u)$, there is a corresponding fuzzy number operation rule:

$$\begin{cases} a \oplus b = \left(a^l + b^l, a^m + b^m, a^u + b^u\right) \\ \lambda \otimes a = \left(\lambda a^l, \lambda a^m, \lambda a^u\right) \quad (\lambda > 0, \lambda \in R) \\ \frac{1}{a} = \left(\frac{1}{a^u}, \frac{1}{a^m}, \frac{1}{a^l}\right) \end{cases}$$

where $\oplus$ and $\otimes$ represent the addition and multiplication of fuzzy numbers, respectively.

Now, we explain the ANP-TOPSIS method using the EMC design of the communication station as an example. The comprehensive evaluation indicators of the communication station are receiver sensitivity, IF rejection ratio, transmitter power, antenna isolation, VSWR of antenna, case shielding effectiveness, and frequency band coupling.

First, the weights of the indicators are calculated using ANP method, as shown in Table 6.7. Next, the seven design indicators are listed as a set $Q = \{Q_1, Q_2, \ldots, Q_7\}$ where in terms of indicator attributes, Q_1 and Q_3 are interval number indicators; Q_2, Q_4, Q_5, and Q_6 are precise number indicator; Q_7 is the fuzzy indicator. Q_6 is a score value, ranging from 1 point (worst) to 10 points (best). Q_7 is a level indicator, which includes seven levels: "no coupling," "not serious at all," "not serious," "common," "serious," "very serious," and "extremely serious." In terms of indicator types, Q_1, Q_2, Q_4, and Q_6 are indicators of benefit (the bigger the better), and the other three are indicators of cost (the smaller the better).

There are four shortwave radio design solutions, and we need to select a good one. These four schemes are defined by the metrics of the seven indicators. For example, in scheme 1, the short-wave receiver has a sensitivity range of 90–105 dBm in various adjustment modes and an IF rejection ratio of 65 dB. The short-wave transmitting power range is 60–85 W. The antenna isolation with a certain piece of core equipment is 55 dB. The VSWR of the antenna is 1.2. The shielding effectiveness of the case is 7 dB. The frequency band coupling with other frequency equipment of the whole aircraft is at a common level. The EMC indicators of these four design schemes are listed in Table 6.7.

Firstly, the fuzzy number of the indicator Q_7 is analyzed and calculated. According to the corresponding relationship between the triangular fuzzy number and the

Table 6.7 EMC performance indicators of the four design schemes

Scheme	EMC indicators						
	Q_1/dBm	Q_2/dB	Q_3/W	Q_4/dB	Q_5	Q_6	Q_7
1	90–105	65	60–85	55	1.2	7	Common
2	95–110	68	70–90	65	1.8	5	Serious
3	85–100	70	80–100	60	1.5	8	Not serious at all
4	100–120	62	75–95	75	1.6	7	Very serious
Indicator weights	0.2204	0.1233	0.1774	0.1742	0.1342	0.112	0.0585

language description variable introduced above, the qualitative indicator is represented by the triangular fuzzy number. The seven levels of the indicator, from low to high, are expressed as (0 0 0.1), (0.1 0.2 0.3), (0.2 0.3 0.4), (0.4 0.5 0.6), (0.6 0.7 0.8), (0.8 0.9 1.0), and (0.9 0.9 1.0). Then, we can derive the decision matrix

$$\overline{\mathbf{A}} = \begin{bmatrix} (90, 105) & 65 & [60, 85] & 55 & 1.2 & 7 & \left(0.4\ 0.5\ 0.6\right) \\ (95, 110) & 68 & [70, 90] & 65 & 1.8 & 5 & \left(0.6\ 0.7\ 0.8\right) \\ (85, 100) & 70 & [80, 100] & 60 & 1.5 & 8 & \left(0.1\ 0.2\ 0.3\right) \\ (100, 120) & 62 & [75, 95] & 75 & 1.6 & 7 & \left(0.8\ 0.9\ 1.0\right) \end{bmatrix} \tag{6.12}$$

Next, we normalize the decision matrix $\overline{\mathbf{A}}$ and obtain

$$\overline{\mathbf{B}} = \begin{bmatrix} (0.4128, 0.5665) & 0.4901 & [0.4119, 0.7666] & 0.4285 & 0.3895 & 0.5119 & \left(0.16\ 0.35\ 0.62\right) \\ (0.4358, 0.5935) & 0.5127 & [0.3889, 0.6571] & 0.5064 & 0.5843 & 0.3656 & \left(0.12\ 0.25\ 0.41\right) \\ (0.3889, 0.5395) & 0.5277 & [0.3501, 0.5749] & 0.4674 & 0.4869 & 0.5850 & \left(0.32\ 0.88\ 0.46\right) \\ (0.4587, 0.6475) & 0.4674 & [0.3685, 0.6133] & 0.5843 & 0.5194 & 0.5119 & \left(0.09\ 0.19\ 0.31\right) \end{bmatrix}$$

The normalized decision matrix is then weighted to obtain

$$\overline{\mathbf{G}} = \begin{bmatrix} (0.0930, 0.1249) & 0.0604 & [0.0731, 0.1360] & 0.0746 & 0.0679 & 0.0687 & \left(0.0094\ 0.0205\ 0.0363\right) \\ (0.0961, 0.1308) & 0.0637 & [0.0690, 0.1166] & 0.0882 & 0.1018 & 0.0491 & \left(0.0070\ 0.0146\ 0.0240\right) \\ (0.0859, 0.1189) & 0.0651 & [0.0621, 0.1020] & 0.0814 & 0.0848 & 0.0785 & \left(0.0187\ 0.0515\ 0.1439\right) \\ (0.1011, 0.1427) & 0.0576 & [0.0654, 0.1088] & 0.1018 & 0.0905 & 0.0687 & \left(0.0053\ 0.0111\ 0.0181\right) \end{bmatrix}$$

According to the above equations, the positive ideal point X^+ and the negative ideal point X^- are calculated as

$$X^+ = [(0.1011, 0.1427), 0.0651, (0.0731, 0.1360), 0.1018, 0.1018, 0.0785, (0.0187, 0.0515, 0.1439)]$$

$$X^- = [(0.0859, 0.1189), 0.0576, (0.0621, 0.1020), 0.0746, 0.0679, 0.0491, (0.0053, 0.0111, 0.0240)]$$

The Euclidean distance d_i^+ from each scheme to the positive ideal point $\mathbf{X}^+$\ and the distance from $\mathbf{X}^-$ to the negative ideal point d_i^- can be determined

$$\begin{aligned} d_1^+ &= 0.1124, d_1^- = 0.0533 \\ d_2^+ &= 0.1339, d_2^- = 0.0483 \\ d_3^+ &= 0.1448, d_3^- = 0.0652 \\ d_4^+ &= 0.1383, d_4^- = 0.0574 \end{aligned} \tag{6.13}$$

Finally, the closeness of the four schemes to the positive ideal point is calculated as

$$\begin{aligned} d_1 &= 0.3217, d_2 = 0.2978 \\ d_3 &= 0.3105, d_4 = 0.2933 \end{aligned} \tag{6.14}$$

According to the above calculation results, the ranking of the advantages and disadvantages of the scheme can be obtained. For the design of a communication station, the comparison results of the four schemes given in Table 6.7 are scheme 1 > scheme 2 > scheme 3 > scheme 4.

Chapter 7
EMC Evaluation and Quality Control Methods

This chapter mainly introduces the methods and principles of EMC evaluation and quality control during system development. The goal of EMC evaluation is to check whether the EMC performance of equipment and subsystems meets the requirement of each milestone during development [70–73]. The evaluation items include the EMC performance of the equipment, EMC performance of the subsystem based on the EMC performance of its consisting equipment, and the EMC performance of the system based on the EMC performance of its consisting equipment and subsystems. The goal of EMC quality control is to propose control requirements for equipment, subsystems, and systems to ensure that based on the evaluation of EMC performance of the equipment, subsystems, and systems at each milestone of the development phase, any deviation will be corrected and any problem will be solved in a timely manner [74].

EMC evaluation and quality control are important for real-time and effective monitoring of EMC of the equipment, subsystems, and systems. They have great significance in solving the existing and potential EMC problems in the early phase.

7.1 The Basis for EMC Evaluation

The EMC evaluation shall be based on the quantitative EMC requirements for the system, subsystems, and equipment from the general requirements for product development (or development specification document and development contract), and the EMC requirement further clarified in the product development review meetings.

7.2 The Scope of EMC Evaluation

The scope of EMC evaluation mainly includes three aspects, namely design, management, and test.

© National Defense Industry Press and Springer Nature Singapore Pte Ltd. 2019

D. Su et al., *Theory and Methods of Quantification Design on System-Level Electromagnetic Compatibility*, https://doi.org/10.1007/978-981-13-3690-4_7

7.2.1 EMC Design

1. Requirement analysis

The EMC requirements of products are usually reflected in the overall requirements and overall technical requirements. Therefore, the scope and keys of design should be formed based on the detailed analysis of the overall EMC requirements and the EMC overall technical requirements.

First, we need to evaluate whether the analysis of the overall EMC requirements is effective, including the overall EMC requirements determined in the general requirements for product development, and the EMC requirements that are further clarified through the product development review meeting. The application requirement and the natural and man-made electromagnetic environment where the product operates are also analyzed.

Then, we can evaluate whether the analysis of the overall technical EMC requirements is effective, which includes (1) analysis of the product's adaptability to the expected electromagnetic environment based on the overall EMC requirement of the product and relevant national military regulations; (2) analysis of the product EMC design and limit values in system, subsystem, and equipment level; (3) analysis of engineering feasibility and technical risks.

2. Design

EMC design should be conducted through the principle design phase, the engineering development phase (including the principle implementation and the system development and demonstration) and the production and deployment phase. Therefore, the evaluation should also cover all of the phases.

The evaluation items during principle design phase include (1) formulate the overall EMC technical requirements for the system and decompose the requirements into subsystems and equipment level; (2) conduct EMC design and modeling of the system, subsystem, and equipment.

The engineering development phase can be divided into two parts: the system level and the subsystem/equipment level:

The evaluation items of the system level include

(1) Determine whether the existing EMC problems in the principle design phase have been comprehensively analyzed, whether the solution is proper, and whether the problem has been solved.
(2) Check the EMC conformity of the supporting nondevelopment item (NDI) and determine whether it is necessary to add supplementary requirement.
(3) Evaluate whether the EMC analysis of the subsystem/equipment principle implementation is valid and sufficient; based on EMC of the subsystem/equipment initial prototype, analyze whether the EMC of the system principle implementation is correct; evaluate whether the EMC design is comprehensive and correct, and whether the design is implemented during the principle implementation phase.

(4) Evaluate whether the EMC problems existing in principle implementation phase are comprehensively analyzed. Assess the problem solutions and check whether the problem has been solved. Determine whether the EMC improvement design of the system's system development and demonstration is effective.
(5) Check and evaluate the rationality and comprehensiveness of the production process and installation measures taken to ensure EMC during the development and demonstration phase of the system.
(6) Evaluate whether the EMC prediction analysis of the system development and demonstration of the subsystem/equipment is effective. Evaluate whether the EMC analysis is and prediction of the system development and demonstration is comprehensive based on the EMC of subsystems/equipment.
(7) Conduct a system joint test to identify EMC problems. Evaluate the comprehensiveness of the work, and check whether the problems have been solved.
(8) Conduct a precompliance test to identify EMC problems, evaluate the sufficiency and effectiveness of analysis and check whether the problems have been solved.

The evaluation items during the engineering development phase of subsystems/equipment include

(1) During principle design phase, evaluate the subsystem/equipment's EMC problems, solutions and check whether the problems have been solved.
(2) Evaluate the implementation of the supplementary requirement from supporting NDIs.
(3) During principle implementation phase, evaluate the EMC design and implementation of the system and subsystem/equipment.
(4) Check whether the EMC problems of the subsystem/equipment's principle implementation has been completely solved and whether the EMC of the system development and demonstration is improved.
(5) In system development and demonstration phase, evaluate the production and installation measures taken to ensure EMC of the subsystem/equipment.
(6) Evaluate whether the EMC problem of the subsystem/equipment of the system development and demonstration is a closed loop.
(7) Before the production and deployment test of the system design, EMC development and demonstration test report of subsystems/equipment should be submitted to the system.

The evaluation items in the production and deployment phase include whether the problems in the EMC test in the production and deployment phase has been fully analyzed and categorized; whether the cause analysis is correct; and whether the improved design is effective.

7.2.2 EMC Management

Product EMC management is reflected in the requirement management and execution, including product EMC organization management and quality control.

The EMC organization management evaluation items mainly include

(1) Whether to set up an EMC management office or to point a person in charge;
(2) Whether to define the EMC management items and milestones;
(3) Whether the operation of the EMC working group is normal;
(4) Whether to formulate an EMC outline, whether the outline is to be reviewed, and whether the outline has been fully implemented;
(5) Whether an EMC control plan has been made and whether there are benefits from the control and management;
(6) Whether an EMC work plan has been formulated, and whether the execution of the plan is normal;
(7) Whether an EMC training plan has been formulated and executed effectively.

The evaluation items of EMC quality control mainly include

(1) Whether to incorporate EMC into the quality management system and operate effectively;
(2) Whether to analyze the EMC control milestones, items and measures of the product and evaluate the control effects;
(3) Assess the quality evaluation of EMC in each phase;
(4) Summarize the benefits of EMC quality management.

7.2.3 EMC Test

EMC tests include research test and production and deployment test, which needs to be evaluated separately.

The evaluation items of the scientific research and test phase mainly include

(1) Scientific research and test result of supporting NDIs;
(2) Scientific research and test result of the principle implementation of subsystems and equipment;
(3) Scientific research and test result of the principle implementation of the system;
(4) Scientific research and test result of the system development and demonstration of subsystems and equipment;
(5) Scientific research and test result of the system development and demonstration of the system.

The evaluation items of the design stereotype test mainly include

(1) Qualification of the test organization, test time, location, environment and conditions, technical status of the object under test, test items, test method, test

layout, and relative spatial position between the sensor and the object under test, number of test frequency points, the operation time sequence of the object under test, limit of the test field strength, the susceptivity criteria and other important test elements, etc.;

(2) Conformity of the evaluation test items, the method and basis of EMC comprehensive evaluation, major problems and the verification results in tests, conclusions of tests in various phases, etc.

7.3 Evaluation Method

The evaluation method depends on the product's characteristics that affect the EMC of the product. EMC evaluation can be conducted in hierarchy and by phase. "Hierarchy" means that the evaluation is performed on system, subsystem, and equipment. "By phases" means that the evaluation is performed throughout the principle design phase, the principle implementation phase, the system development and demonstration development phase, and the production and deployment phase.

7.3.1 The Hierarchical Evaluation Method

1. System-level EMC evaluation

System-level EMC is evaluated based on the electromagnetic environment, field coupling strength, inter-equipment interference strength, shielding effectiveness, harmonic characteristics, layout characteristics, antenna layout, cable, multi-equipment coupling, frequency allocation, and subsystem performance [75].

(1) Electromagnetic environment evaluation: Evaluate the impact of the electromagnetic environment inside and outside the system according to the environment of the intended mission;
(2) Field–circuit coupling strength evaluation: According to the electromagnetic environment of the system and the key circuits of each equipment, analyze the coupling relationship and their degree of influence of field–circuit, circuit—field, and circuit–circuit;
(3) Inter-equipment interference strength evaluation: Quantitative analyze the equipment that forms the interference, and evaluate the degree of the interference;
(4) Shielding effect evaluation: Evaluate the shielding effect of the key part of the equipment;
(5) Evaluation of resonance characteristics: evaluation of electromagnetic resonance characteristics of the key parts of the system;

(6) Evaluation of the layout characteristics: The system layout evaluation mainly includes evaluation of the equipment layout, antenna layout and cable layout;
(7) Antenna layout evaluation: Evaluate the antenna pattern after installation, isolation between antennas, etc.;
(8) Cable evaluation: Evaluate the radiated emission and conducted emission of the cables of high-power electrical equipment in the system, and evaluates the coupling of the cables of mission-critical equipment to EMI;
(9) Multi-equipment coupling evaluation: Based on the radiation emission and susceptibility characteristics of each equipment, evaluate the EMC of the subsystem or equipment set composed of multiple equipments;
(10) Frequency allocation evaluation: Evaluate the occupied frequency and frequency allocation of each subsystem and equipment in the system;
(11) Evaluation of subsystems/equipment's performance degradation: After the equipment is interfered, quantitatively evaluate the degradation of performance indicators such as receiving sensitivity, decoding performance, and functional distance.

2. EMC evaluation of subsystems and equipment

Through the analysis and evaluation of PCB design, filter design, shielding design, cable and connector design, power supply characteristics, etc., we can predict the EMC of subsystems and equipment [76].

(1) PCB design: component selection, clock and bandwidth design, layering, functional partitioning, routing, impedance matching, decoupling and bypass, grounding, interface circuits, etc.;
(2) Filter design: filter design for power supply and signal;
(3) Shielding design: case shielding design, crystal shielding design, etc.;
(4) Cable and connector design: connector filtering, cable selection, cable termination, cable shielding design, etc.;
(5) Power supply characteristics: voltage and current characteristics, ripple and noise, temperature drift characteristics, overshoot amplitude of switching, etc.

The following should also be evaluated for the frequency using equipment:

(1) Emission characteristics: transmission power, harmonic rejection, and out-band emission attenuation;
(2) Receiving characteristics: reception sensitivity, out-band rejection, image suppression, and cross talk suppression.

7.3.2 Evaluation Method by Phase

Four phases, namely principle design phase, principle implementation phase, development and demonstration phase, and production and deployment phase are included.

1. Principle design phase

In the principle design phase, we mainly evaluate the EMC design schemes of systems, subsystems, and equipment.

(1) Analyze the dominant or implicit interference between the subsystems and equipment in the system and evaluate the system EMI prediction model.
(2) Evaluate the antenna layout scheme, focusing on the orientation pattern and antenna coupling after the antenna installation in the aircraft.
(3) Evaluate frequency allocation, equipment layout, and cable layout design.
(4) Analyze and evaluate the system shielding effectiveness, the electromagnetic environment inside and outside the system, the electromagnetic resonance characteristics of key parts of the system, and the performance degradation of the equipment subsystem.
(5) Evaluate the design measures of transmission power, harmonic rejection, out-band emission attenuation, receiving sensitivity, out-band receiving rejection, image rejection, cross talk suppression, shielding, grounding, and power supply filtering.
(6) Analyze the board radiation, line cross talk, signal integrity and other characteristics.

The main supporting documents from the research institute in this phase are:

(1) "EMC Management Requirements" of systems, subsystems, and equipment;
(2) "EMC prediction and evaluation report in principle design phase" of systems, subsystems, equipment;
(3) "EMC quality evaluation report in principle design phase" of systems, subsystems, and equipment;
(4) "Summary report on EMC work in demonstration phase" of systems, subsystems, and equipment;
(5) Other reports related to EMC, such as quality analysis reports, standardization reports.

Based on the above-supporting documents, the EMC management office will then review the following transition documents:

(1) "EMC outline";
(2) "EMC control plan";
(3) "EMC test plan";
(4) "EMC design plan in principle design phase" for systems, subsystems, and equipment;
(5) "Overall technical requirements for EMC" for systems, subsystems, and equipment.

2. Principle implementation phase of subsystems and equipment

(1) Evaluate the EMC design scheme in the principle implementation phase, check whether the evaluation result of the principle design phase is implemented in the principle implementation design, and determine the rationality of the design scheme.
(2) Through the EMC test, evaluate the implementation of the EMC design for subsystems and equipment in principle implementation phase.
(3) Evaluate the EMC of the prototype in principle implementation phase by analyzing the electromagnetic emission and susceptive characteristics of the subsystem and equipment.
(4) Evaluate system EMC performance by predicting and analyzing factors including the electromagnetic environment of the system, the strength of the field coupling, the interference intensity between the equipment, the influence of system EMC oversize.
(5) Evaluate the EMC design indicators of the optimized subsystems and equipment.
(6) Evaluate the improved system integration design scheme by analyzing the optimized system-level EMI prediction model.

The main supporting documents from the research institute in this phase are:

(1) "Summary report on EMC work in principle design phase";
(2) "EMC prediction and evaluation report of principle implementation phase";
(3) "Evaluation report on EMC quality of principle implementation phase";
(4) "EMC test outline of principle implementation phase";
(5) "EMC test result of principle implementation phase";
(6) Other EMC-related reports, such as quality analysis reports, standardization reports.

Based on the above-supporting documents, the EMC management office will then review the following transition documents:

(1) "EMC design scheme of principle implementation phase";
(2) "Summary report on EMC work of principle implementation phase".

3. System development and demonstration phase of subsystems/equipment

(1) Evaluate the EMC design scheme of the system development and demonstration phase. The major evaluation items include whether the EMC evaluation results of subsystems/equipment of principle implementation phase have been reflected in the design of system development and demonstration phase; and whether the method of the EMC process control in the system development and demonstration phase is reasonable.
(2) Through the EMC test, evaluate the implementation of EMC design on subsystems/equipment in system development and demonstration phase.

(3) By analyzing the effect of EMI suppression measures of the prototype in system development and demonstration phase, evaluate the EMC of the system and equipment in system development and demonstration phase.
(4) Evaluate system EMC through the prediction and analysis of the influence from subsystems/equipment's EMC oversize on the system performance, combined with the EMC evaluation results of the subsystems and equipment in system development and demonstration phase.
(5) Through system EMC test, evaluate system EMC and optimize the EMC design.

The major supporting documents from the research organization in this phase are as follows:

(1) "Summary report on EMC work in principle implementation phase";
(2) "EMC prediction and evaluation report in system development and demonstration phase";
(3) "EMC quality evaluation report in system development and demonstration phase";
(4) "EMC test outline in system development and demonstration phase";
(5) "EMC test result system development and demonstration phase";
(6) "EMC joint test report";
(7) "EMC process control requirements";
(8) "Hazard analysis report of excess EMC";
(9) Other reports related to EMC, such as quality analysis reports, standardization reports.

Based on the above-supporting documents, the EMC management office will then review the following transition documents:

(1) "EMC design scheme of system development and demonstration phase";
(2) "Summary report on EMC work of system development and demonstration phase."

4. Stereotype phase

In the production and deployment (verification) phase as well as the application and maintenance phase, we need to evaluate the EMC of the systems, subsystems, and equipment.

(1) Evaluate the EMC of subsystems and equipment through EMC stereotype (verification) test of subsystems and equipment.
(2) Evaluate the system EMC through the system EMC stereotype test.

The major supporting documents input by the research organization in this phase are:

(1) "Summary report on EMC work in the system development and demonstration phase" of subsystems and equipment;
(2) "Summary report on EMC work of engineering development phase" of system;

(3) "EMC quality evaluation report in the production and deployment phase" of system, subsystem, equipment.

Based on the above-supporting documents, the EMC management office will then review the following transition documents

(1) "EMC stereotype (verification) test outline" of subsystems and equipment;
(2) "EMC stereotype test outline" of the system.

7.3.3 Specific Requirement for Documents

1. EMC outline

The EMC outline is used to specify the working requirements and methods for EMC organization, management, design, prediction, evaluation, verification, process control, delivery, training, etc. It is a general guide for EMC development of information systems. It is also the basis for developing EMC work plans and other documents.

The purpose of the EMC outline is to enable managers and engineers to treat EMC as a basic performance requirement throughout the entire life cycle of an information system and meet the EMC requirements.

The EMC outline should be developed at the beginning of the principle design phase. It mainly includes the following aspects:

(1) EMC standards, regulations the system follows and how they have been tailored;
(2) EMC organization and its duty, authority, division of labor, and scope of work;
(3) The objectives, contents, requirements, and methods of EMC management in the development process;
(4) EMC work plan;
(5) EMC training, design and verification, prediction and analysis, evaluation, process control, assessment requirements and methods.

2. EMC control plan

The EMC control plan is formulated to fully explain the planning and technical measures related to EMC, and establish the basis for EMC work in the development process to ensure that the EMC requirements specified in the contract are met.

The contents majorly include management, spectrum protection, design of EMI prevention structure, electronic/electrical wiring design, EMC circuit design. In the actual EMC design, the EMC of the system, subsystem, and equipment is controlled by quantitative design.

3. EMC test plan

To ensure a good EMC performance for equipment, a series of EMC tests are required during the development process. In order to coordinate the EMC with other aspects in engineering development, and determine the content, type, scheme and progress of the EMC test, a special EMC test plan needs to be formulated and submitted to the management office in a timely manner.

The test plan mainly includes the scope of application, references, product technical status description, test items and requirements (including system-level test and equipment/subsystem-level test), test schedule, test facilities and equipment, budget, possible problems and solutions and test plan improvement measures.

4. EMC design scheme

EMC design mainly includes the scope of application, reference, electromagnetic environment, EMC requirements, EMC design guidelines, EMC design schemes, EMC prediction and analysis, and EMC test verification.

EMC design scheme includes subsystem or equipment key categories, performance degradation criteria, safety factors, interference and susceptibility control, wire and cable wiring, power, spikes, lapping and grounding, lightning protection, static protection, personnel protection, damage to flammable and explosive devices, design for external environment adaptation, design and application of devices.

EMC prediction and analysis mainly include internal system, between subsystems/equipment; between systems; among system, subsystem, equipment and electromagnetic environment; spectrum utilization and spectrum allocation.

5. Overall EMC technical requirement

The overall EMC technical requirements are used to stipulate the EMC technical indicators that the system and its subsystems and equipment should meet. The requirement should include a clear specification for the technical measures and test items to be carried out in order to meet the EMC requirements in the development process. The overall EMC technical requirements include the followings:

(1) Susceptive criteria when systems, subsystems, and equipment are subjected to EMI;
(2) EMC categories of subsystems and equipment, and the safety margins they should have;
(3) Test items of subsystems and equipment and the corresponding limit values. Description of the tailoring of the existing regulation should be included;
(4) Power supply characteristic requirements of systems, subsystems, and equipment;
(5) System cable selection, categorized laying, and wiring specifications;
(6) Design specifications for lap joint and grounding of systems, subsystems and equipment;
(7) Filter selection and installation specifications for subsystems and equipment;
(8) The specification of the crystal oscillator frequency of the clock circuit of the subsystem and equipment;
(9) Board layout and wiring specifications for subsystems and equipment;
(10) The digital circuit pulse rising edge specification of subsystems and equipment;
(11) Case structure and shielding design specifications for systems, subsystems, and equipment;

(12) Lightning and electrostatic protection class and measures for systems, subsystems, and equipment;
(13) Spurious rejection, harmonic rejection, cross talk rejection, and adaptability of external RF environment of RF equipment;
(14) Special tests and methods for systems, subsystems, and equipment.

The major supporting documents are:

(1) "EMC management requirements" of systems, subsystems, equipment;
(2) "EMC prediction and evaluation report of the principle design phase" of systems, subsystems, and equipment;
(3) "EMC quality evaluation report of the principle design phase" of systems, subsystems, and equipment;
(4) "Summary report on EMC work of the demonstration phase" of systems, subsystems, and equipment.

6. EMC management requirement

EMC management requirements are formulated according to the development requirements of systems, subsystems, and equipment.

7. Prediction and analysis report

EMC prediction and analysis should be performed to determine the scope and extent of the EMC problem of the engineering system, such that the project management personnel and engineers, as well as the production and maintenance personnel, will have an expectation or even discover the potential EMC problems in advance. EMC prediction and analysis provide basis for decision making for engineering development.

The content of prediction and analysis is as follows:

(1) Scheme demonstration and preliminary design phase. This phase is to assist the determination of the main characteristics and technical conditions of the system, such as modulation type, data rate, information bandwidth, transmission power, receiving sensitivity, antenna gain, and parasitic signal rejection. The general contents to be predicted and analyzed are:

① electromagnetic problems between internal equipment and components of the system;
② electromagnetic problems between systems and subsystems;
③ electromagnetic problems among systems or equipment, components, and electromagnetic environment;
④ spectrum utilization and frequency configuration issues.

(2) Development and trial production phase. In this phase, it is necessary to determine the specific performance parameters and functional level components of the equipment, such as amplifiers, mixers, filters, modulators, detectors, display

or readout devices, power supplies. Common contents of prediction and analysis are:

① Electromagnetic problems caused by external electromagnetic signals coupling to different devices and components in the system;
② Cable coupling;
③ Case coupling;
④ Case shielding effectiveness.

(3) Production, deployment, and application phase. In this stage, the electromagnetic control is analyzed and solved by frequency management and electromagnetic control in the time domain and airspace. The prediction and analysis contents generally include

① Site effect;
② Frequency management;
③ Effective radiated power limitation;
④ Coverage of antenna beam;
⑤ System EMC comprehensive analysis.

8. Quality evaluation report

The evaluation report mainly includes the following contents:

(1) The basic conditions of the product such as product name, research organization, development phase, development type, product function, product composition, and principle;
(2) System EMC such as electromagnetic environment, shielding effectiveness, resonance characteristics, layout characteristics, and frequency allocation.
(3) Installation characteristics of subsystems and equipment:

① Installation environment: Specify the installation position of the product. For equipment with antennas, there should be a clear description of the installation location for the antennas and whether there are special requirements of electromagnetic field environment where the antennas are installed;
② Operating power: Identify the transmission power of the transmitter, receiver sensitivity, voltage and current usage of nonRF transceivers, etc.;
③ Operating frequency: Specify frequency information including the local oscillator, IF, RF on the transmitting link and receiving link of the product;
④ Interfered object: Identify other products that may be interfered by this product;
⑤ Interfering object: Identify other products that may interfere with this product.

(4) EMC requirements for the product:

① Standard items: Select the test items according to the test requirements of existing regulations combined with the characteristics of the systems, subsystems, and equipment;
② Description of the tailoring: Explain the tailoring of existing regulations by the product;
③ Other explanations: In addition to the existing regulations, if there is any other EMC test needed due to the product's own requirement, the research organization should list the additional test requirement.

(5) Existing EMC problems in products: Identify the EMC problems found in product testing. With the existing problems, we either keep the product as it is or solves the EMC problems that have occurred before;
(6) Product EMC design evaluation: Evaluate all EMC design of the products, including spectrum, structure, PCB, grounding, shielding, filter, cable interface, power supply;
(7) Test condition of the product: Identify the test organization, test items that the product fails and describe the test condition;
(8) Result of product improvement: Provide the system EMC performance after improvement;
(9) Quality management: Including the supervision and inspection of EMC issues by the quality system, and whether the technical quality problems have been solved;
(10) Product EMC compliance analysis.

9. Quality analysis report

The quality analysis report mainly includes introduction to the development process, how well the technical indicators are in line with the mission specifications, product technical status compliance, implementation of quality assurance outline; how well the product performance is in line with the technical indicators, product quality, quality issues; and whether the problems have been solved, special review, and product quality conclusions. The items are described in detail as following:

The implementation of the quality assurance outline includes outline preparation, work principles and quality objectives, management responsibilities, documents and records control, quality information management, technical status management, personnel training and qualification assessment, customer communication, quality control of design process, test control, procurement quality control, trial and production process quality control and standardization reports.

The execution of quality problem solving includes problem description, cause analysis, corrective action and final check of whether the problem has been solved.

10. EMC test outline

The main contents of the EMC test outline include scope of application, basis for outline, purpose of the test, quantity and technical status of the test object, accompanying test object and supporting equipment, test items, test contents, test methods, number of major test equipment and key parameters, data processing principles and qualification criteria, test organization, institutes that participate the test and task division, test schedule, test security and requirements.

11. EMC test report

The main contents of the EMC test report include test overview, test items, steps and methods, test data, main technical problems and solutions in the test, test results, conclusions, problems and suggestions for improvement, test photos, real-time audio, and video data of the main test items.

12. EMC comprehensive evaluation report

The main contents of the EMC comprehensive evaluation report include electromagnetic environmental conditions, EMC requirements, EMC design guidelines, EMC design schemes, EMC test contents and results, EMC problems and solutions, and the remaining problems and the suggested solutions.

Among them, the EMC design scheme mainly includes key categories of subsystems and equipment, performance degradation criteria, safety factor, interference and susceptibility control, wire and cable wiring, power supply, spike signal, lap and grounding, lightning protection, static protection, personnel safety protection, flammable and explosive materials and equipment hazards, external electromagnetic environment, and suppression measures taken.

EMC problems and solutions mainly include phenomena, problem causes, solutions, and verification.

7.3.4 Specific Evaluation Methods

The analytic hierarchy process is used to classify the EMC requirements involved in the evaluation and determine the weights. Then, according to the interference correlation relationship and the interference correlation matrix, each scoring criteria is determined, and finally, the EMC factors during product development are rated by experts.

1. Determination of the weight value

(1) Cluster analysis by existing standards

According to the existing standards, the key categories of subsystems and equipment stipulate that all subsystems and equipment installed in the system or related to the system should be classified as one of the key EMC categories. These divisions are based on the effects of EMI, failure rates, or degradation procedures for assigned tasks. There are the following three categories.

(1) Class I: This type of EMC problem may result in shortened life, damage to the vehicle, disruption of tasks, costly transmission delays, or unacceptable system efficiency degradation.

Table 7.1 Scales of level using the hierarchical analysis method

Level of importance	Meaning
1	Indicator p and indicator q are equally important
3	Indicator p is slightly more important than indicator q
5	Indicator p is obviously more important than indicator q
7	Indicator p is strongly more important than indicator q
9	Indicator p is extremely more important than indicator q
2, 4, 6, 8	Indicates the intermediate value of the above judgment
Reciprocal	The ratio of the importance of the indicators p to q is k_{pq} The ratio of the importance of the indicators q to p is $k_{qp} = 1/k_{pq}$

(2) Class II: This type of EMC problem may cause malfunction of vehicle, decrease of system efficiency, such that the task cannot be completed.
(3) Class III: This type of EMC problem may cause noise, slight error, or performance degradation, but will not reduce the effectiveness of the system.

According to the classification of the safety equipment in the existing regulations, all subsystems and equipment are first clustered according to I, II, and III classes.

(2) Constructing weight indicators using hierarchical analysis process.

When the analytic hierarchy process is used to determine the weight of the subsystem and equipment, it is necessary to comprehensively consider various factors, including the probability of problems in the development of information system and equipment in the past, the degree of influence on other equipment and on the system. The metrics of the analytic hierarchy process indicators are shown in Table 7.1.

2. Determine the evaluation criteria

(1) Determine the limit

According to the interference correlation relationship and the interference correlation matrix, we can determine the radiation and susceptibility related limits of each subsystem and equipment. The main limits are as follows:

(1) Susceptibility threshold: refers to the signal level that causes the test object to exhibit a minimum discernible undesired response;
(2) Susceptibility threshold: the interference threshold level at which the system cannot work normally;
(3) Failure interference level: the level of EMI that the system is not allowed to accept, in other words, the interference level that causes permanent failure or permanent malfunction of the receiver;
(4) Allowed interference: the maximum interference that the system allows;

(5) EMI emission value: EMI (including radiation emission and conducted emission) brought to the surrounding environment when the system with electromagnetic transmission is allowed to work, the EMI value must not affect the operation condition of surrounding subsystems;
(6) Safety margin: the difference between the relative threshold and the actual interference signal level in the environment.

(2) Optimization limits

The limit value is optimized through the comparison with the test data, and the system's feasibility, cost-effectiveness, and other factors.

3. Expert scoring system

The expert scoring system is used to evaluate the EMC of the system, subsystems and equipment, and the evaluation results are comprehensively processed [77].

The expert scoring comes in two levels: Firstly, according to whether the planned EMC work has been carried out, a 0–1 score will be filled by computer, that is, 1 point for work being carried out, and 0 point for not carried out. The work carried out is then evaluated in detail from a technical level and scored by experts.

Chapter 8
EMC Engineering Case Analysis

Most products cannot satisfy the existing requirements for CE102, RE102, and RS103 when they undergo equipment/subsystem EMC test. This chapter analyzes the causes of failure in the above tests and the resulting hazard. Finally, we provide solutions for the products to pass the tests.

8.1 Hazard of Failure in CE102, RE102, and RS103 Test Items

The CE102 test is applicable for the input leads (including phase, neutral, and ground) of the equipment under test (EUT) power supply, but it is not applicable to the output leads of the EUT power supply. The low-frequency part of the test is to ensure that the EUT does not deteriorate the quality of the power supply (voltage distortion allowed) on the existing power bus bars of the platform. At higher frequencies, the limit of CE102 has a certain control over the radiation of the power leads of RE102. This radiation may be coupled to a susceptive receiver with an antenna. The CE102 limit value needs to be consistent with the RE102 limit value; i.e., the CE102 emission limit value should not exceed the RE102 radiation limit. For example, in the lab environment, a 2.5-m-long power line is connected to the line impedance stabilization network (LISN). The signal level applied to the line is 60 dBμV, and the level detected by the RE102 rod antenna is approximately 20 dBμV/m. The detected electric field frequency is flat at approximately 10 MHz and is approximately equal to (X-40) dBμV/m, where X is the applied signal level in dBμV.

If the CE102 result of equipment exceeds the limit, the low-frequency part mainly affects the quality of the platform power supply, which will reduce the working efficiency of the electric generator. In severe cases, the electric generator may be damaged and the normal operation of other equipment in the system may be affected; i.e., it might cause computer operation errors and video picture jitter. Moreover, it

© National Defense Industry Press and Springer Nature Singapore Pte Ltd. 2019

D. Su et al., *Theory and Methods of Quantification Design on System-Level Electromagnetic Compatibility*, https://doi.org/10.1007/978-981-13-3690-4_8

may cause radiation emission in high frequency and interfere with the highly sensitive receiving equipment.

RE102 test is applicable for radiation emissions generated by the cases and all interconnect cables of the equipment or subsystems, but it does not apply to the transmitter's fundamental or antenna radiation. The purpose of the RE102 test is to protect the sensitive receiver from antenna coupling. The sensitivities of many tuned receivers are on the order of 1 μV, which require stringent requirements to prevent platform problems caused by performance degradation.

There is no implicit relationship between RE102 test and RS103 test. RE102 test is related to the potential effects of an antenna-connected receiver, while RS103 test simulates the field produced by antenna-connected transmitters. Usually, the same equipment is involved in both requirements. For example, a 30 W-ultralong wave AM radio with antennas operates at 150 MHz. In the receiving state, an electric field of 40 dBμv/m (about −81 dBm at the receiver input) can be detected; in the transmitting state, the same equipment will produce a field of 150 dBμV/m (32 V/m) at a distance of 1 m at the same frequency. The two field levels differ by 110 dB.

The RE102 limit curve is based on the relationship between the level of the platform and antenna-connected receivers and the typical amount of shielding associated with the wiring between the antenna and the equipment. Therefore, RE102 draws the limits for surface ships and submarines, the limits for aircraft and space systems, and the limits for ground equipment.

For internally installed equipment, the air force and navy limit curves are developed for aircraft that are not specifically designed with shield volumes, but have a small amount of shielding within the test frequency range. For externally mounted equipment, even a small amount of shielding is not available, so the curve is 10 dB more stringent. Most of these experiences are from fighter size aircraft. Since the tuned antenna effective aperture size (Gλ2/4π) decreases with the increasing frequency, the limit increases by 20 dB/dec above 100 MHz, leaving the coupled power level from the isotropic tuned antenna constant. The curve breaks at 100 MHz are due to the difficulty in maintaining a tuned antenna, because the actual size of the antenna increases with the increase of the wavelength and the coupling decreases with the increase of the wavelength.

For the low-frequency band of the RE102 limit value curve, we mainly consider that the allowed value of the radiation level included in the quality of the power supply may contradict with the curve. For example, on a 115 V 400 Hz power bus, the voltage RMS value is allowed to be about 0.63 V at 15 kHz. In laboratory tests, the radiation field strength at this level is about 76 dBμV/m, which is higher than the narrowband radiation limit of GJB 151 RE102 (A1 class) by 31 dB higher. If the GJB 152 rod antenna is used to measure from the low end to 400 Hz, the indicator level from the power supply waveform is about 1 V/m. Therefore, in order to avoid a conflict between the power quality and this requirement, the RE102 limit should be considered for the aircraft power supply equipment at lower frequencies.

The failure of RE102 test is mainly due to radiation emission from the cases or cables, which may be received by the high-sensitivity receiver in the platform to form interference. It is also possible that the interference signal inserts into the adjacent cable to form new interference. Moreover, the interference caused by the failure of RE102 is the most difficult to identify. When the equipment is tested in the laboratory, it often uses copper foil or shielding tape to strengthen the shielding of the equipment and cables. However, when the platform is actually installed, there are no additional measures as with testing. Therefore, the interference signal is released and may form real interferences. Under this condition, to find out the real source of interference, it is almost necessary to search the entire platform; i.e., all equipment and cables need to be eliminated one by one, which will waste a lot of human and material resources and time and affect the development progress of the platform.

The failure of RE102 will lead to even worse results—early detection by the enemy's passive detection system and detected information that can support fine identification.

RS103 ensures that the performance of equipment onboard or external to the platform does not degrade in the presence of electromagnetic fields generated by various transmitting antennas. There is no implicit relationship between RS103 and RE102. The RE102 limit is mainly to protect the antenna-connected receiver, while RS0103 is to simulate the field generated by the antenna transmission.

The limit of RS103 is directly related to the platform environment. Therefore, in the regulation, the platform environment is divided into eight cases. The limits of different platforms are specified according to the expected environmental level during the service life cycle of the equipment. Then, the limit value is further selected based on the frequency range of the army, navy, and air force equipment. These limits do not necessarily represent the worst environmental conditions that the equipment may be exposed to, as the RF environment may vary widely, especially from the field generated by the transmitters outside the platform. However, the level set by this limit requirement is sufficient for most situations.

Due to the variability of the ground installation environment, some tailoring may be tightened. For example, equipment installed within or close to a large ground radar station may be interfered by the back lobe of the antenna with a transmission power exceeding 1 MW. Therefore, the limit requirement must be tightened to a suitable level.

For aircraft and ships, different limits are specified depending on whether the equipment is protected by the platform structure. This basis does not apply to the army systems such as tanks, because the equipment used in a certain structure is also often used in places without structure protection. Army aircraft are all required to be 200 V/m, which is not related to equipment installation location or safety critical equipment. This is based on the use of army aircraft. When encountering high-power transmitters, the army aircraft is often closer in distance with a longer duration, and sometimes, some external environments are even higher than 200 V/m. Usually, the qualified level of helicopters is higher than that of other aircraft.

The failure of RS103 test, on the one hand, is due to the shielding performance of the equipment cases and cables and, on the other hand, is due to the circuit design inside the equipment. When an external interference signal is coupled into the equipment through cases or cables, the equipment performance may be degraded or even damaged. When the equipment installed on the platform has radiation-susceptive problems, it is also extremely difficult to find and eliminate because when the equipment was tested, the manufacturers usually have additional shielding measures to pass the test, or in the worst-case condition, the test was not carried out according to the actual installation on the platform. For example, many sensors on an aircraft are not tested, so the signals collected by the sensors are often simulated. As a result, although the equipment passed the 200 V/m test required by RE103, in the actual platform, however, there is often unstable interference which is difficult to reproduce, locate, and eliminate. Such interference may happen throughout the entire service period; in addition, when the platform is in an environment with field strength below 200 V/m, it is obvious that the platform is interfered, which will cause serious flight safety problems.

8.2 The Main Reasons for CE102, RE102, and RS103 Test Failures

The monitoring object of the CE102 test is the input leads of the EUT power supply. The main reason for the test failure is that the power supply filtering of the equipment is not properly designed. Since the internal power consumption of each equipment is not the same, it is necessary that the power input passes multiple power conversion modules to fulfill the requirements for the internal modules of the equipment. In this circumstance, it is required to filter not only at the input leads of the primary power supply, but also at the input end of each secondary power supply. Moreover, each power supply filter must be designed for the specific situation; otherwise, signals exceeding the standard will be generated and it will enter the primary power supply through the secondary power supply, causing the CE102 result of the entire equipment to exceed the standard. A common cause of the excess result of CE102 is the pulse-width modulation (PWM) signal during DC–DC conversion. The signal is usually at tens of kilohertz or hundreds of kilohertz, with strong amplitude and harmonics. It appears as a strong signal with a fixed interval in the CE102 test.

The monitoring object of the RE102 test is the radiation emissions generated by the equipment and subsystem cases and all interconnecting cables. The main reason for the test failure is that the interference signal is radiated from the slots, openings of the case, or connection cables due to poor design of the internal circuit EMC. Generally, the test frequency of RE102 is 10 kHz–18 GHz. According to the frequency characteristics of the signal, the wavelength of the radiated emission signal in the low-frequency band is longer, and the radiation is difficult to leak with respect to the slot of the case. Therefore, the excessive radiation at low frequencies is mostly transmitted through cables. In the high-frequency band, the signal wavelength is shorter, the cable attenuation is larger, and the excess radiation is emitted from the case slots; the boundary between the high and low frequency is related to the equipment size and cable length, and the empirical value is about 200 MHz. Although the radiation exceeding the standard of RE102 emits from the case slots, openings, and cables, the reason for the excessive radiation is the internal circuit design. One of the most important points to be protected in internal circuits is the crystal. The crystal oscillator is a narrowband signal, often with strong amplitude and various harmonics. It appears as continuous multiple peaks with the excess result from RE102 test.

RS103 test is to simulate the field generated by the antenna transmission and use the field to illuminate the equipment case and cable to observe whether the equipment is susceptive. Similar to the RE102, the common test frequency is 10 kHz–40 GHz. According to the frequency characteristics of the signal, the low-frequency coupling channel is usually cables, and the high-frequency coupling channel is the case slots and openings. The susceptibility of the equipment usually goes into the equipment through cables, openings, and slots of the case, causing the internal circuit to respond and form interference. For cable coupling, cables and equipment interconnection connectors are the key parts to be protected against susceptibility. For aperture coupling, we need to focus on the treatment of the openings and the slots of the case. In general, to avoid susceptibility, we need to consider shielding and grounding of the cable; 360° bonding and filtering at the connector; internal filtering and wiring; design of the grounding system of the equipment.

8.2.1 CE102 Test

Example 8.1 A motor-type product uses 28 V DC and 115 V AC power. After the preliminary test, it was found that the 28 V DC passed the CE102 test, while the 115 V AC failed the test. The failure frequencies were mainly concentrated at 10–700 kHz, and the maximum value exceeded the limit for about 30 dB. To investigate this problem, two sets of contrast tests were carried out.

(1) Comparison between rotating and nonrotating motor in CE102 test. With all the other circuits working normally, the product was under CE102 test with rotating motor and nonrotating motor. From the test result (Fig. 8.1), we found that the rotating motor was only about 5 dB larger than the result of the nonrotating motor. The two results had the same trend and the same frequency band that exceed the standard. Therefore, the interference of the motor rotation to the power line is not the main factor of the CE102 failure.
(2) Comparison between the PWM control signal sent and unsent without motor rotation. In the motor driver circuit module, the PWM signal is used to control the rotation of the motor. When the motor was not rotating, the product was subject to CE102 test with the PWM control signal being sent and unsent. The test results (Fig. 8.2) show that when the PWM control signal was not sent, the test results all met the standard limit value requirements, and when the PWM control signal was sent, the test was failed.

The contrast test shows that sending PWM control signal will cause the product's 115 V AC to fail the CE102 test. Now, we will analyze the result with the schematic diagram of the circuit (Fig. 8.3).

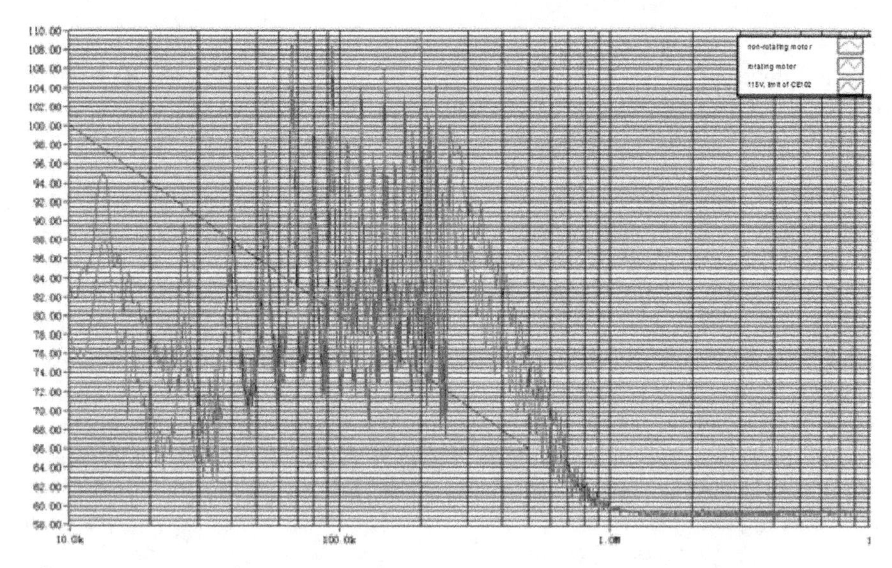

Fig. 8.1 Comparison of CE102 test curves between rotating motor and nonrotating motor

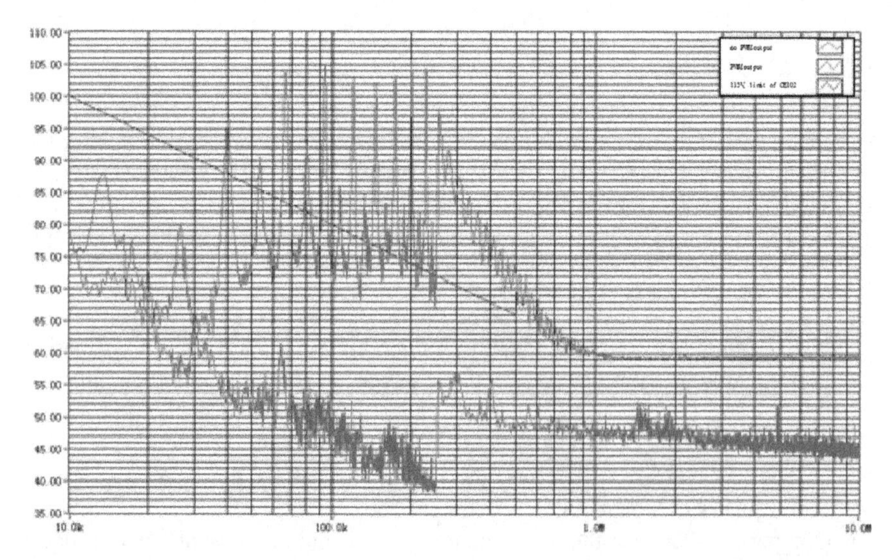

Fig. 8.2 Comparison of the CE102 test curve with PWM control signals sent and unsent

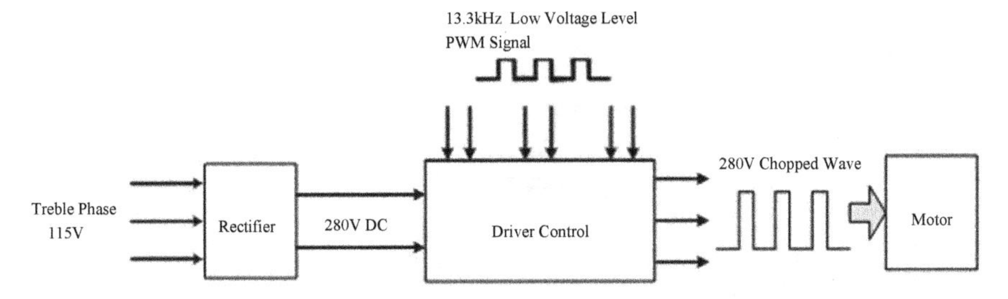

Fig. 8.3 Schematic diagram of the motor driver module

The 115 V three-phase AC power was rectified to generate 280 V DC. The 280 V DC was controlled by PWM signal to form 280 V chopping wave to drive the motor. When the 280 V chopping wave was formed, with the rising and falling of voltage, the burrs was generated to form an EMI source, which went back to the 280 V DC bus bar and affected the 115 V three-phase AC line. At the same time, it can be seen from the CE102 test results that the peak of the failure curve is 3–46 harmonics, and the odd-order harmonic amplitude is greater than the even-order harmonic amplitude. Therefore, when designing the driver circuit, the filtering measures matched with the driver control circuit should be added to the 280 V DC bus to suppress the conduction interference.

8.2.2 RE102 Test

Example 8.2 A product failed RE102 test, and the failure frequency points were concentrated at 2–30 MHz, and the maximum value was 5 dB larger than the limit.

In the problem diagnosis test, it was found that the RE102 test results were inconsistent before and after the product case was opened for simple adjustment. Therefore, it was suspected that the shielding of the case was unstable, resulting in inconsistent test results. Through experiments, it was found that the RE102 test results were very different when the case screws were tightened and untightened (a half turn to lose after tightening). Since the shield design of the upper cover of the product was not good enough, the upper cover of the case was so thin that it deformed after being subjected to force, thereby creating a slot. Therefore, the screws must be tightened firmly; otherwise, the cover on the case would be poorly connected, resulting in leakage of electromagnetic energy. Figure 8.4 shows the bonding situation between the upper cover and the sidewall of the case.

After the old cables were replaced and the screws were tightened, the product passed the RE102 test. The comparison of the RE102 test curve with the screw tightened and untightened is shown in Fig. 8.5.

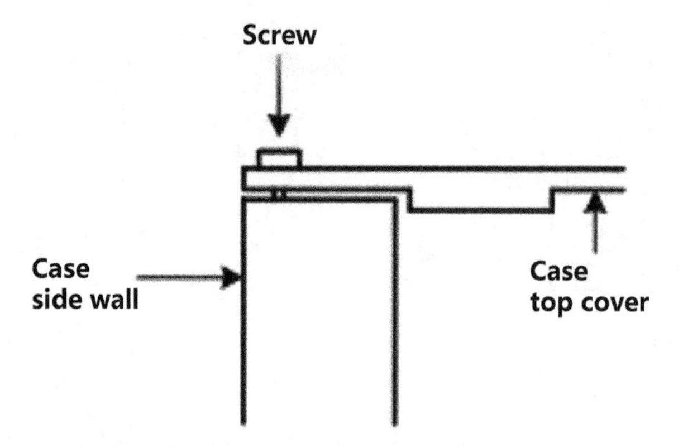

Fig. 8.4 Bonding situation between the top cover and the sidewall of the case

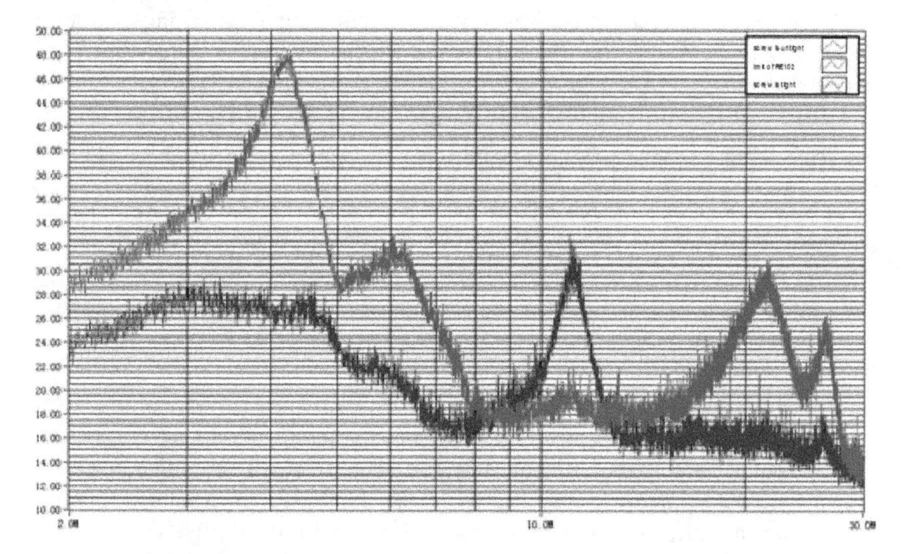

Fig. 8.5 Comparison of RE102 test curve with screw tightened and untightened

Example 8.3 A subsystem for a certain mission consists of 14 pieces of equipment, and it failed the RE102 test. The failure frequencies were concentrated at 30–200 MHz, and the maximum value was 5 dB over the limit.

Necessary accessory equipment was added so that the 14 pieces of equipment could work separately. Each of the equipment went through the RE102 tests, and one piece of equipment was suspected to be the cause of the RE102 failure. Then, a copper mesh was used to cover the connectors, keys, fan power cable and other parts of the equipment, and the RE102 test is performed again. Through the tests, it was found that when the copper mesh covered the fan power cable, the RE102 test could be passed. Figure 8.6 shows the fan power cable and structure of the case.

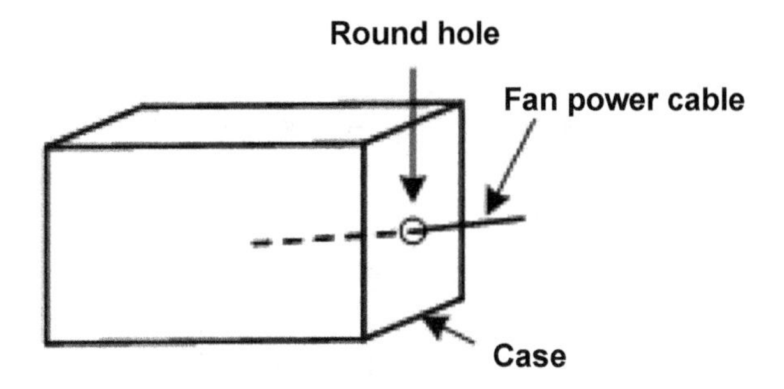

Fig. 8.6 Fan power cable and structure of the case

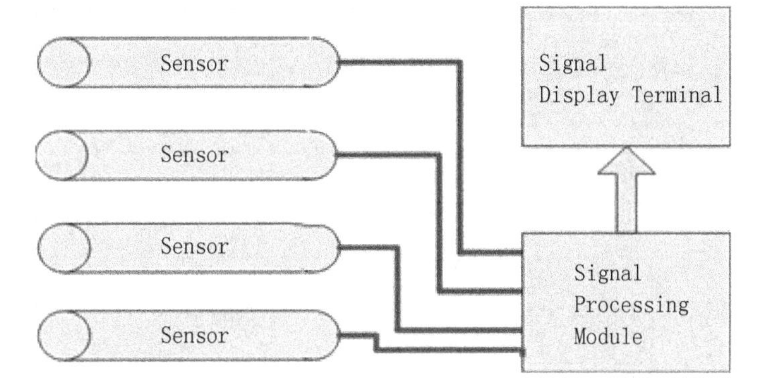

Fig. 8.7 Composition of the sensor equipment

If the shielding of the case is continuous, i.e., there is no opening in the conductive case, then the shielding of the case shall be very effective in theory. However, any opening such as wires and holes can greatly reduce the shielding effectiveness.

8.2.3 RS103 Test

Example 8.4 In the RS103 test of certain sensor-type equipment, the 200 MHz–1 GHz susceptibility test was failed. And the higher the frequency, the more susceptive the equipment. The equipment was used to measure the change of a physical variable through a sensor. The change was then converted into an electrical signal and transmitted to the processor, as shown in Fig. 8.7.

In order to find the susceptive object and the coupling path, the copper mesh was used to cover the display module, the processing module, the interconnection cable, and the sensor of the product one by one, and the RS103 test was then performed. The test results showed that when the copper mesh covered the sensor and the transmission cable, the anti-interference ability of the equipment was improved. Therefore, it could

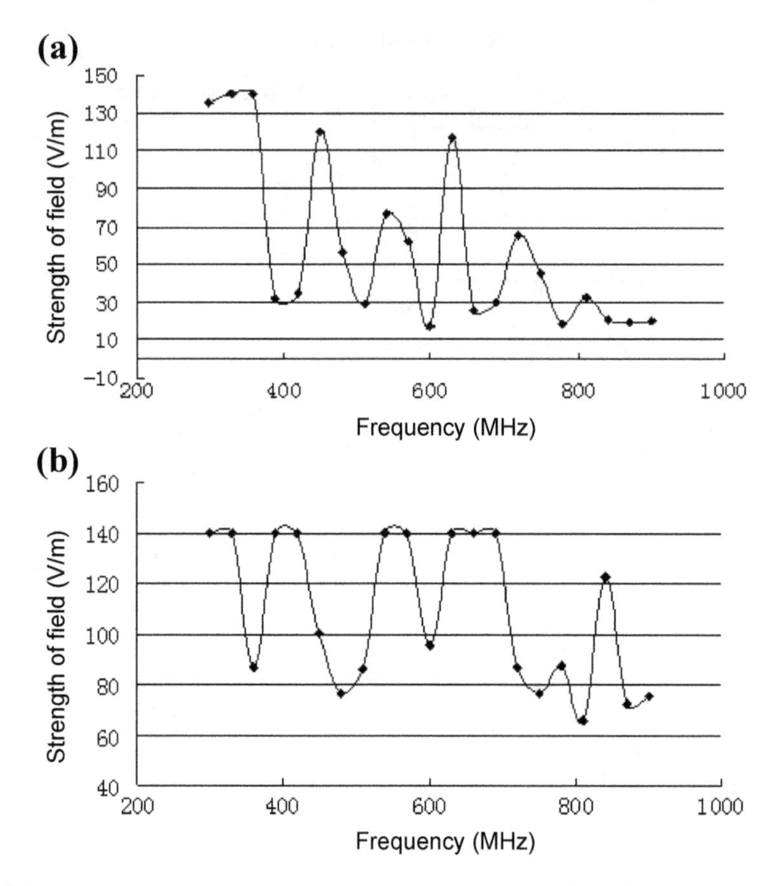

Fig. 8.8 Product susceptibility curve—**a** RS103 susceptibility curve, **b** RS103 susceptibility curve when the sensors and cables were covered by the copper mesh

be preliminarily determined that the sensor of the product, the transmission cable, and the connection portion between the two were the main paths for picking up electromagnetic energy, thereby causing the RS103 to fail. Figure 8.8a, b shows the RS103 susceptibility curves for the equipment before and after using the copper mesh. It can be seen from the susceptibility curve that due to the small size of the sensor, the coupling effect on the high-frequency electromagnetic field is strong, and as the frequency increases, the anti-interference ability of the product increases. The electric field intensity that causes susceptibility increases from a minimum of 15 V/m before the copper mesh covering to 64 V/m after the covering.

8.3 The Solutions to Pass CE102, RE102, and RS103 Tests

8.3.1 The EMC Failure Location

There are four main methods to locate failures in CE102, RE102, and RS103 tests:

(1) Method of exclusion: It is applicable to finding the major cause (which is a certain module) of failure in the equipment/subsystem. Using this method, we start from the equipment, and gradually check the modules and circuits. This method can locate the failure down to the equipment or circuit level.
(2) Method of replacement: Replace a circuit module in the equipment with a simulated load to determine whether the module is the main cause of the test failure. This method is simple to implement and can diagnose at the equipment or circuit level.
(3) Method of interchange: Replace a circuit module in the equipment with a circuit module without EMC problem to determine whether the circuit module is the main cause of the test failure. This method can diagnose the failure at equipment or circuit level.
(4) Method of diagnostic test: When the equipment/subsystem fails a test item, the method of diagnostic test is a convenient and effective way to determine the main factors causing the problem, such that the problem can be solved fast and accurately. Diagnostic tests generally use sensors such as probes, combined with spectrum analyzers and oscilloscopes, to test local electric field strength, magnetic field strength or time-domain waveform, amplitude, etc. Diagnostic tests sometimes do not require standard test equipment and test sites. We can use simple and versatile equipment without a high accuracy requirement. As long as the test is highly targeted, it is easy to find out the problem through testing and determine the effectiveness of the improvement measures. When using the method of diagnostic test, we can get a better result if we also combine other methods described above.

8.3.2 Trouble Shooting Suggestions

1. CE102

In Example 8.1, the failure of CE102 was because that there was no appropriate filtering added toward the power module. Therefore, in order to meet the limit requirement of CE102, it is necessary to include filtering measures such as:

(1) Install internal filter:
An EMI filter shall be installed on the 280 V bus in the case, as shown in Fig. 8.9. However, due to the space limitation in the case, the internal inductance and the capacitor component of the filter are limited in size, such that they cannot be

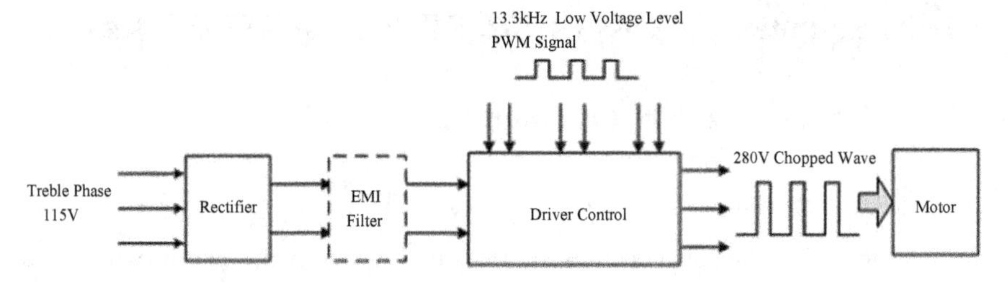

Fig. 8.9 Schematic diagram of EMI filter installation

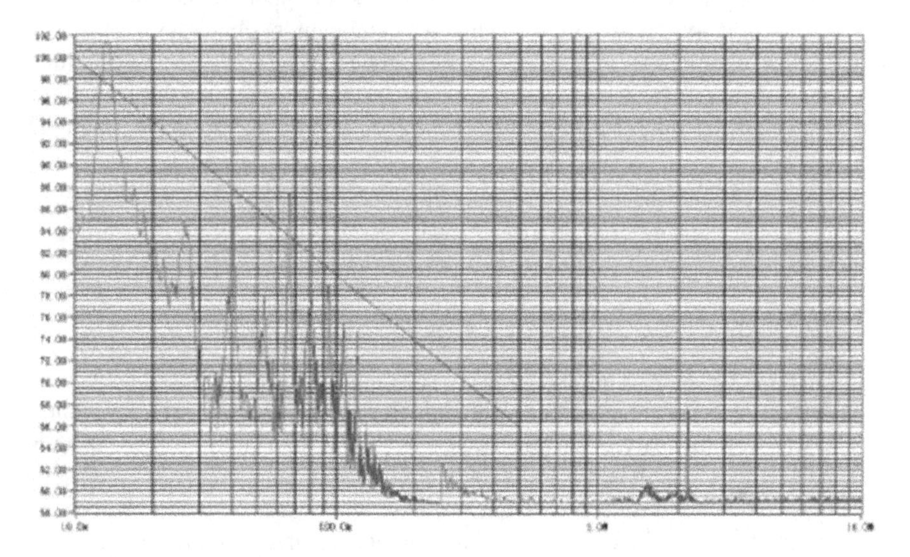

Fig. 8.10 CE102 test result after installing the internal filter

well grounded, causing the performance of the filter not meeting the application requirements. After the internal 280 V DC bus was installed with the filter, there were still three frequency points failing the test, and the maximum value exceeds the limit by 4 dB, as shown in Fig. 8.10. Compared with Fig. 8.2, it can be seen that the result of CE102 has been significantly improved after the filter was installed.

(2) Install power filter:
A power supply filter could be installed at the inlet of the three-phase 115 V power leads of the product. The test results showed that after the external filter was installed, the CE102 test result as shown in Fig. 8.11 can meet the standard limit value. However, the external filter was rather bulky, and the overall weight of the product increased.
Since the internal circuit board of the product has been basically determined, it is difficult to install and connect the internal filter. What's more, it is difficult to optimize the filter parameters. The final measure is to install a power filter.

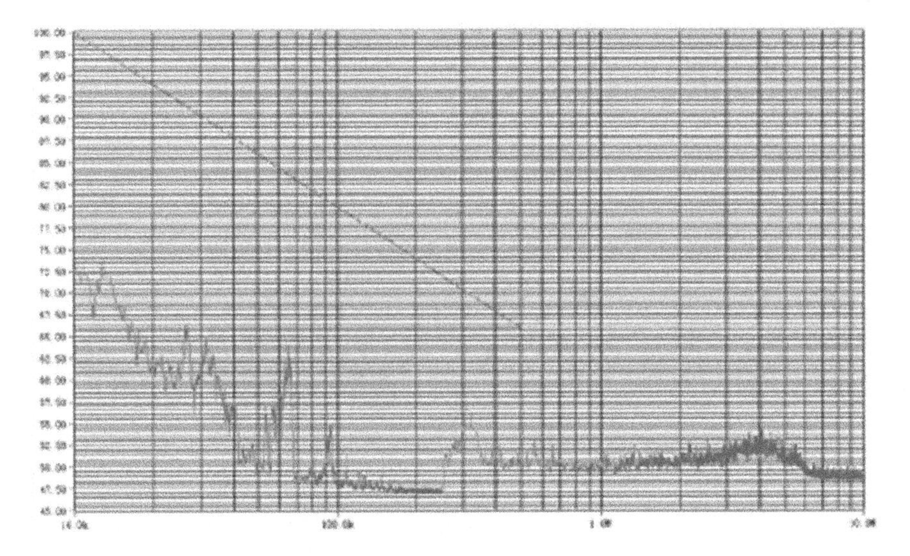

Fig. 8.11 CE102 test result after installing the power filter

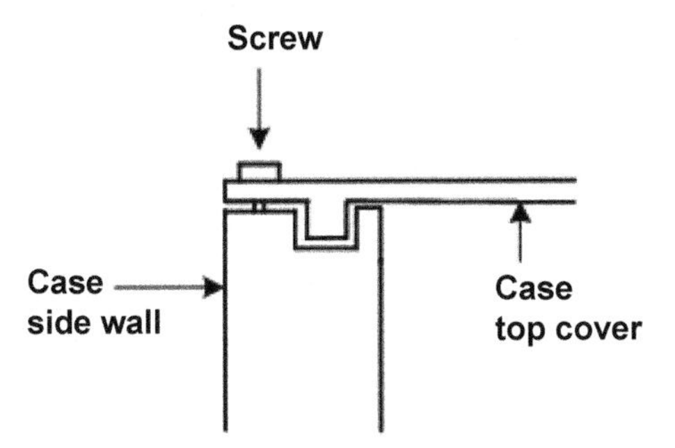

Fig. 8.12 Measures to strengthen the shielding

2. RE102

In Example 8.2, the RE102 test did not satisfy the standard limit value due to the insufficient design of the case cover. Therefore, in order to pass RE102, a more reasonable case shielding structure should be adopted as shown in Fig. 8.12.

However, because the sidewall of the case is thin, it is not suitable for this modification. The final shielding scheme is adding copper reeds at the seam between the upper cover and the sidewall of the case, thereby increasing the electrical continuity of the case.

In Example 8.3, since the fan power cable was taken out from the opening of the case wall, the shielding strength of the case has been greatly reduced, resulting in a failure of the RE102 test. To solve the problem, we can connect a feed-through capacitor in series to the fan power cable and well ground the feed-through capacitor;

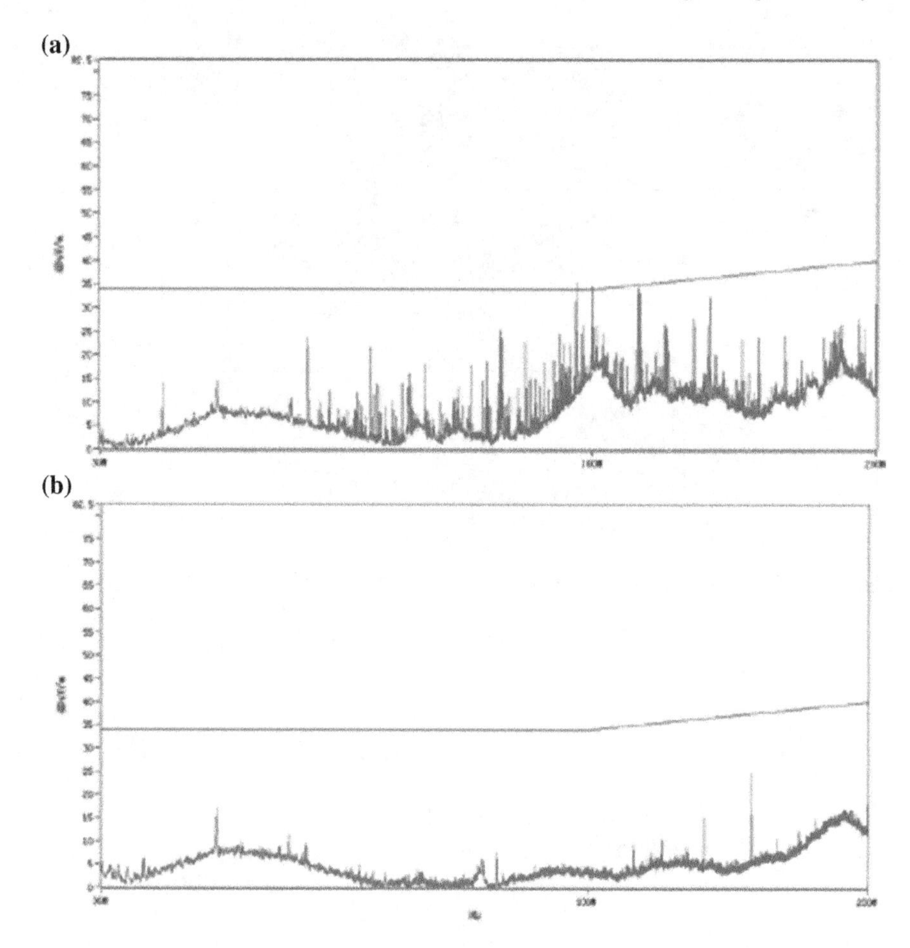

Fig. 8.13 RE102 test results before and after taking measures—**a** the product without any measures; **b** after the feed-through capacitor is added to the power cable of the fan

thus, the radiation emission of the product can be greatly reduced to meet the standard limit requirement of RE102. Figure 8.13a, b shows the RE102 test result of the equipment before and after the series connection of the feed-through capacitor to the fan power cable.

3. RS103

In Example 8.4, the main reason that the product failed the RS103 test was that the sensor, the transmission cable, and the connection location of the product were the main path of picking up electromagnetic energy.

According to the interference mechanism and physical structure of the front-end sensor of the product, the designed shielding structure shall not affect the physical variable that the sensor induced, and at the same time, the external electromagnetic wave shall be effectively shielded and attenuated. Therefore, the sieve structure shield

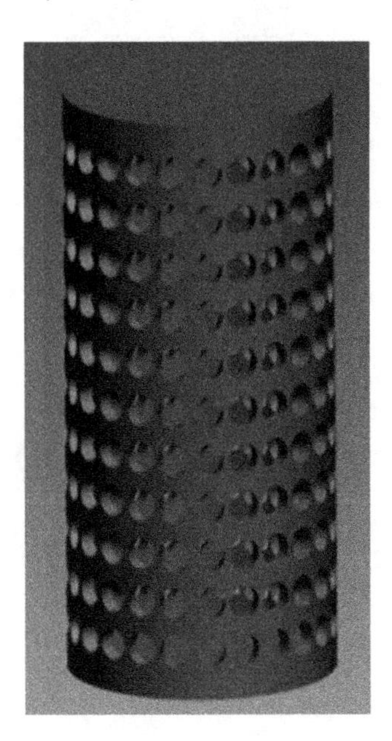

Fig. 8.14 Structure of the shield

is designed based on the actual usage, as shown in Fig. 8.14. The practical application shows that the structure can effectively protect the sensor and the connecting part of the cable from the external electromagnetic environment and thus ensure the normal signal acquisition.

Appendix A
Introduction to System-Level EMC Quantitative Design Software Platform

There is complex mutual EMI between equipment/subsystems in large systems such as airplanes. The key issue in EMC design is how to avoid the mutual interference among equipment/subsystems from the top layer and improve the overall performance and compatibility of the platform. The Institute of EMC Technology of Beihang University has developed a system-level EMC design software (EMCIC). The software can be used for EMC modeling, simulation, prediction, evaluation, indicator decomposition and iterative optimization design of the platform from the perspective of the entire system. The software automatically writes all the design processes into the design documents and stores in the database, such that the EMC design process can be traced back and the design quality can be controlled. At the same time, the development cycle as well as the cost-effectiveness ratio can be reduced.

The EMCIC software has a navigation window. The function of the whole software can be embodied through the navigation window. Different navigation windows can be customized by setting different design indicators, and the function can be controlled and organized this way. The software also has a good version management and process control function. Figure A.1 shows the navigation window of the overall EMC design process of an aircraft.

It can be seen from Fig. A.1 that the software includes six main functional modules: pre-processing, post-processing, program management, EMC evaluation, system-level EMC design and database management. The basic functions of EMCIC are listed in Table A.1.

© National Defense Industry Press and Springer Nature Singapore Pte Ltd. 2019

D. Su et al., *Theory and Methods of Quantification Design on System-Level Electromagnetic Compatibility*, https://doi.org/10.1007/978-981-13-3690-4

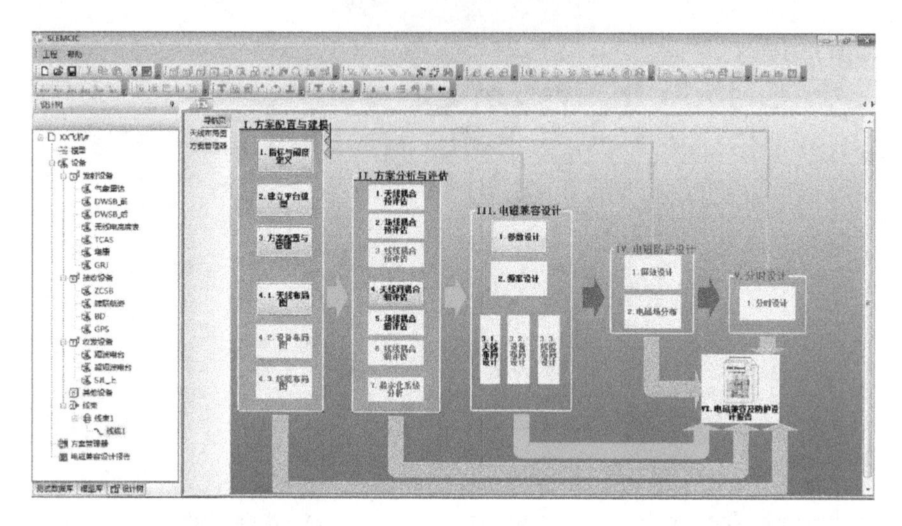

Fig. A.1 Overall software function

Table A.1 Summary of basic functions of EMCIC

Pre-processing	Post-processing
Program management (version control, create/inherit, delete, copy, etc.)	Matlab (Mathworks) visualization
Indicator customization	Visualization of parameter S
Software platform wizard	Visualization of antenna pattern (2D, 3D)
Platform import, extraction and transformation	Visualization of the E/H near field distribution (2D, 3D)
Antenna layout visualization and adjustment	Report generation in the format of Microsoft Word
Program management	EMC evaluation
Transmitter management	EMC pre-analysis
Receiver management	EMC interference correlation matrix
Transceiver management	Simulation and prediction of isolation between antennas
Non-RF equipment management	Simulation and prediction of the near-field distribution
Management of Interference correlation relationship	Simulation and prediction of antenna pattern characteristics
Management of simulation analysis frequencies	Simulation and prediction of the shieling effectiveness
Spectrum browsing	Simulation and prediction of field-line coupling
Behavioral modeling	Simulation and prediction of line-line coupling
System-level EMC design	Database management
Frequency-domain design	Model database
Margin design (power design, susceptibility design, out-of-band emission attenuation design, out-of-band rejection design for receiver, shielding effectiveness design, layout design, protection design)	Test database
Time-domain design	Simulation task release and download

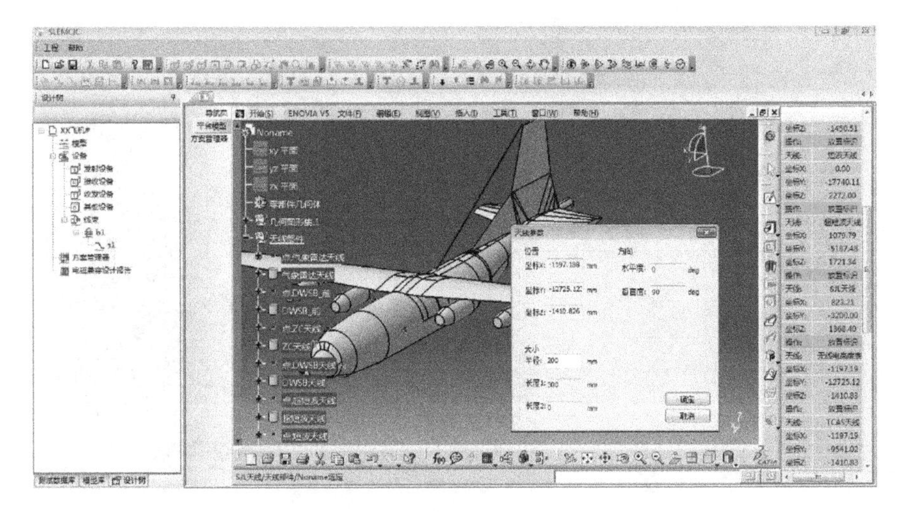

Fig. A.2 Construction of EMC geometric model of the platform

A.1 Pre-processing Function

EMCIC integrates a commercial CATIA environment and AI Environment. It has all the functions required to define complex geometries, and supports import and generation of files in CATIA, IGES, STP, and STL formats. EMCIC also supports meshing. The platform EMC collection model is constructed as shown in Fig. A.2.

A.2 Post-processing Function

Through the post-processing function, the S parameters and the Cartesian and polar plot of antenna patterns (amplitude, phase, main polarization and cross polarization, circular polarization axis ratio, phase center) can be rendered, the field strength is visualized and the EMC design reports can be generated. The visualization of an antenna pattern is shown in Fig. A.3.

A.3 Program Management

The program management function is to model the EMC design scheme, including: basic attributes of transmitter, receiver, transceiver and non-RF equipment, as well as the construction of transmission attributes, reception attributes, case attributes, cable attributes, and interference correlation. Initial layout settings for antennas, cables, and devices on the platform are also covered in the program management function as shown in Figs. A.4, A.5, and A.6.

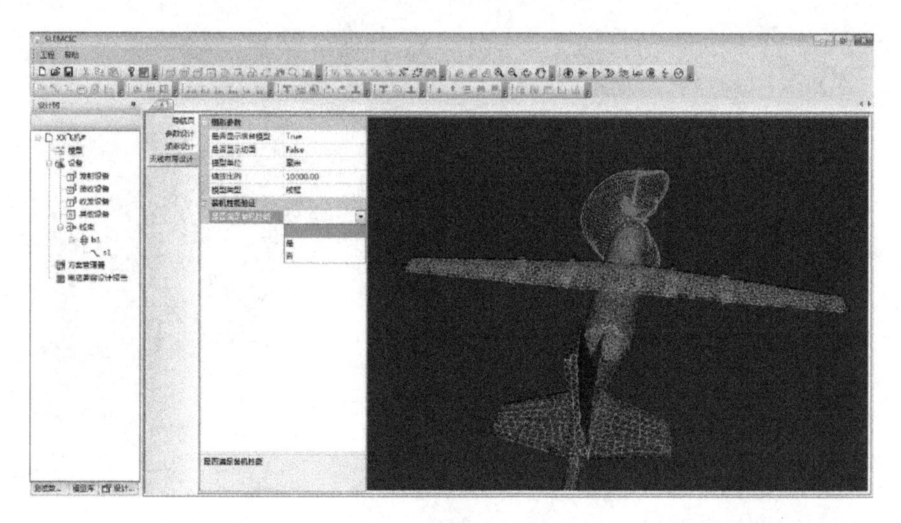

Fig. A.3 Data visualization of an antenna pattern

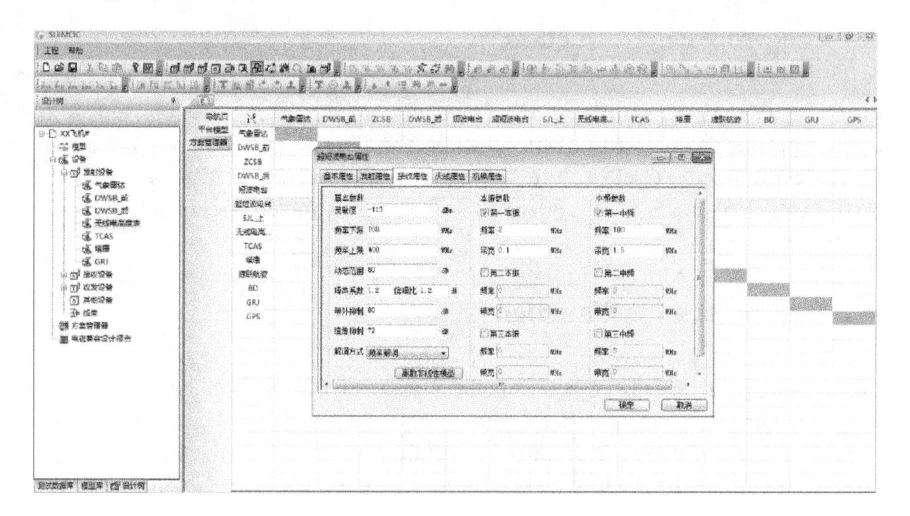

Fig. A.4 Parameterized modeling of equipment/subsystems

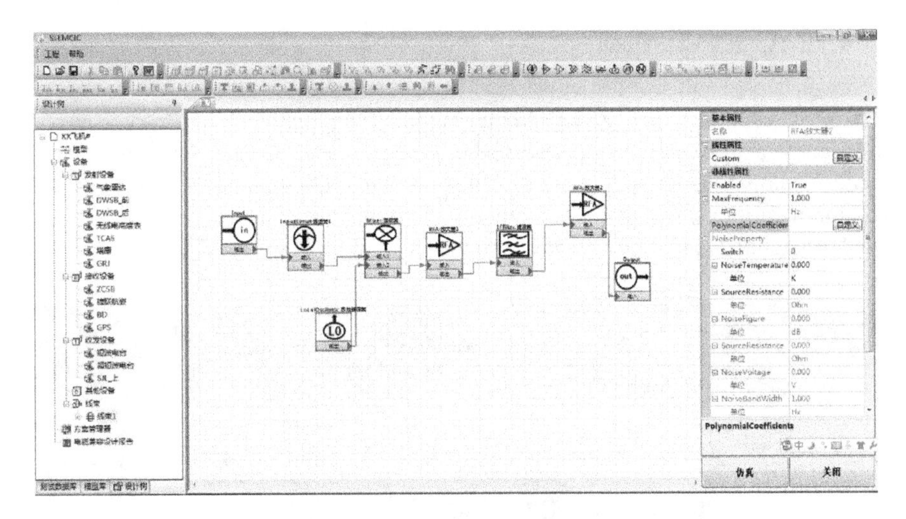

Fig. A.5 Behavioral modeling of equipment/subsystems

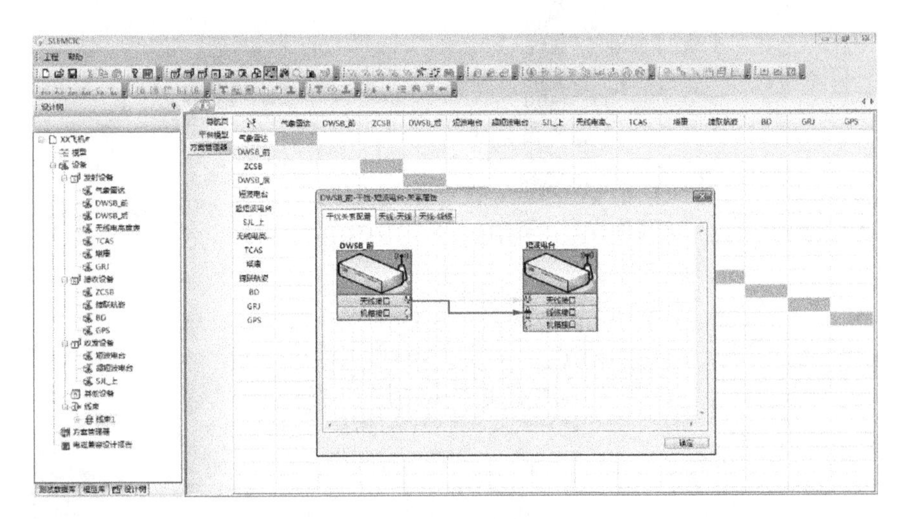

Fig. A.6 Construction of interference correlation relationship between equipment/subsystems

A.4 EMC Evaluation

The EMC evaluation function allows users to do rapid analysis, amplitude analysis, frequency analysis and detailed analysis. For equipment pairs that are electromagnetic incompatible or with potential safety hazard, self-developed simulation prediction algorithms, relevant commercial software or test data are used to obtain the electromagnetic interference correlation matrix as shown in Figs. A.7, A.8, and A.9.

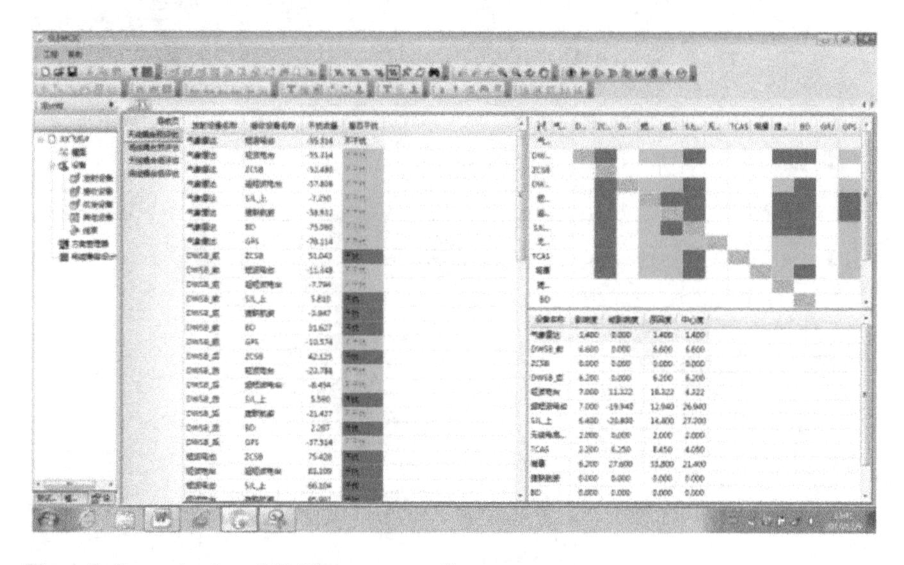

Fig. A.7 Pre-evaluation of EMC antenna coupling

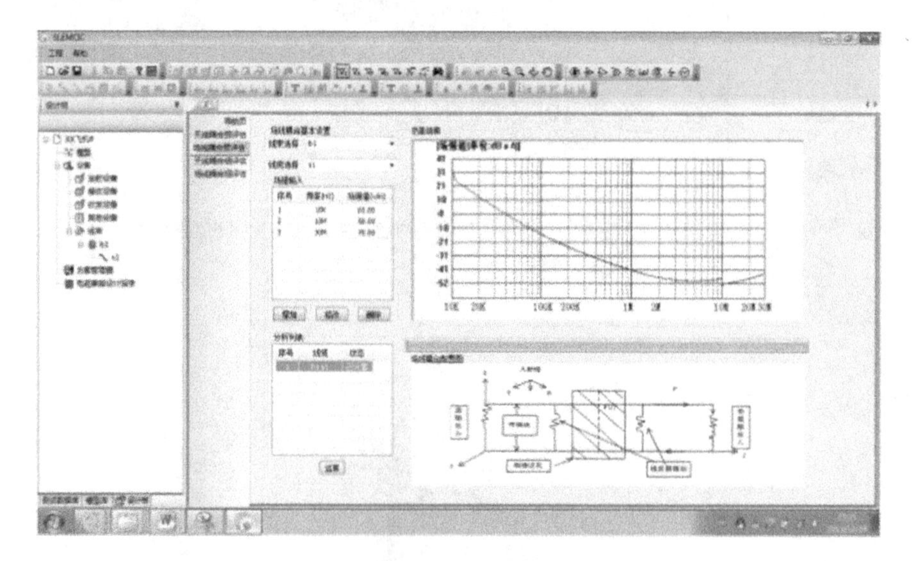

Fig. A.8 Pre-evaluation of EMC line crosstalk

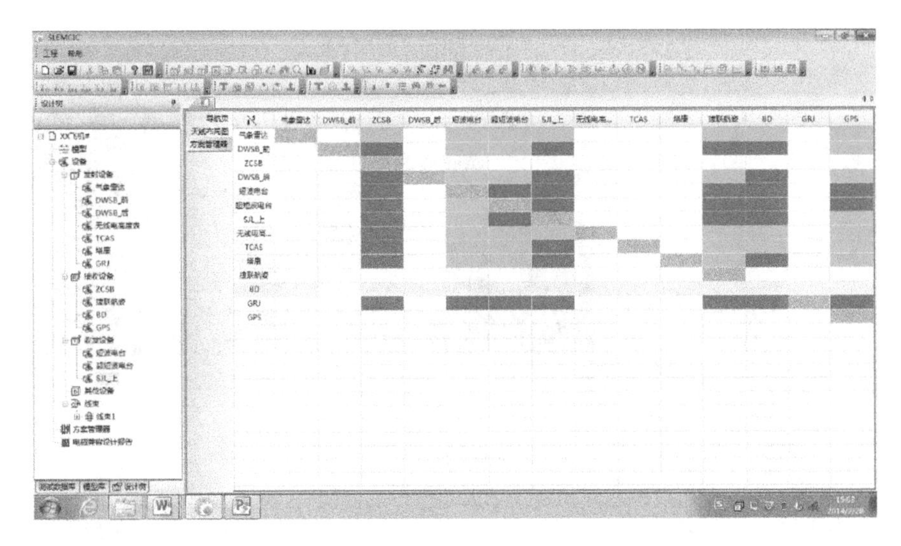

Fig. A.9 EMC Interference correlation matrix

A.5 System-Level EMC Design

This function allows the users to optimize the initial EMC design scheme based on the evaluation results. The optimization involves thirteen items including frequency design, antenna layout design, transmission power design, receiving susceptibility design, out-of-band emission attenuation design, out-of-band receiving rejection design, shielding effectiveness design, and latching design as shown in Figs. A.10, A.11, A.12, and A.13.

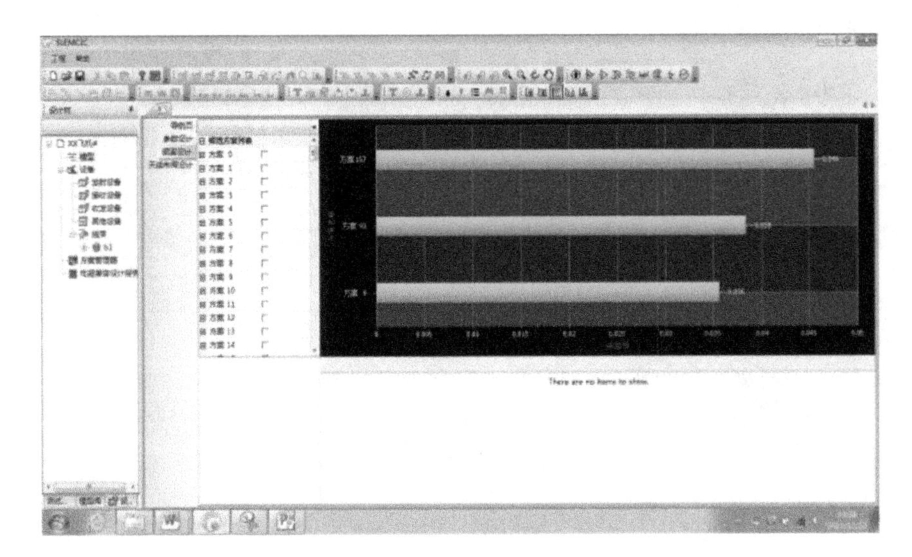

Fig. A.10 Frequency design

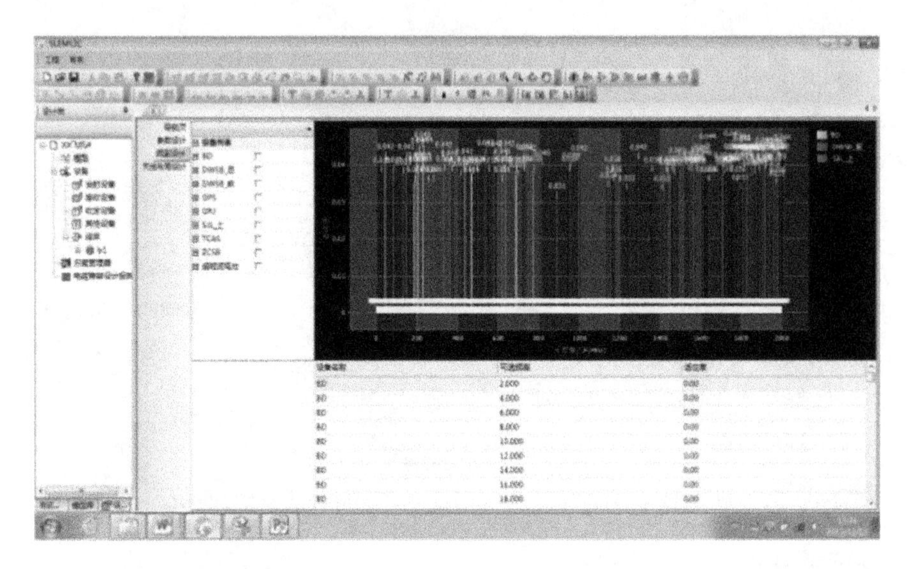

Fig. A.11 Frequency optimization

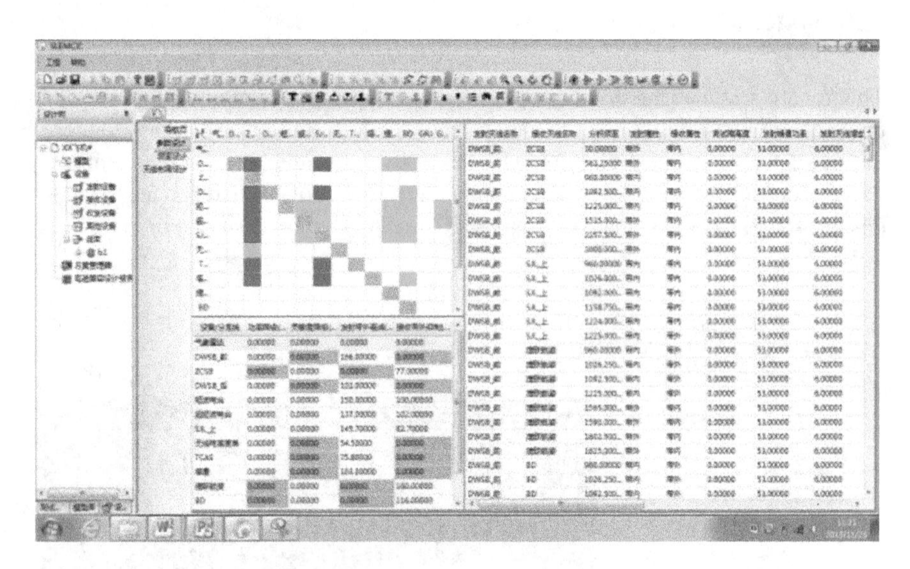

Fig. A.12 Frequency-using equipment parameter design

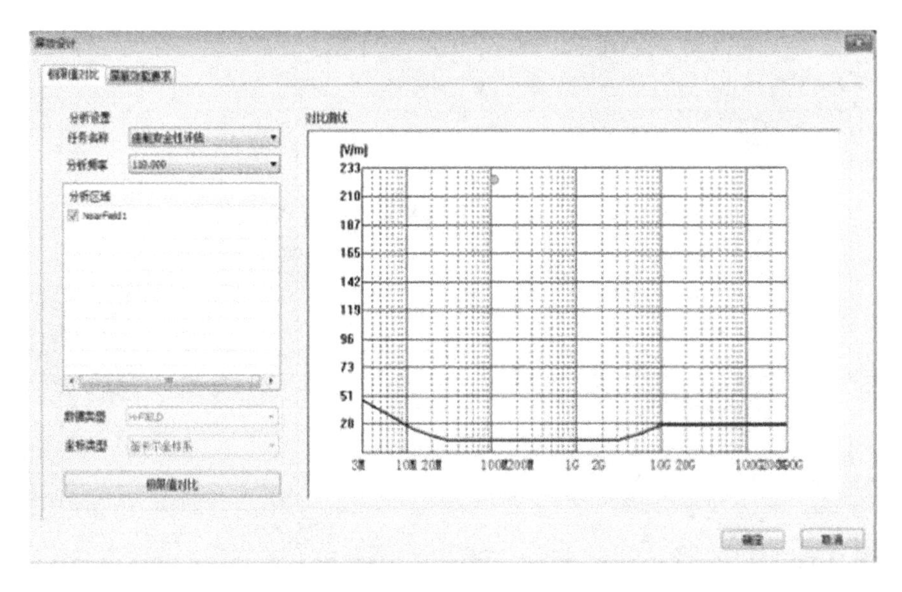

Fig. A.13 Shielding effectiveness design

A.6 Database Management

This function allows users to manage the model data, test data, engineering data and simulation task data. Users can build parametric model of the antenna using EMCIC, and the generated model can be used directly by common electromagnetic

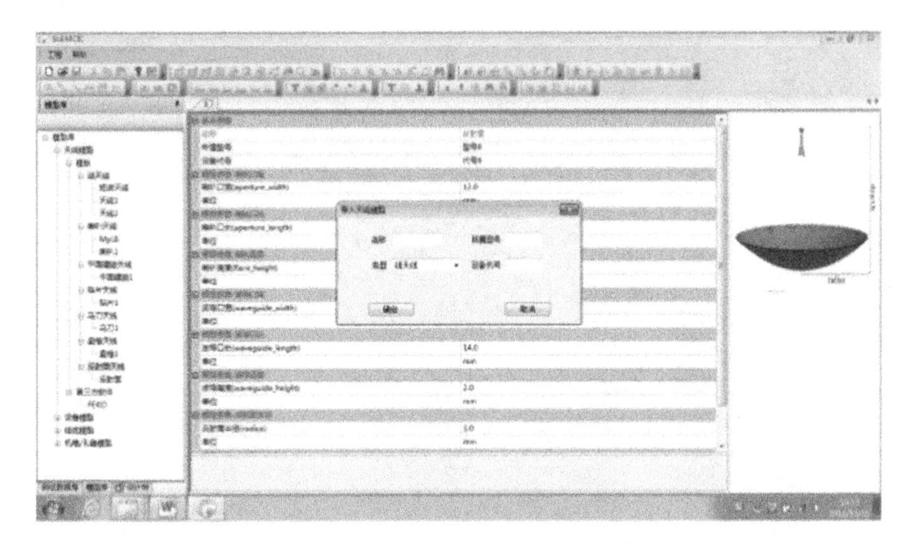

Fig. A.14 EMC Parameterized Model Library

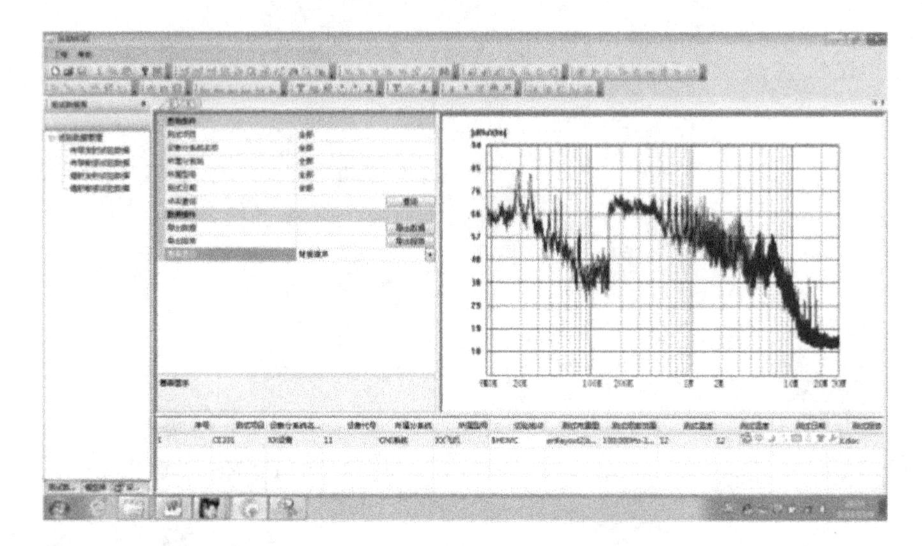

Fig. A.15 EMC Test Database

field simulation software to do electromagnetic calculation. In addition, the software supports the entry and maintenance of EMC test data for each equipment/subsystem on the platform. The test data can be further used for verification and conformance test of the design. Figs. A.14 and A.15 illustrate the parameterized model library and the test database separately.

Appendix B
Software for Aircraft EMC Evaluation in Different Phases

In view of the difficulty in EMC evaluation during product development, based on the engineering experience of several models of aircraft development, and the different EMC requirements of aircraft systems and airborne equipment in different development phases, we developed a software for aircraft EMC evaluation management in different phases, as shown in Fig. B.1.

1. **Main functions of the software**

   (1) Product information entry: The information that can be entered includes basic information, functional principles, product installation characteristics, EMC requirements, EMC problems, EMC design, test conditions and information of the data entry personnel.

   (2) Expert scoring system: Experts will score the EMC performance of the current development phase in accordance with the input information of the product and the EMC indicator system established by the software in different development phases. Experts provide their evaluation opinions and suggestions, and the software automatically calculates the total score of the EMC condition of the product in the current development phase.

   (3) Report generation: The software can generate a report in Microsoft Word format. The report covers the input EMC information of the product, as well as the expert evaluation scores, suggestions, etc.

   (4) Database storage: The input EMC information of the product and expert evaluation scores, suggestions, etc., can be stored as an Microsoft Access database file, which can be called by the supporting "aircraft EMC quality management software" (Fig. B.2).

© National Defense Industry Press and Springer Nature Singapore Pte Ltd. 2019

D. Su et al., *Theory and Methods of Quantification Design on System-Level Electromagnetic Compatibility*, https://doi.org/10.1007/978-981-13-3690-4

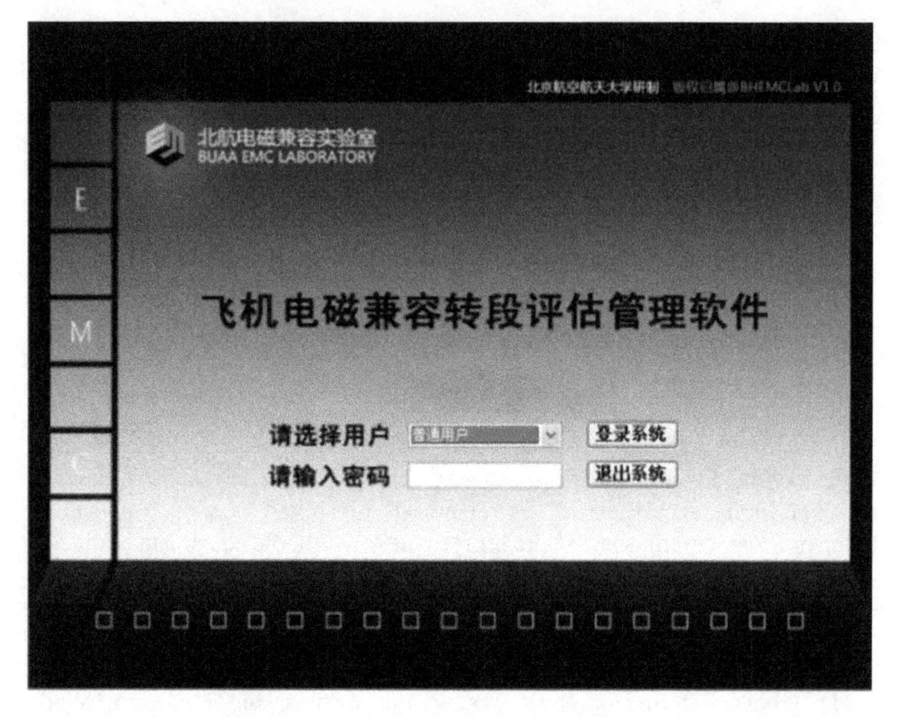

Fig. B.1 Main interface of the software for aircraft EMC evaluation management in different phases

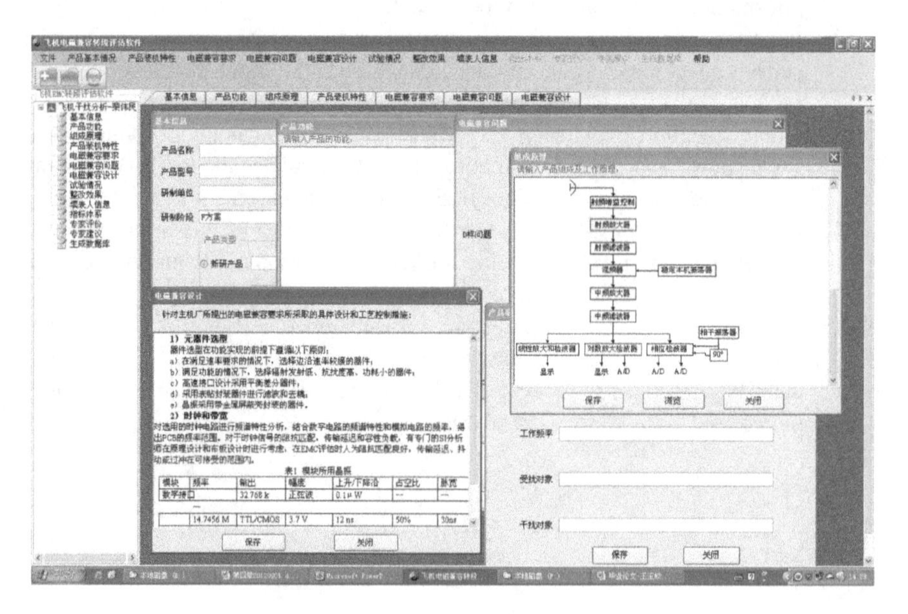

Fig. B.2 Main functions of the software

2. **Software modules**

(1) Engineering operation module: This module allows the users to manage the product information in each phase. The product information in different phases is always stored in the same project folder, so that it is easy to create a new project or open an existing project for viewing and modification. In addition, the function of reports generation and exit are also in this module. Users can click "Generate a Report" in the file to display the information of the current project in a Microsoft Word format. Users can click "Exit" to exit the software.

(2) Product information entry module: This module allows users to enter the information of the product series in each phase. In this module, users can input basic information, product installation characteristics, EMC requirements, EMC problems, EMC design, test conditions, troubleshooting results, and information of the data entry personnel. The information of each sub-item can be modified and saved. Every piece of information entered will appear in the generated report.

(3) Evaluation module: This module allows experts to evaluate and score the product provided by the manufacturer. The software will display different scoring system in accordance with phase of the product. Using the provided scoring system, experts can evaluate the products, decide the weights of each indicator and give score (Fig. B.3); The total score of the product in this phase can then be calculated. In addition, experts can also enter corresponding evaluations and suggestions in text for the product. The text will also appear in the generated report.

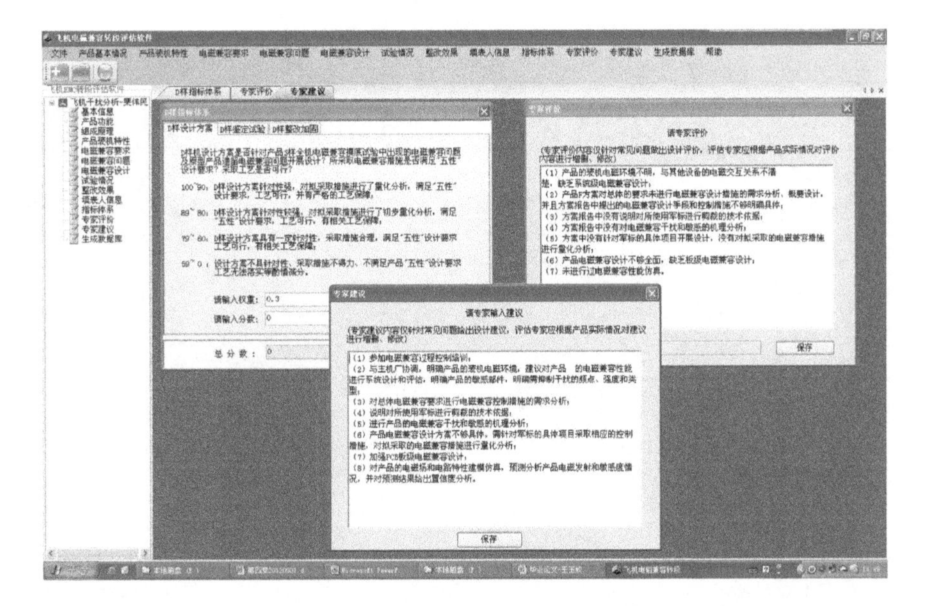

Fig. B.3 Interface of the expert scoring system

(4) Database module: The main function of this module is to store some of the useful information of the product into a Microsoft Access database file, so that the product information in different phases can be called and compared by other database management software.

3. **Applications**

The software has been applied to the development of several pieces of equipment and several aircraft models in different organizations. Using this software, users can collect and analyze product's basic information, product function and working principles, product installation characteristics, product EMC requirements, EMC design of current development phase, EMC test, EMC problems, troubleshooting measures and effect in current development phase. The software then generates EMC requirements and evaluation criteria in accordance with the development phase. Next, the product EMC is evaluated using the expert scoring method and recommendations will be provided to improve product EMC design. Finally, the software generates a comprehensive evaluation report and database files.

The software allows users to evaluate the EMC technology status of the entire aircraft and the airborne electronic equipment in each development phase, so that all users can understand the current EMC technology status of the product, so that the EMC control can be improved during development.

Appendix C
Aircraft EMC Quality Management Software

We have developed the "aircraft EMC quality management software" for the EMC information communication for all users. The software allows users to integrate the EMC status information of each phase of the system or equipment, and add or delete the database files generated by the "software of aircraft EMC evaluation management in different phases" in batches. the EMC status data of the whole aircraft in different phases is integrated into a single database file. This way, it is convenient for users to manage and store the EMC status information of the electronic equipment of the whole aircraft. The software can also generate the EMC quality report of the whole aircraft in PDF format.

The software was developed in Microsoft Visual C++ 6.0 integrated development environment with Microsoft Access database; the overall structure of the software is in framework style. The main framework includes system menu, maximization, minimization and closing the dialogue box; the sub-frame consists of label-type frames. Each project is a sub-frame divided into two display areas. The software allows multiple projects, or product in different phases open at the same time for searching, adding and deleting the EMC equality information (Fig. C.1).

1. **The main function**:

   (1) Database management: This function is to manage the product's EMC status information in Microsoft Access database files. This function allows users to add, delete, search and display the files, so that all users can analyze and manage the EMC status of the aircraft.

   (2) Statistics of quality management information: By gathering the EMC performance score of the whole aircraft, the software will automatically generate statistical result such as the EMC excellent rate, good rate, pass rate and failure rate of the product;

   (3) Quality management report generation: The EMC status information of the whole aircraft, including the EMC quality of all equipment, can be generated into a report in PDF format.

© National Defense Industry Press and Springer Nature Singapore Pte Ltd. 2019

D. Su et al., *Theory and Methods of Quantification Design on System-Level Electromagnetic Compatibility*, https://doi.org/10.1007/978-981-13-3690-4

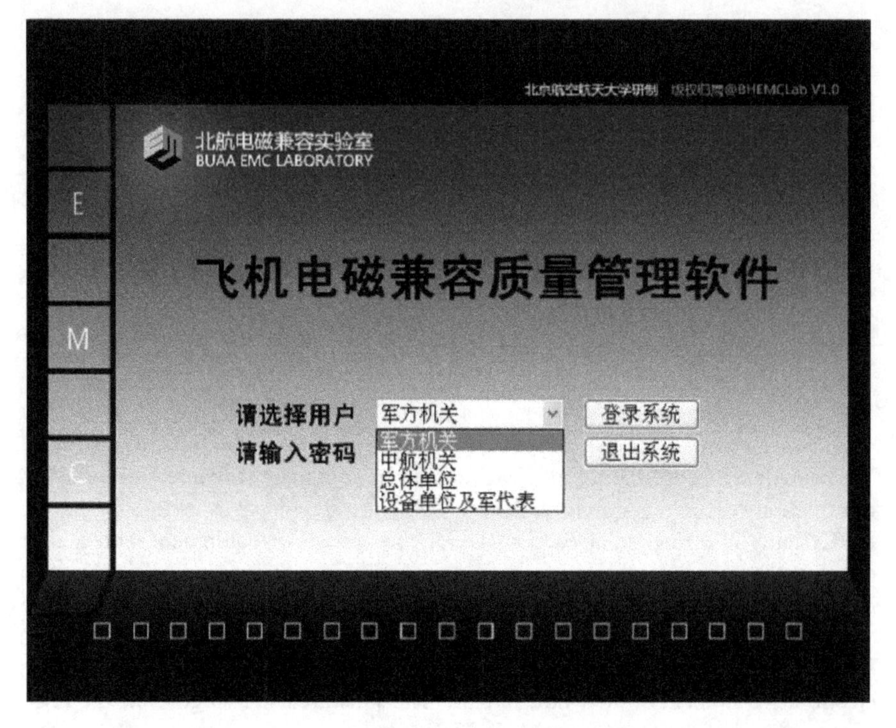

Fig. C.1 Main interface of aircraft EMC quality management software

The "aircraft EMC quality management software" allows users to add or delete information files of different products or products in different phases, such that users can manage the aircraft EMC quality information. The software can display all information entered and highlight the expert score; it can display the manufacturer's test report; at the same time, the database centrally manages the quality information and generates quality reports. The software interface is shown in Fig. C.2.

2. **Software modules**

(1) **Project management module**: This module allows users to perform project management operations, such as create, open, save, and save as. The result of the user's operation of the database file will be saved in the open or saved project. The project management drop-down menu has the "recently opened project" button, which makes it convenient for the user to open the last three projects that have been opened. In addition, the "exit" function is also in this menu. Users can exit the system by clicking the "Exit System" button.

(2) **Report generation module**: This module allows users to generate reports. There are two generation modes, one is to only generate the content of the current table into a PDF report; the other is to generate a report containing all the details on the right side.

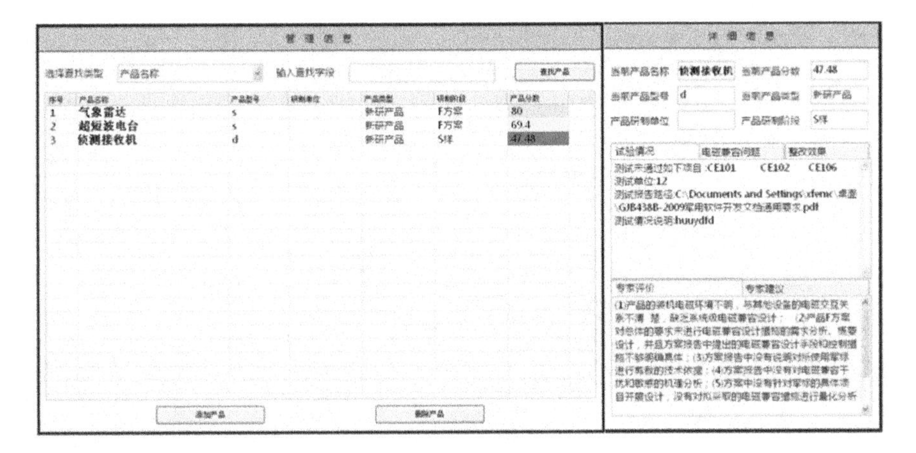

Fig. C.2 Software function interface

(3) **Toolbar module**: This module contains shortcuts of the menu. The shortcuts are linked with: new project, open project, save project, generate report (all information), generate report (table information), help, and about the system.

(4) **Display module**: The left side is the management information, which mainly displays the information list of the Microsoft Access file for current project. When a row in the list is selected, other detailed information of the row will be displayed on the right side. The information management function includes adding, deleting, and finding the Microsoft Access file in the correct format.

3. **Applications**

This software has been applied in the development of several aircraft models in conjunction with the "software of aircraft EMC evaluation of different phases". The software provides a platform for users to manage the product's information, the EMC design and the control work of each phase, such that the EMC technology status and the aircraft EMC quality can be controlled and managed.

References

1. Donglin S, Chen A, Xie S (2009) Electromagnetic field and electromagnetic wave (in Chinese). Higher Education Press, Beijing
2. Cheng'en L, Dazhang C (1979) Foundations of microwave techniques (in Chinese). Beijing: National Defense Industry Press
3. Shen Z (1980) Microwave technology (in Chinese). National Defense Industry Press, Beijing
4. Lv S (1995) Foundations of microwave engineering (in Chinese). Beihang University Press, Beijing
5. Pozar DM (2011) Microwave engineering. Wiley
6. Huibi X (2003) Research and application of field theory of network used in transmission line theory including mutual inductance (in Chinese). Huaqiao University, Fujian
7. Kang X (1993) Antenna principle and design (in Chinese). Beijing Institute of Technology Press, Beijing
8. Xie C, Qiu W (1985) Antenna principle and design (in Chinese). Northwest Telecommunication Engineering Institute Press, Xi'an
9. Rudge, AW, Milne K, et al (1988) The handbook of antenna design (in Chinese). Translate by Yukuan M, et al. PLA Press, Beijing
10. Donglin S, Wang M et al (2002) Electromagnetic compatibility analysis of plane antennas (in Chinese). J Beijing Univ Aeronaut Astronaut 2:20–22
11. GJB1389-92 (1992) Electromagnetic compatibility requirements for systems (in Chinese)
12. Donglin S, Wang B, Jin D et al (2006) EMC pre-design technologies on EW special aircraft (in Chinese). J Beijing Univ Aeronaut Astronaut 10:1241–1245
13. MIL-HDBK-237C (2001) Electromagnetic environmental effects and spectrum certification guidance for the acquisition process
14. Wang Bingqie (2007) Key techniques research of EMC comprehensive design for EW special aircraft (in Chinese). Beihang University, Beijing
15. Chen W (2010) Key techniques research of system-level EMC behavioral simulation method (in Chinese). Beihang University, Beijing
16. Liao Yi (2012) Research on equivalent circuits of wire antennas and behavioral modeling for system-level EMC analysis (in Chinese). Beihang University, Beijing
17. Liu S, Dang Y, Fang Z (2004) Grey system: theory and application (in Chinese). Science Press, Beijing
18. Wang Q, Wang S, Zuo Q et al (1996) Grey mathematic foundation (in Chinese). Huazhong University of Science & Technology, Wuhan
19. Qiong W, Donglin S, Shuguo X, et al (2006) Noise density research in the system-modeling of RF receivers (in Chinese). J Beijing Univ Aeronaut Astronaut 32(4):395–398.

© National Defense Industry Press and Springer Nature Singapore Pte Ltd. 2019

D. Su et al., *Theory and Methods of Quantification Design on System-Level Electromagnetic Compatibility*, https://doi.org/10.1007/978-981-13-3690-4

20. Daqing LBH (2006) Research of an EMC design methodology based on prediction method (in Chinese). Chin J Sci Instrum S3:2499–2500
21. Пьедровский ВИ, шейдерников ЮЁ (1992) Electromagnetic compatibility of radio electronic equipment (in Chinese). Translated by Jiazhen F, et al. Aviation Industrial Publishing House, Beijing
22. Wood J, Root DE, Tufillaro NB (2004) A behavioral modeling approach to nonlinear model-order reduction for RF/microwave ICs and systems. IEEE Trans Microw Theory Tech 52(9):2274–2284
23. Bogdanor JL, et al. Intrasystem electromagnetic compatibility analysis program. McDonnell Aircraft Company, 1974: AD-A008526.
24. Bogdanor JL, et al. Intrasystem electromagnetic compatibility analysis program. McDonnell Aircraft Company, 1974: AD-A008527.
25. GJB786-89 (1989) Preclusion of ordnance hazards in electromagnetic fields, general requirements for (in Chinese)
26. MIL-STD-464A (1997) Electromagnetic environmental effects requirements for systems
27. GJB5313-2004 (2004) Limits and test methods for exposure to electromagnetic fields (in Chinese)
28. Qiong W, Donglin S (2006) Prediction design of the noise and gain specifications of the RF receiver. In: 17th international Zurich symposium on electromagnetic compatibility. EMC—Zurich
29. Balanis CA, Peters L. Aperture radiation from an axially slotted elliptical conducting cylinder using geometrical theory of diffraction. IEEE Trans Antennas Propag 17(4):507–513
30. Yu C, Burnside WD et al (1978) Volumetric pattern analysis of airborne antennas. IEEE Trans Antennas Propag 26(5):636–641
31. Burnside WD, Wang N, Pelton EL (1980) Near-field pattern analysis of airborne antennas. IEEE Trans Antennas Propag V28(3):318–327
32. Chung H, Burnside WD (1984) Analysis of airborne antenna radiation patterns using spheroid/plates model. In: Antennas and propagation society international symposium
33. Yamamoto K, Yamada K (1994) A method to improve computation efficiency for the radiation pattern of an airborne antenna. In: Second international conference on computation in electromagnetics
34. Lin Z, Mou S, Zhao Y (1995) The analysis and calculation for airborne antenna's radiation pattern (in Chinese). Modern Radar 17(3):71–76
35. Xue Z-H, Gao B-Q, Liu R-X, et al (1998) Calculation and analysis of near field of airborne short wave antenna. J Beijing Inst Technol V7(3):286–292
36. Cao X, Xiang T, Ma F et al (2002) Study of the airborne antenna pattern (in Chinese). J Microw 18(1):15–19
37. Wen H, Guo P, Feng X et al (2003) Calculation of geometric diffraction method for airborne antenna pattern (in Chinese). Flight Test 19(2):18–26
38. Chen J (1998) The theory analysis and application for antenna polarization mismatch (in Chinese). Space Electron Technol 4:48–63
39. Liu J (1996) Research on antenna layout and electromagnetic compatibility of spacecraft (in Chinese). Beijing University of Posts and Telecommunications, Beijing
40. Tsinghua University (1976) Microstrip circuit (in Chinese). Posts & Telecom Press, Beijing
41. Chen Z, Donglin S, Liu Y et al (2013) Ray path tracing on discrete surface (in Chinese). J Beijing Univ Aeronaut Astronaut 5:665–669
42. Zong W, Liang C, Cao X, et al (2002) Diffraction ray tracing of GTD on the cone and cylinder (in Chinese). J Xidian Univ (4):482–485, 489
43. Ping W, Xianliang W (2003) Ray path tracing on convex surface with applications to creeping wave (in Chinese). Microcomput Dev 9:60–62
44. Wang N, Liang C, Zhang Y, et al (2007) Study on the creeping ray-tracing algorithm of NURBS-UTD (in Chinese). J Xidian Univ (4):600–603,668

45. Taylor C, Satterwhite S, Harrison CW, et al (1965) The response of a terminated two-wire transmission line excited by a nonuniform electromagnetic field. IEEE Trans Antennas Propag 13(6):987–989.
46. Cai R (1997) Principle, design and prediction technology of EMC. Beihang University Press, Beijing
47. Saleh AAM (1981) Frequency-independent and frequency-dependent nonlinear models of TWT amplifiers. IEEE Trans Commun 29(11):1715–1720
48. Bosch W, Gatti G (1989) Measurement and simulation of memory effects in predistortion linearizers. IEEE Trans Microw Theory Tech 37(12):1885–1890
49. Vuong XT, Guibord AF (1984) Modeling of nonlinear elements exhibiting frequency-dependent AM/AM and AM/PM transfer characteristics. Can J Electr Comput Eng 9:112–116
50. Ku H, Kenney JS (2003) Behavioral modeling of nonlinear RF power amplifiers considering memory effects. IEEE Trans Microw Theory Tech 51(12):2495–2503
51. Wen O, Han C (2001) A solution to the curse of dimensionality in Volterra functional series identification (in Chinese). J Xi'an Jiaotong Univ 6:658–660
52. Zhu A, Brazil TJ (2004) Behavioral modeling of RF power amplifiers based on pruned Volterra series. IEEE Microw Wirel Compon Lett 14(12):563–565
53. Wang J. Geophysical inversion theory (in Chinese). China University of Geosciences Press, Wuhan
54. Zhu A, Pedro JC, Brazil TJ (2006) Dynamic deviation reduction-based Volterra behavioral modeling of RF power amplifiers. IEEE Trans Microw Theory Tech 52(3):4323–4332
55. Nan J, Gao M, Liu Y et al (2008) Analysis and comparison of behavioral models for nonlinear RF power amplifier (in Chinese). J Microw S1:170–175
56. Zhu A, Wren M, Brazil TJ (2003) An efficient Volterra-based behavioral model for wideband RF power amplifiers. In: IEEE MTT-S international microwave symposium digest
57. Chen Q, Jiang Q, Zhou K et al (1993) Electromagnetic compatibility engineering design manual (in Chinese). National Defense Industry Press, Beijing
58. Jianchang O, Lin S, Lv Y (2010) Theory and practice of EMC design for electronic equipment (in Chinese). Publishing House of Electronics Industry, Beijing
59. Zhang Z, Zhou L, Wu D (2006) Microwave engineering, 3rd edn (in Chinese). Beijing: Publishing House of Electronics Industry
60. Wang B, Donglin S, Zeng G et al (2007) Intermodulation suppression design and measurement for FH receiver (in Chinese). J Wuhan Univ Technol 3:385–388
61. Feng Y (2007) Research on predistortion RF power amplifier. Beijing University of Posts and Telecommunications, Beijing
62. Lee S, Lee Y, Hong S et al (2005) An adaptive predistortion RF power amplifier with a spectrum monitor for multicarrier WCDMA applications. IEEE Trans Microw Theory Tech 53(2):786–793
63. Wei Y (2013) Research on the quantization method for the typical factors of the electrical equipment (in Chinese). Beihang University, Beijing
64. Tan G (2000) The structure method and application of background value in grey system GM (1, 1) model (in Chinese). Syst Eng Theory Pract 4:98–103
65. Zhu L (2012) Electromagnetic compatibility design and countermeasures and case analysis (in Chinese). Publishing House of Electronics Industry, Beijing
66. Mardiguian M (2008) 著.陈爱新,译.辐射发射控制设计技术[M].北京: 科学出版社
67. Mardiguian M (2008) Controlling radiated emissions by design (in Chinese). Translated by Aixin C. Science Press, Beijing
68. Chengbin F (2013) Research on method of EMC prediction and analysis for airborne RF equipments (in Chinese). Beihang University, Beijing
69. Meng C (2013) The research based on ANP of the aircraft EMC Indicator quantity analysis (in Chinese). Beihang University, Beijing

70. Xia Y, Qizong W (2004) A technique of order preference by similarity to ideal solution for hybrid multiple attribute decision making problems (in Chinese). J Syst Eng 6:630–634
71. Yazar MN (1979) Civilian EMC standards and regulations. IEEE Trans Electromag Compat 21(1):2–8
72. Gao J, Su D, Dai F, Wang B, Cao C (2007) Method of systemic EMC evaluation. In: 2007 international symposium on electromagnetic compatibility, pp 251–254
73. Shiva SR, Pande DC (2008) System level EMC evaluation of radars [C]. In: 2008 10th international conference on electromagnetic interference & compatibility, pp 313–316
74. Wang B, Wang L, Du Q (2013) A method of systemic EMC evaluation for missile. In: 2013 international conference on information technology and applications, pp 432–435
75. Rao KR (1999) EMC quality and its approach. In: Proceedings of the international conference on electromagnetic interference and compatibility, pp 168–171
76. Kudyan HM (2007) A hierarchical model for prescribing EMC design targets to the components of a system. In: 2007 IEEE international conference on electromagnetic compatibility, pp 1–6
77. Chen W, Su D, Li J, Cai D (2007) The EMC analysis and estimation method for complicated electromagnetic radiation system. In: 2007 IEEE international conference on electromagnetic compatibility, pp 247–250
78. Rao KN, Ramana PV, Krishnamurthy MV, Srinvas K (1995) EMC analysis in PCB designs using an expert system. In: 1995 IEEE international conference on electromagnetic compatibility, pp 59–62

CPSIA information can be obtained
at www.ICGtesting.com
Printed in the USA
LVHW061124260319
611865LV00002B/69/P

9 789811 336898